TIME REGAINED

Time Regained
Symmetry and Evolution in
Classical Mechanics

Volume 1

Sean Gryb
University of Groningen

Karim P. Y. Thébault
University of Bristol

OXFORD
UNIVERSITY PRESS

OXFORD
UNIVERSITY PRESS

Great Clarendon Street, Oxford, OX2 6DP,
United Kingdom

Oxford University Press is a department of the University of Oxford.
It furthers the University's objective of excellence in research, scholarship,
and education by publishing worldwide. Oxford is a registered trade mark of
Oxford University Press in the UK and in certain other countries

Published in the United States of America by Oxford University Press
198 Madison Avenue, New York, NY 10016, United States of America

British Library Cataloguing in Publication Data

Data available

Library of Congress Control Number: 2023946171

ISBN 9780198822066

DOI: 10.1093/oso/9780198822066.001.0001

Printed and bound by
CPI Group (UK) Ltd, Croydon, CR0 4YY

MIX
Paper | Supporting
responsible forestry
FSC
www.fsc.org   FSC® C013604

Preface

This book is the first volume of a two-part project, *Time Regained*, articulating a new approach to the analysis of time in modern physical theory. Our particular concern is with the Problem of Time as it occurs in theories which feature dynamical laws that are invariant under time reparameterization symmetries. The class of such theories includes both simple mechanical theories and the general theory of relativity and prospective theories of quantum gravity. Our goal is to provide a solution to the problem of time such that determinate time-ordering structure can be reconciled with reparameterization symmetry. The present volume is focused on analytical formulations of classical mechanical theory. This restricted context will allow us to develop a new framework for the analysis of symmetry and structure in physical theory and provide a fully general and explicit resolution of the classical mechanical problem of time. In the second volume, the framework will be extended to the context of classical and quantum gravitational theories towards both formal resolution of the gravitational problem of time in general terms and an explicit solution realized in the context of simple cosmological models. This cosmological solution will be found to have attractive novel consequences including unitarity and singularity resolution.

In the most general sense our project is one concerned with the *natural philosophy of time*. As such, we intend our work as a contribution to both physics and philosophy. Ours is not, however, a project regarding the *interpretation* of physical theory or the *metaphysics* of time. Rather, our principal goal is to motivate and articulate a novel proposal for the *development* of physical theory. The main focus of the present volume is the problem of time as it occurs in classical mechanical theories and our proposal is designed specifically as a response to the problem in that context. First and foremost, therefore, this book is intended as a contribution to mathematical physics. In order to formulate and implement our solution to the classical mechanical problem of time new work on the geometric foundations of mechanical theories will be required. In particular, we will provide a formalism for the comprehensive and rigorous analysis of reparameterization symmetries that supersedes that provided by the standard approach due to Dirac.

Our formal analysis will be built upon the sustained analysis of a tangled knot of historical and philosophical problems relating to time and symmetry. Most significantly, we will consider philosophical arguments regarding time, as found in the work of Newton, Leibniz, and Mach, and philosophical arguments regarding symmetry, as found in the work of various contemporary authors in philosophy and foundations of physics. In this respect, this book is also a research monograph in the philosophy of physics which engages with core issues in the history of natural philosophy. We do not take the mathematical physics, philosophical, and historical aspects to be disconnected. Rather, we will adopt the methodological outlook of what might be called a 'three-field' approach to the foundations of physics: an interdisciplinary approach simultaneously applying methods and insights from mathematical physics, the history of natural philosophy, and the philosophy of physics.

Our ultimate goal is not, however, merely to provide a unified physical, philosophical, and historical analysis of the problem of time taken in isolation. Rather, we find the problem of time to be both a domain of enquiry in its own right and a 'resource', providing tools that can be productively applied to other domains, cf. Rickles (2020). On the one hand, we will view the problem of time as a physical, philosophical, and historical subject in its own right, with attendant problems, questions and formal challenges. But on the other, we will view the problem as an arena for developing new concepts, tools, and techniques, which find applications in other domains. When seen from this perspective, the problem of time gains considerable value as a topic of study since it brings into relief key issues in the study of time in classical mechanical theories.

Acknowledgements

Support and Funding. The genesis of this project can be traced to a a precise spatiotemporal coincidence in the terms of the first discussion of its topic in the first meeting of its authors at a conference in Brisbane in 2011, and its development has been spread across space and time encompassing Australia, Canada, the Channel Islands, the Netherlands, Germany, the United Kingdom, and Romania. The debts we have jointly accrued in completing this book are difficult to exhaustively express. Of particular note are early discussions regarding the problem of time which one or both authors profited deeply from with Julian Barbour, Oliver Pooley, and Hans Westman. Parts of the book have greatly profited from further interactions with Edward Anderson, Ric Arthur, Alex Blum, Martin Bojowald, Harvey Brown, Jeremy Butterfield, Adam Caulton, David Cobb, Erik Curiel, Ana-Maria Crețu, Karen Crowther, Radin Dardashti, Richard Dawid, Pete Evans, Sam Fletcher, Simone Friedrich, Steffen Gielen, Henrique Gomes, Jonathan Halliwell, Stephan Hartmann, Philipp Höhn, Nick Huggett, Lucy James, Tim Koslowski, James Ladyman, Huw Price, James Read, Dean Rickles, Bryan Roberts, Donald Salisbury, Simon Saunders, Steven Savitt, Lee Smolin, Tony Short, Dave Sloan, Kurt Sundermeyer, Tzu Chien Tho, Jim Weatherall, and Chris Wüthrich. We are particularly appreciative to Ric Arthur and Erik Curiel for detailed comments on draft chapters of the manuscript. Material developed towards this book has also received feedback from a wide variety of audiences in presentations and from various anonymous reviewers of research articles. KT is grateful to University of Bristol students on the courses *Space, Time and Matter* and *Advanced Philosophy of Physics* for feedback on material that has used in the present book.

Initial work on this project was supported by the Perimeter Institute for Theoretical Physics, the Centre for Time (University of Sydney), Radboud University, Utrecht University, and the Munich Center for Mathematical Philosophy (Ludwig-Maximilians-Universität München). We are grateful to Huw Price and Stephan Hartmann for facilitating this early institutional support as well as to the Alexander von Humboldt Foundation and the Netherlands Organisation for Scientific Research (NWO) (Project No. 620.01.784) for financial support. KT is particularly grateful to colleagues, students, and academic visitors at the Bristol Department of Philosophy. This work could not have been completed without the rich scholarly and collegial atmosphere that Bristol has provided over the last eight years. Significant work towards this project was completed by KT during a University Research Fellowship in 2018/19, supported by the University of Bristol. Thanks is also due to Oxford University Press, and in particular Sonke Adlung and Giulia Lipparini, for patience and support notwithstanding the somewhat glacial progress of this project.

A final and most significant note of thanks is due to Ana and Nienke for their constant support and encouragement.

Previously Published Material. The vast majority of the material presented here is new. That said, various elements build in different ways on existing published works of the authors. When we are drawing explicitly on published material, this is indicated below. There are two substantial tranches of material that are reproduced from previous work. First, we are particularly grateful to Elsevier for permission to republish parts of the article 'On Mach on Time' which was published in *Studies in History and Philosophy of Science Part A*, 89, 84–102 which is included in Chapter 4. Second, Chapter 12 features material from the presubmission, preprint version of the article 'Schrödinger Evolution for the Universe: Reparameterization', philsci-archive.pitt.edu/11299/. Finally, at a more minor level, we are grateful to Oxford University Press for permission to republish several paragraphs from the article 'Time Remains', *British Journal for the Philosophy of Science*, 67(3), 663–705, which are reproduced in Chapter 14.

Contents

1

Introduction

1.1 The Objects of Our Enquiry

At once possessing a unity of nature and an infinite variegation of form. Singularly constitutive of human experience and the subject of inexhaustibly diverse artistic, philosophical, and scientific discourses. Suffused through everything and yet nothing in-and-of itself. Time is problematic. We know precisely what it is, until we begin to discuss it. It is this combination of unity and multiplicity, familiarities and intangibility, that should perhaps be designated the master problem of time, of which all others are mere shadows. Yet, there can be no substantive progress in any philosophical enquiry into the nature of time without first setting a firm delimitation of its remit. What will we mean by the problem of time? What approach will we take towards its investigation? What would constitute a solution?

Let us start by providing a brief comment on our methodological outlook, echoing the remarks of the Preface. The primary restriction we will make in what follows is the assumption of a contemporary scientific perspective in which our outlook on both the problem and its solution will be anchored in methods of modern theoretical physics. Our methodology will be an inherently interdisciplinary one, drawing, in addition, upon tools from analytic philosophy, the philosophy of physics, and the history of natural philosophy. The overriding approach we will adopt, however, will be that of mathematical physics. Our goal will be to precisely formulate and solve a specific problem in the articulation of classical mechanical theory, understood from a modern geometric perspective. In doing this we will be required to construct new formal approaches to the geometric analysis of symmetry and evolution in mechanical theory. There is no scope for a purely interpretative solution to the problem of time in classical mechanics. An approach founded upon the methodological outlook of mathematical physics is thus unavoidable.

We can further constrain the scope of our investigations by delimiting two dimensions of analysis. The first dimension is the aspects of time or, as we shall more precisely frame them, the *temporal structures* that we will consider. Our enquiry is a foundational one in the sense that the target of our formal excavations shall be the *bare substratum* of time as represented in mechanical theory. We will focus our sustained attention upon the provision of a maximally precise physico-mathematical characterization of the basic structural characteristics of time. The limitation in the scope of our enquiry to basic temporal structure allows us to centre our analysis upon the temporal structure of mechanical theories in which the function of time and temporality is at its most minimal in terms of providing *undirected time orderings* and *durations*. The first we will call *chronordinal structure*, the second *chronometric structure*. A third key temporal structure relates to the representation of *empirical features* of time or temporality. This we will call *chronobservable* structure. A major preoccupation in what follows shall be the explication and analysis of chronordinal, chronometric and chronobservable temporal structures as found within classical mechanical theories.

The second dimension of analysis that will act as a frame for our investigations is the identification of the problem itself. That is, our endeavours will be focused upon a specific chain of interrelated issues in the analysis of particular aspects of time in classical mechanical theory which can together be identified as *the* problem of time. In order to better understand the problem as it occurs in modern physical theory it will serve us particularly well to consider two *precursor problems*, each of which can be identified in historical debates regarding the nature of time. In order to present these problems we must first make two crucial further terminological specifications, each of which will be explicated in more detail shortly. First, temporal structure that is unchanged under variation between symmetry related dynamical models in a theory is *invariant* structure. Second, temporal structure that is changed under variation between dynamically related models in a theory is *dynamically relative* structure. The *chronordinal problem* is then to fix *determinate chronordinal structure* as both *invariant and dynamically relative*. The *chronometric problem* is to fix *determinate chronometric structure* that is *invariant and dynamically relative*.

Each of these problems comes from a desire to represent temporal structure as invariant and dynamically relative, in the sense we will define below, combined with a requirement for two key types of such structure, chronometric structure and chronordinal structure, to be determinate. The requirement for determinate chronordinal structure will be understood as a *pragmatic necessity*, while the requirement for determinate chronometric structure will be understood as an *empirical necessity*. Thus in each case, these are more than interpretative problems; rather they are physico-mathematical problems, relating to the temporal structures which are required to be mathematically well-defined in order to provide an adequate mechanical representation of dynamical motions.

A novel achievement in what follows will be the identification of the precursor problems at two crucial junctures within historical discussions of the natural philosophy of time. The precursor problems we will identify are found within the early modern debates regarding the 'absolute' and 'relative' status of time in mechanical theory, in particular, within the classic absolute vs relative time dialectic, as expressed in the correspondence between Leibniz and Clarke in the early 1700s. Still in the historical natural philosophical context, we will also find these problems perspicuously expressed in the work of Ernst Mach towards the end of the 1800s. Our discussion of the Machian view of time will take place not only in the particular context of a dialectic with the original Newtonian treatment but also in comparison with work by Mach's contemporaries on the reframing of Newtonian absolute time in terms of inertial clocks. In each case, we will argue, the informal problem is at the heart of key debates, disputes, and ambiguities. Newton, Leibniz, and Mach all grappled with the precursor problems of time, and valuable lessons can be learned from their struggles. It will prove of invaluable profit to our constructive project to deepen our understand the foundations of the progenitor problems of time as found in the writings of these three thinkers.

Discussion of the precursor problems in the context of the history of natural philosophy of time and the views of Newton, Leibniz, and Mach will be the focus of Part I of this monograph. This will lead into a more formal and general analysis of symmetry and structure in physical theory that will be developed in Part II, the main ideas of which will be summarized shortly. The formalized modern problem of time is focus of Part III. It will be worthwhile to give a brief overview of how we propose to formulate and solve this problem now.

This problem of time occurs within all mechanical theories in which the laws are invariant under local time relabelling. Such *reparameterization invariant theories* are the context in which the modern classical mechanical problem of time can be formally stated and in which our solution shall be articulated. The problem can be stated as follows: *the problem of time in classical mechanics is that the symmetry properties of reparameterization-invariant theories with regard to chronometric and chronobservable structure enforce indeterminacy with regard to such structure under the standard approach to gauge symmetry.*

Our solution will be first to demonstrate that the standard approach to gauge symmetry is inadequate to the treatment of reparameterization invariant theories and second to formulate a general and fully adequate alternative based upon a more perspicuous analysis of temporal symmetry and structure. Our approach is such that reparameterization invariance is demonstrated to be fully compatible with determinate chronordinal structure. As such, the novel approach to symmetry and evolution in classical mechanics that we will establish in this monograph will provide a resolution of the problem of time in classical mechanics. We will provide a formal analysis of the symmetry and structure of reparameterization invariant theories that allows for determinate and well-behaved chronometric, chronobservable, and chronordinal temporal structures.

This work will provide a formal basis to reject alternative views on the problem of time in classical mechanics. Most significantly, we will demonstrate the formal inadequacy of approaches based upon supposed equivalence between dynamics and gauge symmetry, which is often taken to lead towards the outright denial of temporal evolution. This view is often only implicitly expressed in the classical mechanical context but is a straightforward implication of the standard interpretation of Hamiltonian constraints when applied to reparameterization invariant theories. The essence of this view has been most vividly expressed in terms of the slogan that 'motion is the unfolding of a gauge transformation' (Henneaux and Teitelboim, 1992, p. 103). That such paradoxical statements can be found within textbook treatments of contemporary physical theory is without doubt remarkable. One might plausibly classify such an understanding as the pseudo-problem of time. One major goal of our project will be to demonstrate that the pseudo-problem is precisely that. Supposed equivalence between motion and gauge symmetry rests upon a formal conflation. In particular, we will implement a new formal analysis that allows one to explicitly differentiate the generators of evolution and 'gauge' symmetry in reparameterization invariant theories. Fundamental timelessness thus rests upon a mistake. Motion is not the unfolding of a gauge transformation. With the correct mathematical tools one can clearly distinguish the fundamentally distinct roles of symmetry and evolution in reparameterization invariant theories of classical mechanics.

On a more constructive note, we should mention a second alternative approach to the problem of time that works towards a deflation rather than a solution. The essence of this view is idea that a functionally adequate mechanics can be recovered based upon a relativized and underdetermined chronordinal structure. Under this view, chronordinal structure can be built internally to a mechanical system by reference to particular degrees of freedom that play the role of internal clocks. This view has been advocated consistently in the context of reparameterization invariant classical mechanical theories by Rovelli.[1] Formally, this approach will be shown to be a complementary, yet in a certain specific sense more limited, than our own analysis.

1.2 Symmetry and Evolution

Our analysis of symmetry and evolution in classical mechanics is designed to be of independent value outside the analysis of reparameterization invariant theories and the problem of time. We construct a new framework for the analysis of symmetry and structure in mechanical theory which allows for the analysis of the full set of transformation properties of any structure that can be defined on the model space of a theory. This framework allows for a fine-grained and unambiguous identification of 'gauge symmetries' in terms of the existence, or not, of initial value constrains. Moreover, our framework will also allow for the identification of core heuristics for theory extension and re-articulation. The core aspects of our framework can be outlined informally as follows. Full formal definitions and proofs will be provided in Part II.

1.2.1 Three Levels of Theoretical Structure

A physical theory can be analysed in terms of three levels of structure, each with different forms and functions. At the most basic level, *constitutive structure* is the most basic structure of a theory. This is the structure that one must assume in order to build the space of kinematically possible models. That is, the space which represents the basic 'pre-nomic' set of possibilities admitted by the theory. Each of these models can be thought of as something like a bare universe, stripped of laws and dynamics. The constitutive structures are those required to formulate the space, K, of *kinematically possible models* (KPM) of the theory. For the most part we will assume that the constitutive structures consist of: i) a manifold structure used to characterize physical events, and represented by a differential manifold; ii) geometric structures, used to characterize relations of ordering, distance and orientation between the events; and iii) matter structures, used to characterize the non-geometric material content.

Models with different tokens of the same type of constitutive matter and geometric structure are typically constitutively distinct KPMs of the same theory despite the fact that they share the same constitutive structure. Thus, for example, two- and three-body particle models can be understood as having the distinct tokens of a common Newtonian constitutive structure.

---

[1] See in particular Rovelli (2002, 2004, 2007).

The token-type distinction will allow us to distinguish between the following two cases. First we have structure common between all KPMs which share the same *type* of structure. Such structure is constitutively fixed. Constants of nature can typically be understood as an example of constitutively fixed structures of a theory. Second we have structure that is common between all KPMs which share the same *token* of structure but which varies between at least two distinct tokens of the same type of constitutive structure. Constants of motion can typically be understood as an example of constitutively fixed structures of a theory. This distinction allows for differentiation between the modal status of two different types of structure. That is, structure that is the same in all KPMs of the same model type, and structure that can vary between kinematically possible models that instantiate different tokens of that type.

With regard to KPMs, our framework takes a form similar to that of what we shall call the *Standard Formalism* used to characterize the model space of a physical theory. The standard formalism is articulated in terms of a specification of the space of KPMs together with a set of solution-independent 'fixed' (or 'non-dynamical') fields common to all KPMs, and a set of 'non-fixed' (or 'dynamical') fields not common between all KPMs. Constitutively fixed in our sense thus coincides with fixed within the standard formalism. We will discuss this relationship at length in Chapter 5 and return briefly to the concept of fixed structure shortly.

Nomic structure or law-like structure is the structure that represents the laws of a particular theory. In our framework nomic structure will have two important and distinct functions. The first function of nomic structure is to pick out a partition in the space of KPMs: that is, to tell us which models are dynamically possible and which are not dynamically possible. We can formalize this partitioning function in terms of a nomic structure that partitions the space of KPMs into the proper sub-space of dynamically possible models, or DPMs, $D \subset K$. The second function of nomic structure is to provide us with an equivalence relation between DPMs. The equivalence relation function provides a methodology for determining which DPMs are dynamically distinct and which are dynamically identical. The relevant notion of distinctness and identity is here a nomic one rather than an ontic one; that is, a distinction based upon a difference that the laws pick out between two models and not a distinction that is necessarily equivalent to a strong metaphysical notion of distinctness and identity.

We can explicitly formalize the equivalence relation function of nomic structure in terms of a *Projection Map*, π_N. This is some nomic structure that serves to 'project out' distinctions between models according to some dynamical equivalence principle. We can designate the space of equivalence classes $\tilde{D}$ of D the space of Distinct Dynamically Possible Models (DDPMs); this allows us to consider the projection from the space of DPMs to the space of DDPMs: $\pi_N : D \to \tilde{D}$. The role of the nomic structure can be given a schematic representation as per Fig. 1.1 where we have introduced the terminology of a 'fibre' for the equivalence class of DPMs.

1.2.2   Symmetry and Equivalence

The terminology we have introduced thus far already affords us resources to demarcate a number of concepts of symmetry and equivalence. In particular, we are now in a

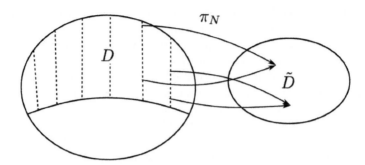

Fig. 1.1: Schematic representation of the nomic structure. D is the partition of dynamically possible models. The dotted lines are 'fibres' that represent dynamically equivalent models which the projection map, π_N, maps into single points in the space of distinct dynamically possible models, $\tilde{D}$.

position to disambiguate the idea of a symmetry as a transformation which 'maps solutions into solutions' via a threefold distinction.

First we can distinguish *non-symmetry maps*. These are maps between DPMs of a theory that are constitutively distinct at the token level; that is, maps between different tokens of the same constitutive structure. These maps take solutions into solutions but are not symmetries in any relevant sense since they do not preserve constitutive structure at the token level. Simple examples include mapping between two-body and three-body solutions in Newtonian mechanics and mapping between Schwarzschild and de Sitter solutions in general relativity.

Second, we can distinguish *broad symmetry maps*. These are maps which transform between DPMs that are constitutively equivalent at the token level. Broad symmetry maps transform between DPMs with the same token of the constitutive structure; for example, maps that transform between the different initial conditions of a three-body Newtonian gravitational system.

Third, we can distinguish *narrow symmetry maps*. A narrow symmetry is a sub-set of the broad symmetries which correspond to maps between DPMs which are constitutively equivalent and dynamically equivalent. The intuitive and vaguely expressed idea of 'symmetry transformation' can be precisified as our concept of narrow symmetry.

Following Belot (2013), it is worth mentioning at this stage of our analysis that the standard approaches to the dynamical formalisation of symmetries in terms of variational symmetries, divergence symmetries, or Hamiltonian symmetries will not on their own furnish a reliable notion of narrow symmetry. This is because, as will be discussed in more detail later, there are always cases where such formalisations of dynamical symmetries lead us to: i) designate as equivalent DPMs which we want to think of as dynamically distinct; and ii) designate distinct DPMs which we want to think of as dynamically equivalent. Our methodology will be to approach the resolution of such problematic ambiguities in stages. The first stage is the formulation of a deliberately too weak preliminary stance towards the classification of narrow symmetries built upon an appropriate refinement of one of the standard formal symmetry definitions. The second stage will then be to modify the nomic structure such that the further physically well-motivated narrow symmetries can be stipulated.

We can specify our weak notion of narrow symmetry via a sufficient but not necessary condition as follows: *preliminary stance on narrow symmetries: all broad symmetries which are variational symmetries (i.e. transformations that leave the variational problem invariant) of the full system under study are narrow symmetries.* Here we are using the word 'stance' following a partial inspiration from van Fraasen's epistemic voluntarism in the context of physical reasoning regarding symmetries, laws, and theoretical models. In particular, we take there to be significant room for 'operation of the will' when it comes to our choice of narrow symmetry principles (van Fraassen, 1989; van Fraassen, 2008; Okruhlik, 2014). Further, the stance is by design a preliminary one, open to revision and modification but constituting a solid working hypothesis that can be used as a first step in the analysis of the narrow symmetries of a physical theory.

The preliminary stance allows us to classify as narrow symmetries the most basic and natural examples of maps between dynamically equivalent models. For example, time-independent rigid Euclidean transformations of a set of gravitating Newtonian point particles are variational symmetries and thus we will recover the expectation that re-embedding the same set of Newtonian gravitational motions in a uniformly spatially shifted space will lead to a dynamically equivalent model of the theory.

The obvious complication for even this simple case are time-dependent transformations such as Galilean boosts. These are not variational symmetries of a set of gravitating Newtonian point particles and thus, under the preliminary stance, models related by such transformations are taken to be dynamically distinct. However, the physical problem at hand may indicate that the relevant transformations *should* be narrow symmetries. This is where the second stage comes in. On our approach, such transformations may be promoted to narrow symmetries by a modification of the variational principle. The narrow symmetries of a theory can be adapted to the particular context in which the theory is being applied, but at the cost of explicitly changing the variational principle used to define the solutions of the theory. Explicit consideration of this case will be provided in §6.5 and §8.5.

1.2.3 Irregular Nomic Structure

The physical significance of the ambiguities in the dynamical definition of narrow symmetry should not be overstated. The primary role of dynamical equivalence principles, and thus the definition of narrow symmetries, within mechanical theory is in the specification of the dynamical evolution of a given system in terms of well-posed equations of motion. From the perspective of the variational problem, the most fundamental subdivision is between those theories in which the nomic structure *enforces* a projection as a requirement for well-posedness and those theories in which the nomic structure which allows for independent well-posed equations of motion for all models which are DPMs.

In the context of Lagrangian action principles this is a distinction which is (almost) entirely between what are usually called 'regular' Lagrangians and 'irregular' Lagrangians. With this in mind we can introduce the distinction between regular and irregular nomic structure. *Regular Nomic Structure* is nomic structure that enforces a partition but not a projection. For a theory with regular nomic structure it is in principle possible to treat all DPMs as prime facie dynamically distinct. However, there are typically good physical reasons to apply a projection based on preliminary stance. It is, however, in principle possible to treat all DPMs as prime facie dynamically distinct, notwithstanding the preliminary stance.

By contrast, *Irregular Nomic Structure* is nomic structure that enforces a partition and a projection. For a theory with irregular nomic structure it is mandatory to classify at least some DPMs as dynamically equivalent in order to construct a well-posed initial value problem. It is in principle then possible to treat all DPMs which are independently well-posed as dynamically distinct. However, again there are typically good physical reasons to apply a projection based on the preliminary stance.

The crucial physical idea is that in theories with regular nomic structure, all degrees of freedom have independently well-posed equations of motion associated with them. For a theory with regular nomic structure, there is an important sense in which there is *formally* no redundancy or underdetermined equations. The class of theories with irregular nomic structure is importantly different. In particular, with a few important exceptions, these theories are such that there exist degrees of freedom which do not have independently well-posed equations of motion associated with them. In such circumstances, we typically encounter a pernicious form of *dynamical redundancy* wherein there exist transformations of the dynamical variables that lead to an underdetermination in the equations of motion. Such redundancy *must* be interpreted as narrow symmetry since otherwise the initial value problem will not be well-posed. The standard terminology for such transformations is 'gauge transformations' and the underdetermined dynamical variables are standardly understood as 'gauge' degrees of freedom.

The interconnected problems of underdetermination, gauge symmetries, and redundancy have been the subject of a large amount of discussion in both physics and the philosophy of physics, both in general terms and in the specific canonical context that is the focus of this book.[2] None of the extant accounts are fully satisfactory, however. Furthermore, as already noted, we take the problem of finding such an account to be the key challenge to the correct diagnosis and resolution of the problem of time in classical mechanics.

The standard treatment of the relationship between symmetry and redundancy derives from the Dirac analysis of constrained Hamiltonian systems.[3] In particular, the key diagnostic tool comes from a prescription due to Dirac for the identification of 'gauge transformations' and the associated redundancy in systems with irregular nomic structure. This approach is built upon the identification of *canonical constraints as generators of gauge transformations*. In brief, the idea is that certain constraint functions that occur in the derivation of the canonical momenta are taken together

---

[2] Extensive reference to the physics literature will be provided in what follows. Key contributions from philosophers are (Belot and Earman, 2001; Belot, 2003; Earman, 2003; Rickles, 2004; Rickles, 2007; Pitts, 2013; Pooley and Wallace, 2022). Our own analysis provided below is based upon an extension of that provided in (Gryb and Thébault, 2014; Gryb and Thébault, 2016a).

[3] The textbook treatments are provided in Dirac (1964), Sudarshan and Mukunda (1974), Sundermeyer (1982), Gitman and Tyutin (1990), and Henneaux and Teitelboim (1992).

with relevant consistency conditions to form an algebra that expresses the failure of the canonical system to be independently well-posed. The solution is then to treat as dynamically distinct only those solutions that are independent of the constraint algebra in terms of the directions of the associated flows on phase space. Conversely, according to the Dirac prescription, phase space points that lie along the orbit of a constraint should be taken to represent physically identical states of affairs. Thus the transformation between such points, as generated by the constraint, is mere gauge.

There are however various general considerations that call into question the Dirac prescription. The first is that the prescription, both in our sketch and in the detailed presentation both by Dirac and by some later authors, elides the significant difference between transformations of instantaneous states and transformations of dynamical histories. Clearly, one would like to be able to move back and forth between the symmetries defined at the level of histories and the dynamical redundancy which exists at the level of the initial value problem. However, considerable care is needed in setting up such correspondences, and on its own the Dirac approach proves too coarse-toothed for the relevant purpose. One goal of our approach is to provide a geometric framework that allows for the disambiguation of the connections between constraints, redundancy, and 'gauge' symmetries at both the level of histories and the level of instantaneous states.

The second more fundamental issue with the Dirac prescription is its implication for theories which are invariant under time reparameterizations. A general feature of such theories is that the Hamiltonian function is itself necessarily a constraint. This then means that the Dirac prescription implies that the Hamiltonian function is a generator of gauge transformations. However, we of course apply the Hamiltonian function also in the role of the generator of dynamical motions. We then arrive at our paradox: motion is the unfolding of a gauge transformation! Here the lines of interpretation become rather criss-crossed. Dirac himself in fact never endorsed such a paradoxical conclusion. Moreover, the argument that he himself provided for the prescription clearly does not support the conclusion either. What is often called Dirac's theorem will be explicitly reconstructed in §7.3. Following on from the discussion of Barbour and Foster (2008), it shall be made explicit in our rendition of the theorem that the result that is proved by Dirac is limited in scope by the assumption that constrained Hamiltonian theory in question contains an external time variable. Dirac's theorem thus simply does not apply to reparameterization invariant theories.

At this point it might be hoped that the ambiguities in this situation can be resolved by appeal to Noether's second theorem. This is because the second theorem establishes a connection between invariance of an action under infinitesimal transformations of an infinite continuous group parameterized by arbitrary functions and the existence of generalized 'gauge' identities which hold irrespective of the satisfaction of the Euler–Lagrange equations. Since reparameterization invariance is precisely an invariance of the action under an infinitesimal transformation of an infinite continuous group parameterized by arbitrary functions, one is thus guaranteed the existence of a gauge identity in this case also. One might expect, then, that the relevant gauge identities allow one to identify dynamical redundancy within the initial value problem and thus remove the ambiguity in the Dirac prescription for the case of reparameterization invariant theories.

The situation is not, however, so straightforward. In fact, new work will be required to establish in general terms the connection between irregular nomic structure, Noether's second theorem, and the existence of *initial value constraints*. It is the latter which are indicative of the specific species of underdetermination problem associated with dynamical redundancy and which are required for the isolation of gauge degrees of freedom. The major formal achievement of this monograph is the establishment of a first-order geometric formalism that provides general criteria for isolating dynamical redundancy. These criteria generalize and supersede those provided by either Noether's second theorem or the Dirac approach. They are also important in identifying the crucial ambiguity regarding the preliminary stance in the context of the distinction between variational symmetries-over-histories and variational symmetries-at-an-instant. As we shall see, reparameterizations should be understood as variational symmetries-over-histories and thus in that sense are narrow symmetries. However, reparameterizations are not variational symmetries-at-an-instant and therefore are not narrow symmetries in that sense.

Our project is to provide new and more fundamental diagnostic tools for interpreting gauge theories. This challenge shall be taken up in two stages. The first in Chapter 8 will be to set out our first-order formalism for the case of theories with a fixed time parameterization. The problem of extending our analysis to reparameterization invariant theories will be the subject of an extended analysis in Part III. Crucial to interpreting the implications of this new formalism will be the new framework for the analysis of symmetry and structure that we will summarize below.

1.3   Transforming Structures

1.3.1   Absolute and Relative Structures

Distinct concepts of 'absolute' and 'relative' are often conflated in discussions of the status of space and time in physical theory. It will prove essential to our discussion to differentiate *three* basic contrasts in the context of our discussions of spatiotemporal structures in physical theory. Here we are taking inspiration from Friedman (1983, II.3). These are: i) absolute vs relational spatiotemporal structure, which is an ontological distinction relating to dependency relations; ii) absolute vs relative spatiotemporal structure, which is a formal distinction relating to non-dynamical transformation properties; and iii) absolute vs dynamical spatiotemporal structure, which is a physical distinction relating to dynamical transformation properties. For clarity, we will make a terminological refinement a rename the trio of distinctions as *substantival* vs. *relational* (ontological distinction), *invariant* vs. *surplus* (formal distinction), and *dynamically absolute* vs *dynamically relative* (physical distinction). With this terminology in hand let us then consider each contrast in a little more detail.[4]

---

[4] There is a vast literature that discusses in different ways and using different terminologies each of these distinctions. Our account is most closely related to that of Friedman (1983, II.3). Further general discussions can be found in Sklar 1974; Earman 1989; Rickles 2007; Maudlin 2012; Pooley 2013; Dasgupta 2015. The literature specifically dealing with absolute and relative spacetime structure in the context of analytical mechanics will be introduced in the context of our specific detailed discussions in Parts II and III.

The first contrast between absolute and relational rest upon a distinction with regard to the 'ontic dependency' between material and spatiotemporal entities. Following Pooley (2013, p. 522), a substantivalist about spatiotemporal structure will maintain that a complete catalogue of the fundamental objects in the universe lists, in addition to the elementary constituents of material entities, the basic parts of the relevant spatiotemporal structure. Relationalists maintain that spatiotemporal structure does not enjoy a basic, non-derivative existence. According to the relationalist, claims apparently about spatiotemporal structure are ultimately to be understood as claims about material entities and the possible patterns of spatiotemporal relations that they can instantiate.

The contrast between relationalism and substantivalism can be connected to different accounts of how one counts distinct ontological possibilities. This is on the basis of the relationalist admitting fewer ontic possibilities than the substantivalist precisely because the relationalist, but not the substantivalist, denies that two states of affairs that are identical in their relational spatiotemporal structure can be ontologically distinct. A direct equation of the views with strategies for possibility counting rest upon an over simplification however, since some prominent substantival views that count possibilities in the same way as relationalists.[5] As we shall explain in more detail shortly, we will seek to 'bracket' the ontological distinction between relationalism and substantivalism in our discussions whenever possible. That the connection between possibility counting and ontology proves to be so underdetermined is arguably a vindication of such a strategy. In any case, the significance for our current discussion is that any ontological distinction between relationalism and substantivalism *does not* have a correlate in the transformation behaviour of spatiotemporal structures under maps between DPMs as characterized in the framework introduced in §1.2.

The second contrast can be characterized precisely in our framework via the behaviour of a spatiotemporal structure under narrow symmetries transformations. As noted above, for clarity, we will refer to the 'absolute' spatiotemporal structures that do not change under narrow symmetry transformations as *invariant* structures and the 'relative' spatiotemporal structures that do change in a well-behaved manner as *surplus* structures. This classification will, of course, only be defined in the context of a definition of narrow symmetries for a particular theory. It will thus inherit all the complexities and qualifications in the definition of narrow symmetries as per our earlier discussions.

The important point is that for any given specification of nomic structure in terms of a partition and projection, one has at hand an entirely unambiguous definition of the maps that transform between DPMs which are identical according to the relevant nomic standard of equivalence. These transformations within an equivalence class are graphically represented as moving along the 'fibres' given by the vertical dotted lines in Fig. 1.1. Invariant spatiotemporal structure is then structure with transformations that are entirely trivial as one moves along the fibre. Surplus spatiotemporal structure

---

[5] Application of this general strategy for substantivalism with reduced possibility counting can be found for instance in Maudlin 1988; Butterfield 1989; Brighouse 1994. See also Rickles (2007) and Pooley (2013).

by contrast is structure that has non-trivial, but well-defined transformation properties as one moves along the fibre.

Finally, we can consider the third contrast that relates to the transformation behaviour of a spatiotemporal structure under 'dynamical transformations', defined as the broad symmetries which are not narrow symmetries. We will refer to spatiotemporal structures that do not change under dynamical transformations as *dynamically absolute* and temporal structures that do change in a well-behaved manner as *dynamically relative*.[6] It is important note here that by our definition the dynamical transformations are maps which transforms between DPMs that are constitutively equivalent at the token level and yet dynamically distinct. This is the complement of the narrow symmetries within the set of broad symmetries.

In parallel to the second 'formal' distinction, in the case of this third 'physical' distinction, it is important that for any given specification of nomic structure in terms of a partition and projection, one has at hand an entirely unambiguous definition of the maps that transform between DPMs which are distinct according to the relevant nomic standard of distinctness. These transformations between equivalence classes are graphically represented as moving between the 'fibres' given by the vertical dotted lines in Fig. 1.1. Dynamically absolute spatiotemporal structure is then structure that is such that its transformations are entirely trivial between the fibres. Dynamically relative spatiotemporal structure by contrast is structure that has non-trivial, but well-defined, transformation properties as one moves between the fibres. Moving between the fibres in space of DPMs D is of course equivalent to moving between points in the space of Distinct DPMs $\tilde{D}$, given by the projection $\pi_N : D \to \tilde{D}$. Thus by our definitions dynamically absolute structure does not change between any point in $\tilde{D}$ but dynamically relative structure displayed well-defined and non-trivial transformation properties under transformations in $\tilde{D}$.

As a brief aside, it is worth noting that dynamical transformations do not constitute the full set of transformations between distinct DPMs. This is because we are also excluding from our definition of dynamical transformations maps between DPMs of a theory which are constitutively distinct; that is, maps between different tokens of the same constitutive structure. These maps take solutions into solutions but are not symmetries since they do not preserve constitutive structure at the token level. Returning to our favourite examples, consider mapping between two-body and three-body solutions in Newtonian mechanics or mapping between Schwarzschild and de Sitter in general relativity. Whilst the invariance or not of spatiotemporal structure in the context of such constitutively distinct models is of great philosophical and physical significance, especially in the context of the problem of providing a substantive definition of background independence, the formal structure of transformations in this wider context is not amenable to the analysis we will provide and will thus be treated as a distinct case from dynamical transformations when required.

To recap, we have introduced three binary contrasts: an ontological contrast between substantival and relational; a formal distinction between invariant and surplus;

---

[6] Dynamically absolute structures in our sense are analogues, in some ways, to so-called Absolute Objects. See Pooley (2017, §7) and Curiel (2019).

	Absolute	**Relative**
Ontological	Substantival	Relational
Formal	Invariant	Surplus
Physical	Dynamically Absolute	Dynamically Relative

Table 1.1 Table illustrating three binary contrasts regarding senses in which a spatiotemporal structure may be absolute or relative. Under the assumption of independence three binary choices yields eight possibilities.

and a physical distinction between dynamically absolute and dynamically relative. For ease of reference and understanding we have set out these distinctions in Table 1.1.

1.3.2   Bracketing Ontology

In principle it is possible for a given spatiotemporal structure to be understood as realized any of the eight possibilities given by arbitrary combinations of the three binary distinctions. For example, in principle, one might hold that a given structure is substantival, surplus, and dynamically relative or relational, invariant, and dynamically absolute. Within the context of our positive project we will make the significant methodological choice to treat the three distinctions, substantival vs relational, invariant vs surplus, and dynamically absolute vs dynamically relative as mutually independent and to focus our analysis on the second and third to the exclusion of the first which will be 'bracketed' in what follows.

This is not to say that we do not recognize that a deep conceptual alliance exists between, for example, relationalism and dynamical relativity. Such connections will be considered at length in our discussion of the views of Newton and Leibniz in Part I. Moreover, in general terms, it is plausibly the case that at least some key arguments are at least in part overlapping. For example, epistemological arguments in favour of relational and invariant structure. However, in our view the relationalist/substantivalist ontological dispute can in fact be largely detached from what we take to the *physically significant* issues regarding the *formal representation* of spatiotemporal structure in physical theory. These issues, in our view, are always issues regarding the invariant/surplus and dynamically absolute/relative distinctions. In line with our programmatic aim of resolving the problem of time in classical mechanics, rather than intervening in debates regarding the metaphysics of time, we will seek to 'bracket' the ontological dispute and focus our energies on the second and third physico-mathematical disputes.

One specific means that we will use to achieve this bracketing is to leave open the representational relationship between distinct DPMs and the ontology they stand in for in the world; that is, we will refrain from assuming an *interpretation* that fixes either a model-world representation relation or standard of representational equivalence between models. It is only given a choice of a particular interpretative attitude towards representation that our notion of dynamically distinct models can be taken to fix a relevant notion of *ontological* distinct models. In particular, so far as the debate between relationalists and substantivalists reduces to the correct ontic possibility, this debate is transformed into a choice of attitude towards representation that we refrain from making. The important point, from our perspective, is that it is perfectly possible for all parties to a dispute regarding ontology and representation to agree on the appropriate criterion of dynamical equivalence whilst disagreeing with regard to the further interpretational move that fixes a notion of ontological equivalence.

1.3.3   Incomplete and Inhomogenous Transformations

A further important point relates to how we will implement the invariant/surplus and dynamically absolute/relative distinctions. It will in fact prove of great utility to understand these as tripartite rather than bipartite distinctions by allowing for the case in which the structure in question do not vary in a well behaved way under the relevant transformation. That is, as well as having structures that do not change (i.e. are invariant/dynamically absolute) and do change (i.e. are surplus/dynamically relative) under the relevant transformations, we will consider structures that are 'incomplete' and thus are 'broken' by the relevant transformation.

As already noted, for any given specification of nomic structure in terms of a partition and projection, one can consider the maps that transform between DPMs which are distinct according to the relevant nomic standard of distinctness and those maps that transform between DPMs which are equivalent according to the relevant nomic standard of equivalence. Recalling Fig. 1.1 once more, the first set of transformations move between the fibres and the second set move along the fibres. We can distinguish *dynamically incomplete* as structure that has ill-defined transformation properties induced by the transformations between dynamically distinct models, corresponding to movement between the fibres (or equivalently between different points in $\tilde{D}$). *Kinematically incomplete structure* which is structure that has ill-defined transformation properties induced by the transformations between dynamically equivalent models, corresponding to movement along the fibres.

Once we bracket the ontological distinction but add the two possibilities for incomplete structure we are left with *two trinary* distinctions. We can think of these six combinations as picking out *invariant* structure that does not vary under narrow symmetries transformations, *surplus* structure that is covariant under narrow symmetries, *dynamically absolute* structure that does not vary in a well defined way under dynamical transformations, *dynamically relative* structure that does vary under dynamical transformations, and then *kinematically incomplete* and *dynamically incomplete* structures that have ill-defined transformation properties under the relevant transformations.

We can now consider combinations of these combinations: structure that is both invariant dynamically and relative or both surplus and dynamically absolute. Since we have two trinary distinctions there are six possibilities. For easy of reference and understanding we have set out these distinctions in Table 1.2. Most significantly within the six possible choices is the combination that we will label *fixed structure*. This is structure which is both invariant and dynamically absolute. Such structure is identical for all DPMs. Recalling our definitions above, clearly all structure that is the same for all KPMs, and thus constitutively fixed or contingently fixed, will be the same for all DPMs, and thus fixed structure. The converse need not hold of course, since structure

	Absolute	Relative	Incomplete
Formal	Invariant	Surplus	Kinematically Incomplete
Physical	Dynamically Absolute	Dynamically Relative	Dynamically Incomplete

Table 1.2 Table illustrating two trinary contrasts regarding the transformation properties of spatiotemporal structure. Under the assumption of independence two trinary choices equates to six possibilities.

that is the same for all DPMs need not be the same for all KPMs at either the token or type level.

A final feature of our framework is the means by which we differentiate transformations that are 'local' from those that are 'global'. Locality in this sense will be defined in terms of the *homogeneity* of the transformations on the space of DPMs. In particular, *global* transformations are those that act homogeneously across all dynamical models and *local* transformations act homogeneously with respect to particular events. This is an essentially formal distinction and we will take some time in the discussion of §9.3 in setting up the relevant mathematical machinery carefully. We will give a brief informal sketch here since the global vs local distinction is of considerable importance to the problem of time in classical mechanics.

The core idea is as follows. Recall that all models, both kinematically possible and dynamically possible, are built out of three levels of constitutive structures: a manifold structure used to characterize physical events; geometric structures, used to characterize relations between the events; and matter structures, used to characterize the non-geometric material content. Consider a particular transformation between two DPMs, $\phi : D \to D$. This transformation acts on the constitutive structures of the two models, respectively transforming the manifold, matter, and geometric structures. In this context we can consider the *point-wise action* of ϕ on the manifold structure; that is, how the transformation ϕ acts on a given point p in the event space.

Let us then consider some coordination of the manifold structure in terms of a *time-like foliation* where we have a series of spatial slices labelled by a time parameter, t; see §9.3 for the coordinate-free version of these ideas. If a transformation is *spatially homogenous* it is unchanged as we vary between spatially separated points on the same spatial slice. Such a transformation will therefore be such that the partial derivative with respect to our coordinates adapted to the spatial slices are zero. We thus have that $\partial_i \phi = 0$ for spatially homogenous transformations. Such transformations are spatially global. Conversely, if a transformation is *spatially inhomogenous* it changes as we vary between at least some spatially separated points on the same spatial slice. Such a transformation will therefore be such that the partial derivative with respect to our coordinates adapted to the spatial slices are not always zero. We thus have that $\partial_i \phi \neq 0$ for spatially inhomogeneous transformations. Such transformations are spatially local. See Fig. 1.2 for a graphical representation of a spatially inhomogenous transformation

Similarly, if a transformation is *temporally homogenous* it is unchanged as we vary between temporally separated points in the manifold structure. Such a transformation

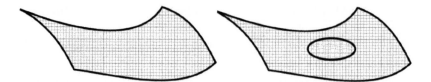

Fig. 1.2: Representation of a spatially inhomogenous transformation of the event space. Left-hand picture is the event space before the transformation. Right-hand picture is the event space after being acted on by the transformation. The transformation acts only on the spatial region inside the 'hole' represented by the oval shape. Inhomogenous transformations formalize the idea of a local transformation.

will therefore be such that the partial derivative with respect to our time parameter is zero. We thus have that $\partial_t \phi = 0$ for temporally homogenous transformations. Such transformations are temporally global. Contrarily, if a transformation is *temporally inhomogeneous* it changes as we vary between at least some temporally separated points in the manifold structure. Such a transformation will therefore be such that the partial derivative with respect to our time parameter is not always zero. We thus have that $\partial_t \phi \neq 0$ for temporally inhomogeneous transformations. Such transformations are temporally local. In general, we will work with a definition such that a transformation ϕ is global if and only if it is both spatially and temporally global and a transformation ϕ is local if it is either spatially or temporally local.

1.3.4 Structure and Heuristics

It is worth saying a few words about the motivation for considering incomplete structures. Why look at spatiotemporal structures within a physical theory that are 'broken' by some elements of the symmetry or dynamical transformations? It is here that the wider context of our project becomes apparent. This is *not* a book about the interpretive analysis of spatiotemporal structure in well-established physical theories. Rather, as noted in the Preface, this is a 'three-field' foundational investigation, where philosophical approaches are integrated with those of history of science and theoretical physics. Ultimately, one of our principal aims is to apply philosophical and historical methods towards the development of novel approaches to outstanding problems of theory construction. Incomplete structures are important because they provide key heuristics for identifying novel routes towards modifying and extending theories. Once more, here we are drawing inspiration from Friedman (2001). In particular, we take identification and modification of incomplete structures to be part of a *regulative metaframework for theory construction* (p.66). We will return to this connection in §10.1.

In general terms, our goal in developing the new framework has been to construct general heuristics for theory extension and re-articulation. Perhaps the most novel of these will be precisely the *completion heuristic* drawn from the analysis of incomplete structures. Further heuristics, which are more familiar, can be identified and systematically categorized within the framework. Of particular importance are the elimination of surplus structure via a *reduction heuristic*, the conversion of invariant to surplus structure by a *surplusing heuristic* which relies upon expansion of the narrow symmetries of a theory, and the conversion of a dynamically absolute structure to a dynamically relative structure via a *relativization heuristic* of which best matching is a specific form. Articulating the basis for these heuristics, and also the attendant drawbacks, will be the central concern in §10.1.

The final heuristic that we shall consider is drawn from the conversion of a structure with global properties to a structure with local properties; that is, modification of the theory and structure such that the transformations on the space of DPMs that the structure is well behaved under are converted from homogenous to inhomogeneous. This is the *localization heuristic* that will provide the final tool that will prove crucial to our analysis of the problem of time.

1.4 Time Regained

1.4.1 The Hierarchy of Structures

A basic presupposition of our approach is that temporal and spatiotemporal structures are defined within the *dynamical* models of a physical theory. Defining things in this manner might seem to indicate that we assume nomic structure to be more basic than temporal. And this might appear problematic since plausibly *some* temporal structure is necessary to formulate nomic structure. However, on our view one should not think of there being a general hierarchy between the nomic and temporal structure. Rather, we take the relationship between nomic and temporal structure as a matter to be investigated a posteriori.[7]

A further immediate concern regarding our classification of structures is whether the geometric structure that is needed to play the constitutive role in defining the KPMs of the theory will also function as at least part of the relevant temporal structure. This will also often be the case. However, what is significant for our purposes is that the constitutive and temporal roles of geometric structures are distinct. In the constitutive role, geometric objects are providing structure to a purely kinematical representation of the universe. As such, there is no necessary connection between the relevant geometric structure and the representation of time within physically possible universes. In contrast, in the temporal role, the geometric structures are, by assumption, representing temporal features of physically possible universes. As with much else, these subtleties will become clear once examples are considered in detail.

1.4.2 The Structure of Time

The primary, and most fundamental, motivation behind our analysis is to provide a suitably precise formal and conceptual representation of the different concepts of time in a physical theory. This conceptual refinement will be achieved by introducing formal representations of temporal structures which correspond to core 'properties of time'. For the most part, these temporal structures will be introduced on a case-by-case

---

[7] A rather different way of approaching the relationship between nomic and temporal structure is that espoused by Callender (2017). On this view, time is distinguished by its 'simplifying' functional role with laws, understood as partial differential equations with a well posed Cauchy problem. For critical discussion of Callender's view see Baron and Evans (2018), and James (2022). These interesting debates are tangential to the questions at issue here.

basis by considering the temporal structures of particular physical theories. They will fall into three general categories: *chronordinal, chronometric*, and *chronobservable*. We hope the reader will forgive our two Greco-Latinate hybrid neologisms. In each case, no suitable word capturing the correct senses could be found and we decided to introduce two new terms following the structure of the existing term 'chronometric'.

Chronordinal structures are structures that provide a representation of time orderings of sets of events. For our purposes, this will principally be undirected total or partial orders. Chronordinal structure provides a means to organize events in an *undirected temporal line*. We neglect here the possibility of non-trivial temporal topology which would complicate things somewhat. This is equivalent to assuming that time has a non-cyclic chronordinal structure. Thus we have that given three events, a, b, c in the domain of a chronordinal structure we will be able to specify, for example, that b is *between* a and c but not necessarily whether it is before or after a. Making things more concrete, chronordinal structure on the space of daily meals will allow one to say that lunch is between breakfast and dinner, but not whether it is before dinner and after breakfast, or after dinner and before breakfast. A structure that is chronordinal need not, therefore, be sufficient to establish a chronology or directed ordering. All chronological structures will necessarily be chronordinal, however, since an undirected ordering is of course necessary but not sufficient for a directed ordering.[8] In the context of the classical mechanical theories that will occupy us for the majority of this book, the crucial means of formalizing chronordinal structure will be in terms of the idea of a *monotonic parameterization*. That is a labelling of the states of a mechanical system that takes the form of a function that is varying consistently so that it either never increases or never decreases. Finally, chronordinal structure as we define it can include the ordering of both local and distant events. This aspect will not be a major focus in the present volume but will of course become highly significant in the full relativistic context taken up in the Volume II of this work.

Chronometric structures are structures that provide a representation of temporal distance relations between pairs of events; that is, they give a real valued assignment of the *duration* between pairs of events. The idea of chronometric structure is a natural generalization of the notion of a spatial distance metric, although of course there are a range of important disanalogies between the idea of a 'distance' and that or a 'duration'. For every (unordered) pair $\{a, b\}$ in the domain of the chronometric structure there is a unique (since we consider non-cyclic time) non-negative real number, the *duration* of the interval between them, τ_{ab} (with $\tau_{ab} = 0$ for $a = b$). To be a properly constituted time metric, this function must satisfy the 'triangle equality': for any triple of events a, b, c, the sum of two of the durations equals the third, $\tau_{ac} = \tau_{ab} + \tau_{bc}$, say. In this case, if the events are distinct, we define b to be between a and c, so that chronometric structure entails chronordinal structure.[9] Significantly, a temporal structure is a chronometric if it provides sufficient structure to define temporal distance relations

---

[8] Our notion of chronordinal structure is closely related to the idea of 'C' theoretic structure of time introduced by McTaggart (1908) and more recently discussed by Farr (2012, 2016). We will comment further on this connection shortly. For discussion of an explicit geometric example of a chronological structure in the context of simple mechanical systems analysed via contact geometry see Sloan (2018).

[9] Thanks to Nick Huggett for input on the definition of chronometric structure.

between events. Thus, a temporal structure may be chronometric whilst also providing further structure, such as a privileged temporal coordinate chart. In general terms we can expect that chronordinal structure is necessary but not sufficient for chronometric structure since the latter will implicitly define the former in anything but the most exotic implementations. It is important for our purposes to note that the notion of chronometric structure is a formal one and need not be connected or even connectable to actual temporal measurements. That is, we will use chronometric structure as a term of art for time-*metric* structure rather than time-*measurement* structure. This brings us to our final notion of temporal structure, chronobservable structure.

Chronobservable structures are structures that function as representations of some empirically accessible aspect of time. This will typically be in terms of a representation of chronometric structure and, as noted above, this means that chronordinal structure will be encoded also. Although in principle one might consider some qualitative chronobservable that represents chronordinal structure without chronometric structure, in the context of mechanical theories such considerations will prove irrelevant. It is important to note here that we are using the term 'observables' as per the standard usage in physics rather than as the term has developed in the literature in philosophy of science. That is, in our usage an observable of a theory is a *empirically quantity that the theory predicts a value for*, rather than some phenomena that is scrutable to aided or unaided human perception. The contrast here can be made by distinguishing our usage from that famously considered by van Fraassen (1980). The classic example of an observable in the sense of van Fraassen is visually accessible phenomena; for example, the phases of Venus themselves seen through a telescope by Galileo. In general, there being no need to privilege sight above the other senses, we can think of observables in the sense of van Fraassen as physical quantities whose value can be directly discerned via the (possibly aided) senses. By contrast, observables in the physicists sense simply correspond to empirically quantities whose value is constrained by the theory given suitable dynamics and initial conditions. This is typically an unambiguous concept in well-formulated physical theories as is vividly realized in Hamiltonian formulations of Newtonian mechanics as the algebra of real valued functions on the phase space and in quantum mechanics as the algebra of self-adjoint operators on the Hilbert space.[10] In the contexts we will consider in this book, the chronobservable structures of a mechanical theory will correspond to sets of temporal quantities that are *observables* by the lights of the theory and will, therefore, usually function as representations of idealized clocks. Typically they will be a sub-set of the observables that implement certain stable relationships with the dynamics such that they 'keep track' of time. It would be naïve, of course, to identify chronobservable structures of a theory with the measurement outcomes of actual physical clocks. This holds just as true for clocks constructed via regular motions under gravity (e.g. pendulums or astronomical clocks) as it does for atoms clocks. In §4.3 we will consider the idea of inertial clocks as an illustration of how approximations and idealizations

---

[10] For further discussion of the seemingly straightforward, but oddly confused notions of 'observable' and 'phenomena' as they occur in the philosophy of science see Massimi (2007) and Evans and Thébault (2020).

accompany even the simplest physically embodied clock system. See Thébault (2021 *a*), for further discussion.

These three categories will in many cases be cross-cutting. That is, in many physical theories we will consider mathematical structures that fit into more than one category. For example, often the same structure will function as a chronometric and chronobservable by providing an observable measure of duration.

1.4.3   Time in Newtonian Mechanics

In order for the reader to have in mind the upshot of the different ideas of temporal structure and how they will interact with the transformation properties we have introduced, we will now briefly sketch the temporal structure of analytical formulations of Newtonian mechanics. This will be an informal overview of the ideas presented in formal detail in §10.2. Our analysis of time Newtonian in mechanical theory will characterize three basic temporal structures.

First we have an undirected chronordinal structure provided by a temporal-ordering structure $>$. This structure is the same for all DPMs. This means that the time-ordering structure is *fixed structure* which does not vary either between dynamically identical models (i.e. it is invariant structure) nor between dynamically distinct models (it is dynamically absolute structure). Furthermore, since in Newtonian mechanics, as we will formulate it, inhomogeneous transformations have been excluded from the set of possible transformations, we automatically have that ordering structure is globally fixed structure, both temporally and spatially.

Second, we will distinguish two forms of chronometric structure. These will be given by a privileged temporal coordinate, t_N, and a Newtonian absolute time, T_N, given by a pairing of t_N and a specific value of an offset that we will label a_0. Let us focus on t_N first. The only transformation that acts on t_N will be the time translations or temporal Leibniz shifts that take the form $t_N \rightarrow t_N + a_0$. However, by definition t_N is invariant under such a translation so we immediately have that the Newtonian absolute time is also fixed structure. Furthermore, recall we differentiated two forms of fixed structure. Structure that is identical in all models that share the same type of constitutive structure is constitutively fixed. Structure that is identical in all models that share the same token of constitutive structure but that varies between at least some tokens, is contingently fixed. The privileged temporal coordinate is constitutively fixed since there is no basis for it to vary between different tokens of the Newtonian constitutive structure. Let us then consider the Newtonian absolute time, T_N. In contrast to t_N we know that changing the offset $t_N \rightarrow t_N + a_0$ corresponds to moving to a different Newtonian absolute time $T_N \rightarrow T'_N$. Thus this structure varies under the relevant narrow symmetry transformation and is surplus structure. Furthermore, since such a transformation is dynamically irrelevant, we know that the structure is also dynamically absolute. Thus the Newtonian absolute time, T_N, is a dynamically absolute surplus structure. Significant for what follows is that both the chronometric structures t_N and T_N have their respective status (i.e. dynamically absolute invariant and dynamically absolute surplus) temporally and spatially globally. Again, this was guaranteed since inhomogeneous transformations have been excluded from our formulation.

Third and finally, we will pick out the chronobservable structures given by *clock observables structure*. Clock observables are a set of phase space functions that satisfy a particular set of relations with the dynamics of the theory. They will be defined such that transformations that leave the nomic structure invariant will also leave the clock observable structure invariant. This will mean that clock observables will be invariant structures in Newtonian mechanics. Clock observables need not, however, be invariant under transformations between dynamically distinct models. We thus will have that the general class of clock observables in Newtonian mechanics are dynamically relative invariant structure. Once more, the clock observables are automatically dynamically relative invariant structure spatially and temporally globally since inhomogeneous transformations have been excluded from in our formulation.

1.4.4   Time in Jacobi Theories

A theory is reparameterization invariant when its equations of motion are invariant under the temporal relabelling symmetry given by locally rescaling a monotonic and smooth time function. We can understand the transition from Newtonian mechanics to Jacobi theory in terms of the localization heuristic discussed above. As noted above, reparameterization invariant theories can be shown in general to be such that their nomic structure is irregular, and, moreover, to be such that they have Hamiltonian constraints. A special class of reparameterization invariant theories are Jacobi theories in which the action generates a geodesic principle. The correct identification of the temporal structure of Jacobi theories requires a solution to the problem of time in that we must find a precise methodology for identifying determinate chronometric and chronobservable structure which is consistent with the symmetry properties of reparameterization invariant theories. Here we will provide an informal summary of the results of our analysis. The formal treatment can be found in §13.3.3.

Let us again start with the undirected chronordinal structure given by a total time-ordering structure $>$. On our approach this structure can be identified as a determinate structure that is invariant under both narrow symmetry transformations and dynamical transformations of the theory. The time-ordering structure is thus *fixed structure* as in the case of Newtonian mechanics. Unlike in that case, however, since inhomogeneous transformations have not been excluded from the set of possible transformations, since we include local rescallings, we automatically have that ordering structure is locally fixed structure, temporally. It is still globally fixed structure spatially.

The next type of structure we will consider highlights a more direct contrast with Newtonian mechanics. This is chronometric structure given by the privileged temporal coordinate t_N. Significantly, we find in Jacobi theories that the chronometric structure of the theory given by the privileged temporal coordinate is incomplete structure. This is hardly surprising since this is not a temporal structure that plays a physical role in the theory.

The crucial conclusion that we reach is then that *the only chronometric structures which are complete in the context of Jacobi theories are those provided by chronobservable structures.* In particular, we can identify clock observables in terms of phase space functions which are automatically invariant under the narrow symmetries associated with reparameterization transformations and are thus complete structures. As with the Newtonian description, clock observables will also be defined such that transformations that leave the nomic structure invariant will also leave the clock observable structure invariant. This will mean that clock observables will be invariant structure in Jacobi mechanics also. Furthermore, also as in Newtonian theory, clock observables in Jacobi theories need not be invariant under transformations between dynamically distinct models. We thus will have that the general class of clock observables in Jacobi theories are dynamically relative invariant structure.This is the same status as in Newtonian mechanics. The crucial difference is that in Jacobi theories this status is temporally local. This marks a precise sense in which Jacobi theory clock observables are more closely analogous to proper time in relativistic theories than their Newtonian counterparts.

1.5   Directed Time Ordering and Time's Arrow

This monograph concerns the analysis of time as represented within the formal structure of classical mechanical theories. One road that we will leave untravelled is the investigation of *directed* chronordinal structures; that is, orderings of events that provide not just a 'betweenness' relation, but also a 'before' or 'after' relation. Directed chronordinal structures are not represented *generically* within the theories that we will concern ourselves with. This leads to another 'problem of time': the problem of the arrow of time, which relates to the recovery of temporally directed chronordinal structure from undirected nomic structure.[11] There is a long tradition of attempts to solve the problem of the arrow of time within physical science by deriving an arrow of time either from special initial conditions or via additional nomic structures (Callender, 2021).

The first approach is most popular and follows a line of reasoning that stretches back to Ludwig Boltzmann. Under Boltzmann's proposal the time symmetry found in nomic structure of a theory is broken by the asymmetrical restriction to possible models that have highly atypical initial but not final states. This kind of explanation is reasonably adequate for simple thermodynamic systems; however, complications arise when attempting to apply this strategy to the universe as a whole. In particular, although one can postulate a highly atypical past state for the entire universe, via what is called the *Past Hypothesis* it is not clear that such a thesis is in fact amenable to a rigorous and physically plausible formalization in the context of actual models of cosmology (Earman, 2006; Schiffrin and Wald, 2012; Curiel, 2015; Gryb, 2021).[12]

The second approach has in the past been conceived in terms of a requirement for a new fundamentally directed law of nature. More recent work, however, points towards the possibility that the apparent time symmetry of general relativity may be an artefact of a representational redundancy and thus towards the possibility that time asymmetry is built into the character of an empirically equivalent nomic structure

---

[11] A related issue is whether the nomic structure of a mechanical theory is time reversal invariant. For highly rigorous and insightful analysis that demonstrates such a property certainly does hold for the nomic structure of analytical mechanics see Roberts (2022, §3.3)

[12] For more positive discussion of the past hypothesis see Albert 2009; Penrose 1979; Penrose 1994; Lebowitz 1993; Goldstein 2001; Goldstein and Lebowitz 2004; Price 1997; Price 2002; Price 2004; Callender 2017. It is worth nothing that responses to, or even substantive discussions of, the problems for the past hypothesis are not provided in this literature.

'hidden' within the equations of motion of general relativity (Gryb, 2021; Sloan, 2021). Such an approach would provide a new basis for providing an explanation for the arrow of time based upon fundamentally directed temporal structure. Furthermore, such a strategy would parallel and further develop the approach suggested by Barbour, Koslowski and Mercati (2014). Whilst there are certain specific formal connections between this approach to the problem of the arrow of time and some of the ideas that we will discuss for the purposes of our current project, we will operate under the working hypothesis that solving the problem of time in classical mechanics is a challenge independent of the problem of the arrow of time, and thus the two classical mechanical problems should be pursued separately.[13]

In more general terms, our discussion will not include consideration of directed 'B-theoretic' temporal structure within the category of chronordinal structures. Rather, our focus when discussing chronordinal structure will always be upon undirected 'C-theoretic' temporal structure. This distinction comes from the classic discussion of McTaggart (1908). Whereas A-theoretic notions of time relate to dynamic notions of time, usually understood in terms of the idea of a privileged temporal *flow* or process of *becoming*, B-theoretic notions of time are fundamentally directed or tensed conceptions of time, and C-theoretic notions of time are bare temporal orderings. Whether or not the metaphysical project of articulating 'A-theoretic' notions of time is a coherent one, it is not relevant to our current concerns which are principally formal and pragmatic. For discussion of the problem of making sense of temporal 'flow' in the context of modern physics see Price (2011), Callender (2017, §5.1 & §11), and Arthur (2019). For a fascinating and wide-ranging philosophical investigation of temporal structure, which includes both B-theoretic ordering structure and more exotic ideas such as branching times, we highly recommend Newton-Smith (1982). For more general discussion on the status of time-ordering structure we strongly recommend the work of Farr (2012, 2016, 2020). Building upon the ideas of McTaggart (1908), Reichenbach (1956), and Black (1959), Farr advocates for a 'C theory of time' in which there exists an undirected causal relation sufficient to define a partial ordering on the space of events. We take our approach to chronordinal structure in mechanical theory to be broadly equivalent to that of Farr. In particular, Farr argues convincingly that C-theoretic structure is sufficient to frame the problem of the arrow of time. Thus, even in the context where one wishes to connect the two problems, at the level of constitutive structure at least, undirected chronordinal structure is the appropriate framework.

---

[13] A notable example of explicit connection between the two in the context of quantum cosmology is due to Kiefer and Zeh (1995).

Part I

NATURAL PHILOSOPHY OF TIME

In Part I we will consider the natural philosophical analysis of time according to Newton, Leibniz, and Mach. We will do this with a view to deepening our understanding of the conceptually complexity of time in mechanical theory. Our Principial aim throughout will be to isolate the challenge of disambiguating a *unique quantitative representation* of time within a Newtonian universe. It is for this reason that we choose to focus on the view of time espoused by Newton himself, and the two critics, Leibniz and Mach, whose influence is most conspicuous in modern critical discussions of the Newtonian natural philosophy of time.

We begin our analysis with consideration of the absolute conception of time defended by Isaac Newton in the famous *Scholium* section that follows the eight opening definitions of the *Principia* (hereafter simply referred to as the Scholium). Discussion of the proper definition of uniform motion pre-dates Newton to some considerable degree (Barbour, 2001; Gaukroger, 2006). However, what can plausibly be taken to be original to Newton is the forging of a close connection between the problem of defining uniform motion and a particular absolute conception of time within a quantitative system of mechanics.[1]

Furthermore, what could also plausibly be taken to be original to Newton is the establishment of an uneasy balance between *two* different conceptions of absolute time. The first is a *dynamically absolute, surplus structure* that encodes a 'true time' of which the sensible measures of our chronometry are mere shadows. The second is a *uniquely defined* yet *dynamically relative* and *invariant structure* corresponding to a privileged organizing abstraction, which in principle could be based upon a time derived from the relative motions of bodies being 'corrected' towards a distinguished, and maximally 'accurate', Newtonian time metric.

Which of the two views Newton is advocating is a matter of dispute within the literature, and there is a high degree of sensitivity to the differences regarding Latin to English translation that we shall make clear shortly. Key points of critical response

---

[1] This is, of course, not to say that the notion of absolute time is original to Newton. Rather, as set out in meticulous detail by Thomas (2018), absolutism about time is far from being a unitary doctrine and has complex and rich historical origins that cross between metaphysics, natural philosophy, and theology.

to Newtonianism about time then rely upon the distinction. In particular, plausibly in their famous correspondence both Leibniz's criticisms and Clarke's defences of Newtonian time make sense only in the context of the dynamically absolute, surplus structure reading. Moreover, reading things the other way round, we shall argue that it is the failure of the Leibnizian conception of time to match the Newtonian sense of time as an organizing abstraction that proves to be the key shortcoming of non-absolute notions of time.

This notwithstanding, from a functional mechanical perspective our two conceptions of 'absolute time' are indistinguishable within the early modern context. It is only within later articulations of the Newtonian formalism during the nineteenth century in terms of inertial clocks that a differing articulation of Newtonian mechanical theory along these two almost parallel lines can be constructed. Ernst Mach, as we shall see, played a crucial role in these debates and the evolution of his response to Newtonianism regarding time closely matches the development of a critical appraisal that focuses on the first and then the second conceptions of how we might interpret Newton's absolute time within a mechanical context. This in turn points to the start of the process of destabilization of the balance between the two conceptions, which we will take ultimately to reach a final point of crisis in the problem of time.

For our purpose, then, the ambiguity as to how we should interpret Newtonian time is more valuable when left unresolved. The competition between the two notions of time provides the essential tension that our analysis will aim to unwind. And plausibly the tension is not in ourselves, but in our stars. It genuinely is ambiguous how we should best understand the absoluteness of time in mechanical theory, and Newton, being well aware of this ambiguity, left open two spaces of distinct interpretations, while his critics such as Leibniz and Mach only pick up on one side of his view. Ultimately, both what Newton precisely thought about time and what Leibniz and Mach thought about Newton, not beyond all conjecture, are matters that will inevitably prove impossible to completely resolve. However, in grappling with them for the next few chapters we will find conceptual responses which will prove invaluable when we eventually return our discussion to the context of the analysis of symmetry and evolution in formalised modern formulations of mechanical theory.

2
On Newton On Time

According to Newton, time is 'absolute, true, and mathematical'. There is a good basis to interpret 'true' here as involving a sense of universality or being defined globally throughout space. The assertion of a true time can be taken to be predicated upon the assumption that the ordering or chronordinal structure of time is dynamically absolute in the sense of not varying between universes that instantiate different sets of dynamical motions, and invariant in the sense of being not varying between universes that differ solely in virtue of the situation in time of the same set of dynamical motions. In Newtonian time, chronordinal structure is fixed structure. The posit of a 'mathematical' time can in turn be taken to indicate that there is a single time parameter with uniform metric structure: a chronometric structure of time. This temporal structure can then be assumed to be dynamically absolute but not invariant.

2.1 On Newton's Methodology in the Scholium

Before we consider Newton's view of time itself in detail it will prove instructive to first consider briefly a more general question regarding Newton's methodology. In particular, the relationship between his metaphysics and his physics as seen in the context of the Scholium.[1] The Scholium contains definitions, arguments, and discussions relating to time, space, place, and motion.[2] Key throughout is the distinction between the 'absolute' and 'relative' versions of each of these concepts.

On one reading, Newton can be taken to be arguing *from* the existence of centrifugal forces (i.e. the famous examples of the rotating bucket and rotating globes) *to* the metaphysical doctrine of his absolute space. Notably, on such a metaphysical reading the treatments of temporal and spatial concepts look rather different. Whereas the definitions of time and space (paragraphs 1 and 2 respectively) have a very similar structure, the arguments towards these conceptions are very different. On this interpretation, Newton's argument towards absolute space ultimately rests upon the distinction of absolute and relative motion, which in turn is justified by reference to the difference between the two as to their 'properties, causes, and effects' (paragraph 8) as demonstrated by the examples of the bucket and the globes (paragraphs 11–14). In contrast, absolute time is not directly argued for. Rather, in the relevant passage (paragraph 5, quoted in full below), Newton first notes that astronomy distinguished

---

[1] All references are to Newton (1999) with the paragraph numbering following that given by Newton—i.e. the first paragraph is unnumbered, paragraph 1 is that numbered 1, the next paragraph is then numbered 2, and so on to 14.

[2] It is important to note that these 'definitions' are not intended as formal definitions as occur elsewhere in the *Principia*.

between absolute and relative time. He then gives general grounds for maintaining the distinction based upon the possibility that there should be no uniform motion that can play the functional role of marking out equal measures of time.

We will discuss the specific implications for this argument regarding time shortly. What is significant for our current purpose is the *style* of argument. With regard to time, this is evidently not an argument towards a particular metaphysics of time based upon empirical evidence or inference to the best explanation. Rather, Newton can be understood as offering a defence of his conception of time as a necessary presupposition for a successful theory of dynamics. Given the evident trouble that Newton has gone to in making the treatments of time and space parallel to one another, this seeming mismatch in style of argument gives us immediate cause to question the metaphysical reading of the arguments regarding space.

What if we attempt to reinterpret the treatment of space and motion in a manner that is more consistent with that of time? That is, we take the discussion of space and motion also to ultimately have as its goal not the establishment of the metaphysical doctrine of absolute space, but rather the necessity of acceptance of Newton's definitions for an adequate theory of mechanics.[3] Considerable plausibility for this interpretation is provided by the analysis of the unpublished *De Gravitatione* (Newton, 1962, pp. 121–156) wherein isolation of specific failures in Descartes' purely relative definitions is used, in part, to motivate their absolute counterparts.

In its strongest form, such a 'dynamical' interpretation of the Scholium takes Newton to be *reducing* the meaning of the concepts of time, space and motion to their successful dynamical application (DiSalle, 2006; DiSalle, 2016). This would be to strip away the metaphysical components of these concepts entirely. More moderately, we could follow a 'weak' dynamical interpretation and take Newton to be offering a *partial definition* of these concepts via their dynamical application, but still leaving open space for richer definitions that include some strictly metaphysical aspects (Huggett, 2012).

The strong dynamical interpretation is not, however, entirely textually adequate, even as a reading of the Scholium taken in isolation. Consider the discussion of paragraph 6:

Just as the order of the parts of time is unchangeable, so, too, is the order of the parts of space. Let the parts of space move from their places, and they will move (so to speak) from themselves. For times and spaces are, as it were, the places of themselves and of all things. All things are placed in time with reference to order of succession and in space with reference to order of position. It is of the essence of spaces to be places, and for primary places to move is absurd. They are therefore absolute places, and it is only changes of position from these places that are absolute motions.

As noted by Huggett (2012), this passage makes sense only in the context of a reading under which Newton attributes to space and time something more than the properties exhibited by mechanical processes.

The absurdity follows because (primary) places are ('as it were') the places of all things including themselves; so if they moved—i.e., changed places—they would move 'from them-

---

[3] This style of reading owes much to the enormously influential paper of Stein (1967). Here and below we have drawn heavily upon Rynasiewicz (1995a, 1995b), Stein (2002), DiSalle (2006, 2016) and Guicciardini (2016). See also Huggett (2012) and Slowik (2016). A fascinating discussion of Newton's methodology with specific reference to gravity and cosmology is Harper (2011)

selves', which is nonsense. It is the idea that parts of space have essences which is relevant to us; for something cannot be said to have an essence unless it is in some sense a thing. So Newton is apparently saying more than that the world in some way has the *structure* of Newtonian spacetime, he is saying that it has it as the structure of some thing—be it accident, substance, or something else.[4]

Following Huggett (2012), the inference is then, *pace* DiSalle, that paragraph 6 indicates that the parts of Newtonian space and time are taken to have *parts* which have *essences*, and thus, as concepts, space and time cannot be reduced to the dynamical structure that is encoded in their application within mechanical theory.[5] Significantly, the issue at dispute here does not depend upon the interpretation of the parts of space and time as *points* but rather the Huggett interpretation of paragraph 6 requires that the parts of space and time can be understood as components which do not depend upon each other—or any mechanical bodies—for their *essence as absolute places*.[6]

The metaphysical reading of the parts of space and time is strongly supported by the way in which Newton talks about 'particles of space' and 'moments of duration' in the theological discussion of space and time within the General Scholium which was added to the 1713 edition of the *Principia*:

He is eternal and infinite, omnipotent and omniscient, that is, he endures from eternity to eternity, and he is present from infinity to infinity; he rules all things, and he knows all things that happen or can happen. He is not eternity and infinity, but eternal and infinite; he is not duration and space, but he endures and is present. He endures always and is present everywhere, and by existing always and everywhere he constitutes duration and space. Since each and every particle of space is *always*, and each and every indivisible moment of duration is *everywhere*, certainly the maker and lord of all things will not be *never* or *nowhere*.[7]

Reading this passage in combination with paragraph 6 makes is difficult to avoid an interpretation in which the parts of space and time not only have essences as 'places of themselves and of all things' but also identity over time (for the particles of space) and identity over space (for the moments of duration).

In this context it is worth considering the countervailing interpretation in which Newton's metaphysics involves merely conceiving of time as an 'order of successions', and the 'parts of time' as being merely individuated solely in virtue of by their place in this order. Such a reading has been most famously advanced by Stein (1967) based on a passage from *De Gravitatione* that runs as follows:

For just as the parts of duration derive their individuality from their order, so that (for example) if yesterday could change places with today and become the later of the two, it would lose its individuality and would no longer be yesterday, but today; so the parts of space derive their character from their positions, so that if any two could change their positions,

---

[4] Huggett (2012), emphasis added, p. 211.

[5] For more on the interpretation of paragraph 6 see Huggett (2008). Although Newton does not explicitly indicate in paragraph 6 that the parts of time have essences, he does indicate that times, like spaces, are the *places of themselves*, and it is thus a plausible inference to take him to believe that the parts of time also have essences. Thanks to Erik Curiel for discussion on this point. An alternative interpretation of this passage, is that Newton, following Gassendi, denies that space is either a substance or an accident. Points cannot be parts of space, since they are not homogeneous with them; but if the order of these parts (assume they are extended) is absolute, and each can be individuated (by God, presumably), then where they are in the order is what they are. Thanks to Ric Arthur for suggesting this reading.

[6] For detailed discussion relevant to the interpretation of the 'parts of space' in Newton see Slowik (2011).

[7] Newton (1999, p. 941), original emphasis.

they would change their character at the same time and each would be converted numerically into the other. The parts of duration and space are only understood to be the same as they really are because of their mutual order and position; nor do they have any hint of the Earth and on the stars, and throughout all the heavens. And just as we understand any moment of duration to be diffused throughout all spaces, according to its kind, without any thought of its parts, so it is no more contradictory that Mind also, according to its kind, can be diffused through space without any thought of its parts' individuality apart from that order and position which consequently cannot be altered.[8]

On Stein's reading, Newton's view of space and time is then 'strikingly similar to Leibniz' and includes elements of what would be understood as a relational view in modern terminology.

Following Arthur (2021, pp. 90–92), a clear line of differentiation between Newtonian and Leibnizian-relationalist views of time can be established as follows. For Newton, unlike Leibniz, time is not *defined* to be order of succession. Moreover, the 'parts of duration' are not equivalent to the locations in time of any 'thing', even if they get their individuality from positions in an order. This is because for Newton duration is a mode of existence of God, and God exists independently of things. Thus, Newton's 'order of successions' obtains among the 'parts of time' *considered prior to any states of things that might individuate them.* This is the key metaphysical feature of Newtonian time which can be identified in the paragraph 6 and the General Scholium. It means that the parts of time, although they may well be individuated by the place in an order, are intrinsically distinguishable and obtain the individuality prior to their being occupied by things or events (Arthur, 2021, pp. 91–92).

More generally, placing what Newton says about space and time in the Scholium, especially paragraph 6, in the context of Newton's wider metaphysical and theological views on space and time provided in *De Gravitatione*, the *Optiks* (Newton, 1952, p. 403), and the rest of the General Scholium, serves to significantly weaken the claim that we can make a clean separation between Newton's 'onto-theology' and his natural philosophy (Ducheyne, 2006; Janiak, 2008; Slowik, 2009; Slowik, 2016).[9] That said, even if it may not be entirely possible to disentangle Newton's onto-theological and dynamical projects, there are good reasons to take the latter, rather than the former, to be the Principial focus of the Scholium. In particular, the *arguments* given in the Scholium do indeed almost entirely (with the noted exception of paragraph 6) rest upon the functional role of absolute space and time in mechanics, astronomy, and chronometry, and can therefore reasonably be assessed independently on that basis. Thus, a 'weak dynamical' interpretation of the Scholium can still plausibly be adopted and will serve us well for our purpose of using the Newtonian natural philosophy of time to better understand the genesis of the problem of time in modern physics.

2.2 The Tripartite Nature of Newtonian Time

With these subtleties of interpretation kept in mind, let us consider in detail the famous passage (paragraph 1) of the Scholium relating to Newtonian time:

---

[8] Newton (1962, pp. 136–7).

[9] For historical roots of Newton's temporal metaphysics see Arthur (1995, 2019) and Thomas (2018).

Absolute, true, and mathematical time, in and of itself and of its own nature, without reference to anything external, flows uniformly and by another name is called duration. Relative, apparent, and common time is any sensible and external measure (exact or nonuniform) of duration by means of motion; such a measure—for example, an hour, a day, a month, a year—is commonly used instead of true time.

First a note on translation. Here we have opted for the literal translation of the phrase 'seu accurata seu inaequabilis' as 'exact or nonuniform' rather than 'precise or imprecise' as per the Cohen–Whitman translation (Newton, 1999). This translation appears more suitable in the context of the discussion of 'nonuniform' motions in paragraph 5 which we will consider shortly. The older Motte–Cajori translation (Newton, 1962) has 'accurate or non-equitable' which captures essentially the same point. What is important here, and in the parallel part of the discussion in paragraph 5, is that Newton is contrasting those motions which can be used to mark out equal time intervals (i.e. inertial motion or rotations with constant angular-momentum) with those non-uniform or non-equitable motions which cannot.[10] We will comment further on a more significant ambiguity of translation shortly.

Then to interpretation. The important thing to note is that the distinctions offered are three-fold; that is, between times that are absolute or relative, true or apparent, and mathematical or common. There are good grounds to treat these as three separate but interdependent distinctions and thus to understand Newton as giving a *tripartite definition* of time (Huggett, 2012; Schliesser, 2013; Brading, 2016; Thomas, 2018). Let us examine, one by one, what Newton means by time being absolute, true, and mathematical, noting the interrelations and implications as we go.

Absolute can be read in *three* importantly different ways. First, and most straightforwardly, we can think of absolute as a *dynamical* term; absolute temporal structure is insensitive to changes in the motions undergone by material bodies. A dynamically absolute temporal structure will not vary between two dynamically distinct universes related by a change in the motions of some particular set of bodies. Second, we can think of absolute as an *invariance* term: absolute temporal structure is invariant under changes in the situation in time of the same set of dynamical motions. An invariant temporal structure will not vary between two dynamically identical universes related by a change in the situation of the same set of motions within time, for instance as induced by a relabelling of all times by moving everything forward in time by a set amount. Third, and finally, we can think of absolute as an *ontological* term: an absolute time is ontologically independent of the changes of material things. In modern terminology this is to hold that time is *substantival*.

Each of these can be contrasted with a 'relative' notion of time. Time being *dynamically relative* would then be sensitivity of temporal structure to changes in the motions undergone by material bodies. A dynamically relative temporal structure will vary between at least two dynamically distinct universes related by a change in the motions of some particular set of bodies. Time being relative in the sense of non-invariant temporal structure would mean we can consider two distinct universes given respectively by the different situation in time of the same set of dynamical motions;

---

[10] Thanks to Tzu Chien Tho both for discussion of this translation issue and for passing on helpful correspondence with Niccolo Guicciardini on this point.

temporal structure would then provide a *surplus structure*.[11] Finally, time being relative in the ontological sense would mean that its existence depends upon changes in material things. This is the view of the ontology of time usually called *relationalism*, and we will use relationalism to indicate the ontological anti-absolutist view henceforth.

The substantival sense of time is of particular significance in the context of the implications for the surplus (i.e. non-invariant) categorization of temporal structure in Newtonian theory; in particular, that time has parts with essences has the plausible implication that it is possible for two universes, consistent with the laws of nature, to differ solely in virtue of the situation in time of the same set of dynamical motions.[12] There is thus a potential for tension between two of the senses of absolute at work here: absolute qua invariant is plausibly in tension with absolute qua substantival. We will return to the path towards reconciling this tension shortly.

Let us now turn to the question of how we should understand what Newton means by time being 'true'. One reading is to take 'true', as it is used in the Scholium with regard to time, motion, and space, as simply synonymous with absolute. Another, is that true time is the time that obtains for an infinite universe (Schliesser, 2013). The final, and in our view, most plausible and most textually well-supported interpretation is that Newton means something like 'privileged', 'unique', or 'universal' (Huggett, 2012; Brading, 2016). The implication would then be that although there could be many different 'apparent' times associated with a physical system, there is only one 'true' time. This final view seems to fit best with the discussion of true motion and will be adopted here.

There is an important connection between time being 'true' in this sense and the idea of time ordering or chronordinal structure. Most directly we have the connection between a time being 'true' and a time ordering obtaining globally throughout the entire universe (i.e. everywhere in space).[13] This is of course indicated explicitly in the quote from the General Scholium we provided above. Moreover, Newton's claim of there existing a single true time plausibly requires chronordinal structure which is both well defined throughout the universe and dynamically absolute in the sense of insensitivity to differences with regard to the dynamical motions. This is because if time were merely indexed to a particular dynamical change, then there is no reason to expect it to provide a unique temporal ordering. In asserting that time is true we can thus understand Newton as assuming that *the chronordinal structure of time is dynamically absolute*. Thus the interaction between the absolute and true aspects of time provides us with the first example of the crucial analytical idea of understanding

---

[11] We use this term to avoid introducing further terminology to an already complex debate. We do not endorse however the negative implication of dispensability that surplus implies. We will return to the crucial questions of the dispensability (or not) of surplus structure in Part III.

[12] Whilst the denial of such a possibility has been shown to be logically reconcilable with the substantialist view within modern debates, in virtue of a denial of merely haecceitistic differences, it would require substantive interpretative work to understand Newton as endorsing such a position, cf. Pooley (2013).

[13] We might also assume such chronordinal structure is globally well defined *throughout time*. Although this idea does not, to our knowledge, appear in early modern discussions it will be of particular significance for our project later.

a view of time in terms of the *properties of temporal structures*. On our analysis, in being true, Newtonian time is such that it requires chronordinal structure that is dynamically absolute.

A similar argument allows us to categorize the chronordinal structure of Newtonian time as invariant structure. That is, if we take time to be true in a modally thick sense, then time-ordering structure will not only be insensitive to differences with regard to the dynamical motions but also with regard to differences between two dynamically identical universes related by a change in the situation of the same set of motions within time. Thus, on our analysis, in being true, Newtonian time is such that it requires chronordinal structure that is invariant. Temporal structure that is dynamically absolute and invariant plays a special role in physical theory and for this reason we will categorize it under the specific name of *fixed structure*.

Finally, we can consider what Newton meant by time being 'mathematical'. According to the interpretation we are following, what Newton has in mind here is that mathematical time, as opposed to common time, has a *uniform metric structure* and can be represented as a single independent parameter in an equation.[14] We can formalize this claim in terms of the existence of a *dynamically absolute chronometric structure*. What this means is that in contrast to common time, mathematical time provides a single quantitative measure of duration that is applicable to all motions. Thus, the quantity of one unit of mathematical time has the same meaning when applied to the suns motion as to the earths. By contrast, one unit of common time, as given by the motion of a particular body (say the day given by the diurnal rotation of a celestial body) does not *even in the case of a uniform motion* provide a uniform measure of time in the straightforward sense that it is quantitatively distinct from the corresponding measure for a different body that is also assumed to be moving uniformly.

It is worth mentioning here, by way of comparison, the discussion of motion in a central potential in Book 1 of the *Principia*. There Newton makes us of an entirely different methodology for quantifying units of time via the Kepler law of equal area.[15] In this context, following Guicciardini (2003, p. 42 and p. 246), it is plausible to think of Newton as exploring the possibility that the role of an independent time variable can be played by a physical time given by the area swept out by a radius drawn from the centre of the force (one focus of the ellipse) to the body that is moving. Time is then a geometric quantity which depends upon the dynamics.

Newton was not, of course, able to identify a geometric quantity able to play this role in general. In particular, for cases such as perturbed Keplerian motion in empty space, or motion in resisting media, no geometric quantity that depends upon the

---

[14] A further possibility, explored at length by Arthur (1995, 2019) is that by mathematical time Newton also is indicating a further ontological structure of time related to 'intrinsic flow', as founded on his method of fluxions, see also Guicciardini (2016). This would be a further metaphysical element to Newtonian time, in addition to, but connected with, the substantival element discussed above. For our purposes it will be safe to take a neutral stance regarding Newton's commitment to flow. For more details see Arthur's fascinating discussion.

[15] For analysis of the parallel discussion in Newton's earlier work *De Motu* see De Gandt (2014, p. 23).

dynamics could be identified as uniformly flowing.[16] This calls for a different strategy in which the independent time variable is not connected to any physical motion or geometric quantity that depends upon the dynamics. Rather, in the *Principia*, Newton assumes equal increments of the abscissa (i.e. distance along the x-axis) to play the role of the independent time variable that flows uniformly. This parallels his mathematical writing where a 'uniform fluxion' provides the uniform independent time variable, and thus a mathematical time.[17]

The crucial point, once more, is that, in virtue of being mathematical, Newtonian time is asserted not only to possess determinate chronometric structure but it must be assumed that such structure is insensitive to dynamical motions. We again have the idea of properties of temporal structures. Mathematical time is not a geometric quantity which depends upon the dynamics but rather a mathematical quantity independent of dynamical motions. On our analysis, Newtonian time is such that chronometric structure is dynamically absolute.

We can also consider whether the structure in question is invariant or surplus. In other words, does the relevant aspect of time vary between two dynamically identical universes related by a change in the situation of the same set of motions within time. This is where the substantival sense of time proves of significance. In particular, that time has parts with essences has the implication that it is possible for two universes, consistent with the laws of nature, to differ solely in virtue of the situation in time of the same set of dynamical motions.[18] Such differences can involve chronometric differences provided the structure used to represent temporal distances is not itself invariant. This is precisely to consider the implications for chronometric structure of what we will later characterize as a Leibniz shift. With this in mind, we will assume that Newtonian time should be taken to be such that it includes chronometric structure, such as a privileged origin, which varies between dynamically identical universes. We thus categorize the chronometric structure of Newtonian time as dynamically absolute surplus structure.

2.3 Was Newton a Functionalist About Time?

Was Newton in fact seeking to *reduce* the meaning of the concepts of time to its dynamical mechanical *function*? This would be to adopt the opposite interpretation that we have offered and take Newtonian 'absolute' time to be dynamically relative and an invariant structure, and not to commit to any form of metaphysical thesis regarding time or its parts at all. The idea that time could be *both* purely functional in the strong dynamical reading sense of DiSalle (2006, 2016) and yet still mathematical and true serves as a tantalizing potential. This is true whether or not we can read Newton as in fact having this as his fully formed view. Further consideration of this dialectic will

---

[16] In fact, according to Guicciardini (2003, p. 246), attempts to follow this physical time strategy in Proposition 10 of the 1687 version of the *Principia* lead Newton into formal problems that he later corrected using a 'mathematical time' approach.

[17] See De Gandt (2014, pp. 209–213) for an extensive discussion.

[18] Whilst the denial of such a possibility has been shown to be logically reconcilable with the substantialist view within modern debates, in virtue of a denial of merely haecceitistic differences, there seem no good grounds to interpret Newton as endorsing such a position, cf. Pooley (2013).

thus serve a crucial argumentative purpose within the project of this book. What will prove of particular significance in this context are contested questions of *interpretation* of paragraph 5 of the Scholium.

Let us consider the rendering of paragraph 5 in full in the two principal scholarly translations. First in the Cohen–Whitman translation:

In astronomy, absolute time is distinguished from relative time *by the equation of common time*. For natural days, which are commonly considered equal for the purpose of measuring time, are actually unequal. Astronomers correct this inequality in order to measure celestial motions *on the basis of a truer time*. It is possible that there is no uniform motion by which time may have an exact measure. All motions can be accelerated and retarded, but the flow of absolute time cannot be changed. The duration or perseverance of the existence of things is the same, whether their motions are rapid slow or null; accordingly, *duration is rightly distinguished from its sensible measures and is gathered from them by means of an astronomical equation*. Moreover, the need for using this equation in determining when phenomena occur is proved by experience with a pendulum clock and also by eclipses of the satellites of Jupiter.[19]

Passages have been marked with emphasis to point to the contrast with the older Motte–Cajori translation which renders paragraph 5 as:

Absolute time, in astronomy, is distinguished from relative, *by the equation or correction of the apparent time*. For the natural days are truly unequal, though they are commonly considered as equal, and used for a measure of time; astronomers correct this inequality that they may measure the celestial motions *by a more accurate time*. It may be, that there is no such thing as an equable motion, whereby time may be accurately measured. All motions may be accelerated and retarded, but the flowing of absolute time is not liable to any change. The duration or perseverance of the existence of things remains the same, whether the motions are swift or slow, or none at all: and therefore this *duration ought to be distinguished from what are only sensible measures thereof; and from which we deduce it, by means of the astronomical equation*. The necessity of this equation, for determining the times of a phenomenon, is evinced as well from the experiments of the pendulum clock, as by eclipses of the satellites of Jupiter.[20]

Common to both translations is the crucial *empirical* requirement of determinate chronometric structure in allowing one to deduce/determine when a phenomenon occurs. This is the connection between chronometric structure and chronobservable structure, the distance measures of time within equations and the durations as represented by physical clock systems. This is an argument that without a formal means to aggregate the sensible measures of duration based upon the motion of different bodies, we will not be able to save the relevant temporal phenomena. The substance of this argument stands independently of questions of translation and interpretation and we will return to its implications in the context of the correspondence between Leibniz and Clarke in §3.3. The interpretative question relates not to the determinacy of the chronometric structure but rather to its status as dynamically absolute and surplus.

The Cohen–Whitman translation lends itself more naturally to a reading in which chronometric structure is dynamically absolute and surplus. In particular, the reference to 'truer time' and the 'gathering' of duration from its sensible measures is consistent with the idea of time as dynamically absolute, surplus, and substantival that we argued for above. Moreover, the Cohen–Whitman rendering of paragraph 5 is not in explicit

[19] Newton (1999), emphasis added.
[20] Newton (1962).

tension with time being independent from the motion of bodies nor it being possible for two universes, consistent with the laws of nature, to differ solely in virtue of the situation in time of the same set of dynamical motions. When read together with paragraph 6, and the full context of Newton's onto-theology, the Cohen–Whitman translation of paragraph 5 creates no tension for an interpretation of Newtonian time as articulated in the last section.

In contrast, our reading is at least in tension with the Motte–Cajori translation of paragraph 5. In particular, the reference to 'correction' and 'more accurate time' makes good sense in the context of the strong dynamical reading. As emphasized by Julian Barbour (personal communication), the Motte–Cajori translation allows for a reading in which Newton is at least cognizant of the possibility of building a maximally accurate 'corrected' time purely based upon relative motions. Such a measure of duration would then be a chronometric structure that was dynamically relative and yet functionally equivalent to a dynamically absolute time. It is worth noting here that the use of 'deduced' in the Motte–Cajori should be read in the context of a wider meaning of 'deduction', in common usage at the time, which includes inductive inference, and thus not restricted to logically valid inferences as in modern usage (Harper, 2011, p. 38). This is significant since it further supports the reading of Newtonian time in which the assertion of absolute time is a claim regarding the limit of a sequence of increasingly refined abstractions, rather that then 'deduction' of an ontological posit.

Significantly, the functionalist reading of paragraph 5 in the Motte–Cajori translation also allows us to draw close parallels between Newton's conceptualization of time and the view of time we will attribute to the 'mature' phase of the thought of Newton's great nineteenth-century critic, Ernst Mach. This is with some irony since, as we shall see, the German translation of paragraph 5 that Mach himself worked with does not allow such a reading. In particular, the rendering of paragraph 5 (which might plausibly be categorized as a mistranslation) makes reference to a *non-comparative* 'true' time that all but rules out the functionalist reading.

These issues will take on particular relevance in the context of our discussion of inertial clocks. As emphasized by DiSalle (2006, 2016), cf. Friedman (2001, p. 76), there is a natural connection between the functionalist reading of Newton and the temporal metric given by the motion of an idealized force-free particle moving at a constant velocity. Such a reading motivates us to anchor the conception of time put forward in the Scholium firmly within the physics of the *Principia* more generally, in particular in the laws of motion and the crucial distinction between inertia and non-inertial motion. It also importantly foreshadows work on the definition of absolute time via inertial clocks in the nineteenth century that the mature Mach was responding to. We will consider these ideas in the context of our discussion of Mach in Chapter 4.

2.3.1   Shadows of the Problem of Time

Our project here does not depend on settling these subtle issues of translation and interpretation. In fact, the tension that exists within the differing translations and differing interpretations of paragraph 5 will prove highly congenial to our argumentative ends. It seems plausible, surely, to interpret Newton himself to have had a degree of tension *within his own mind* regarding whether his absolute time need be considered, in our terminology, a dynamically absolute, surplus, and substantival posit or rather a unique but dynamically relative and invariant structure. Ultimately the crucial connection here is to the *pragmatic necessity* of determinate chronordinal structure and the *empirical necessity* of determinate chronometric structures. Both forms of temporal structures are required for viable mechanical theory.

As we saw at the end of the last section, there is certainly good evidence that Newton wanted to find a way for the independent time variable in his equations to be connected to a physical motion or geometric quantity that depends upon the dynamics. However, such a representation of time would be required to be determinate in terms of the induced chronordinal structure, or else the practical problem of organizing motions of different systems would be insoluble. It would also be required to be determinate in terms of the induced chronometric structure, or else we would not have a common increments of time, commensurable between different mechanical motions. However, such a conception of time was not formally realized in the *Principia*, and it is simply unclear the extent to which Newton believed that a *uniquely well-defined* 'correction' procedure exists which allowed one in principle to deduce a single absolute time from mechanical motions. Certainly he does not provide an explicit characterization of such a procedure. Our settled view is that these difficulties in translation and interpretation reflect a genuine ambiguity that is best left unresolved. And ultimately we will take the tension between the two sides of Newtonian time to be the first indication of a longstanding problem for the natural philosophy of time, of which the problem of time in contemporary physics provides the final crisis point.

A further important point of interpretation, which further evidences the dynamically absolute reading of Newtonian time, relates to an important *modal* claim at the heart of the argument of paragraph 5. The key issue is to understand what Newton means by the idea that it is *possible* that there is no uniform motion by which time may have an exact measure. The modality in question here is most plausibly understood to be physical possibility rather than epistemic uncertainty since in Proposition 17 (Theorem 15) of Book III of the *Principia* Newton asserts that the diurnal rotations of the planets around their axes are examples of uniform motion.[21]

The modal claim can then be read in two different ways. First, we could take Newton here to be arguing that since the Earth's diurnal motion is non-uniform and must be corrected 'on the basis of a truer time', it is possible that the same *could have been* true of all the celestial motions, even though, in actual fact, it is not. Read in this manner this is rather a weak argument. Unlike in the case of space, Newton is not providing examples of physical situations for which a purely relative account is deficient. Rather, he would be simply inferring from the fact that there are possible scenarios in which there is no physical motion that can be used to give an exact measure of time that the assumption of an absolute, true, mathematical time is a necessary component of an adequate system of mechanics.

---

[21] This crucial observation is due to Rynasiewicz (1995*a*), footnote 19. In contrast, Newton's great contemporary, John Locke, does appear to advance the epistemic form of the same claim (Locke, 2008, II XIV §21). For more on the subtle relationship between Newton and Locke's views on time see Thomas (2018, §7).

A second, more plausible, reading of the modal claim runs as follows: the problematic situation that Newton is foreseeing for the dynamically relative notion of time is based not on the existence of no uniform motions at all, but rather that of constructing functionally adequate measures of time in a context which combines uniform and non-uniform motion. Such a reading evidently makes better sense in the context of Proposition 17 of Book III and leads to a precise negative claim that dynamically relative accounts do not provide a general means by which to equate different common times when there is the need to correct the times given by non-uniform motions. Although they may have access to local uniform motions that are functionally adequate for providing a temporal metric for local change, in the context where at least some motions are non-uniform, a dynamically relative account is taken to be unable to provide a basis to make the relevant corrections that allows for equation of the uniform and non-uniform common times.

To use Julian Barbour's vivid phrase, Newton is suggesting that under a dynamically relative account of time, it is not guaranteed that all of the various common times, given by particular motions, will 'march in step'. Newton is thus pointing to the possibility of a generic failure to deal with the time aggregation problem rather than a specific failure to deal with a scenario in which no motions whatsoever are uniform. This way of thinking about the argument that Newton is making in the modal claim chimes particularly well with a 'weak dynamical' interpretation. In particular, Newton is concerned with the practical astronomical problem of the 'equation of times' and the correction to the time given by the pendulum clock[22] read in parallel with an acceptance that we require at least some further non-dynamical and metaphysical attributes for our conception of time to be fully functionally adequate. We will return to this crucial aspect of the absolute vs. relative time dialectic in terms of what we will call the 'chronometric problem' in our discussion of the Leibniz–Clarke correspondence in the next chapter.

2.4 Chapter Summary

Newtonian time is a complex and powerful concept. In this chapter we have argued that we should interpret Newton in the Scholium as wasting no words, and take seriously the tripartite definition of time as 'absolute, true, and mathematical'. We interpreted 'true' as involving a sense of universality or being defined globally throughout space. We took this assertion to be predicated upon the assumption that the ordering or chronordinal structure of time is dynamically absolute in the sense of not varying between universes that instantiate different sets of dynamical motions and invariant in the sense of not varying between universes that differ solely in virtue of the situation in time of the same set of dynamical motions. In the context of a Newtonian time we find that chronordinal structure is a fixed structure. The posit of a 'mathematical' time we took to indicate that there is a single time parameter with uniform metric structure: a chronometric structure of time. This temporal structure we took also to be dynamically absolute but not invariant. This is the sense in which Newtonian time according to our

---

[22] Of great relevance here are the various relations to Newton's work of the ideas of Gassendi, Flamsteed, Huygens, and Barrow. For more details see Arthur 1995; Arthur 2019; Barbour 2001; Brading 2016.

analysis is understood as surplus structure: it is consistent with Newton's discussion for there to be two universes that differ solely in virtue of the situation in time of the same set of dynamical motions. Putting everything together, according to our interpretation, Newtonian time is constituted by a single fixed chronordinal structure, which is dynamically absolute and invariant, and a surplus chronometric structure, which is dynamically absolute but not invariant.

3
On Leibniz On Time

The basic desiderata of a Leibnizian account of time are that chronometric and chronordinal structure will be determinate, invariant, and dynamically relative. This is because his metaphysics does not allow for the possibility for two universes to differ solely in virtue of the situation 'in time' of the same set of dynamical motions or two universes that involve the same sets of dynamical motions should involve the realization of distinct temporal structures. The Leibnizian metaphysics of time runs into two immediate problems, however. First, what determines the order of succession needed to fix the chronordinal, and second, what determines the duration measure needed to fix the 'quantity of time' between successive actual phenomenal states as needed for a functional system of mechanics.

In the context of Leibniz's views on space, time, and motion, much focus is usually put upon the contrast with the views of Newton; in particular, the critical response to the Newtonian framework articulated in the exchange with Clarke.[1] Whilst the parts of the correspondence relating to time will indeed be the major focus of the second half of this chapter, before we bring our discussion to that point it will be important to first investigate in some depth Leibniz's positive views on time. This preliminary investigation will prove of particular importance since the details of Leibniz's positive view are necessary to unpack the crucial idea, articulated in the correspondence, that time is an 'order of successions'.

3.1 Leibniz's Metaphysics of Time

In parallel with our discussion of Newton, it is worthwhile to start by considering the relationships between Leibniz's metaphysics of space and time and his natural philosophy. It is difficult to contest that the two are much more closely entwined in Leibniz than in Newton. Whereas for Newton, the proposition that the two can be separated is at least plausible enough to be defended, the parallel thesis for Leibniz would certainly not be tenable. Leibnizian philosophy of space and time has at its root his metaphysics, and in particular his notion of simple substances, the famously perplexing monads.

---

[1] The famous and wide-ranging correspondence between Leibniz and Clarke took place between 1715 and 1716, terminated by Leibniz's death. Both the direct role of Newton in the drafting of Clarke's letters and the relation between their respective views of space and time is controversial; see e.g. Vailati (1997) and Thomas (2018, §9). For what it is worth, the reading of Newton above would seem to suggest he played a more minimal role, since the dynamical strand of argument in favour of absolute time is not directly advanced by Clarke.

In the philosophy of the mature Leibniz,[2] monads are the fundamental basis of reality. They are non-spatiotemporal, causally isolated and each contain within themselves a representation or perception of everything else; albeit with different degrees of clarity. Despite this disconnection, the representations of the monads are correlated in a consistent manner via a 'pre-established harmony' (this feature will prove important later).

Setting out a detailed account of how Leibniz conceived the relationship between his monadic metaphysics and his natural philosophy would be a lengtly task, ill-fitted to the current work.[3] What is key for our current purpose is the situation of Leibniz's metaphysics within both the mechanical philosophy of his contemporaries and the earlier philosophical traditions from which he drew extensively. In describing this context we can follow the wonderfully clear summary of Arthur (2021, p. 4):

> [Leibniz's] commitment to the mechanical philosophy was early and lasting, but it was overlaid on certain principles, both Platonic and Scholastic, that privileged minds as sources of activity in the world. On the one hand, Leibniz was as convinced a mechanist as Pierre Gassendi and Robert Boyle, holding that all natural phenomena are explicable in principle, without appeal to substantial forms, by efficient causal explanations involving the motions of bodies; on the other, he subscribed to a view emanating from the Scholastic doctrine of the plurality of forms, whereby the seeds of all living things were created at the beginning of the world, each seed consisting in an immaterial form or active principle dominating the organic body containing it, with the body containing within itself other bodies activated by their own subordinate forms. . . Seen in this light Leibniz's famous doctrine of preestablished harmony was intended as a solution not just to the mind–body problem bequeathed by Descartes, but to the deeper problem of how the teleological activity of forms, leading to the increased perfection of the world, could be reconciled with the impossibility of the action of immaterial forms on matter.

As we shall make clear in what follows, in order to understand Leibniz's philosophy of time, including his criticisms of Newtonianism in the correspondence with Clarke, we need to appreciate both the mechanist and scholastic side of Leibniz, and thus both the role of mechanical change of bodies and the active principles of simple substances.

What will prove most significant is that, again following Arthur (2021, p. 1), Leibniz's monads are understood as partless but to possess 'internal qualities and actions' in terms of their *perceptions* ('representations of the composite or external in the simple') and *appetitions* ('principles of change by the action of which one perception passes continually into another'). The internal monadic perception and appetition are then respectively associated with the real relations of *situation* and *succession* respectively. The relations of situation and succession are then in turn the basis for space and time considered either as supported by actual relations or possible relations. Thus the view of space and time that Leibniz endorses is inseparable from his monadic metaphysics and attempts to understand his criticisms of Newtonian time without adequate framing are inevitably going to be at best partial and at worst incoherent.

A useful schema that provides an intuitive crux for better understanding Leibniz's metaphysics is to refer to three 'levels' or 'realms' (Winterbourne, 1982; Hartz and Cover, 1988). At the most basic level are the monads with their intrinsic properties of

---

[2] The mature period of Leibniz is usually taken to be from the late 1680s onwards. For discussion in the development of Leibniz's views see Ariew (1994) (as well as the other essays in that volume) and Arthur (2014).

[3] See Garber (1994), De Risi (2007), and Arthur (2014).

perception and appetition. Next, we have the 'phenomenal level' that is made up of *phenomena bene funda*—well founded phenomena—that, due to pre-established harmony, are accurate reflections of the real and actual monadic states. Finally, we have the ideal level which, by contrast, is made up of *entia rationis*–abstract or fictional things—that include 'phenomena' founded upon possible but non-actual monadic states. Crucially, although both the phenomenal and the ideal levels can include things which are infinite, all concepts that depend upon the continuum are only applicable to the ideal realm. Phenomenal things for Leibniz can only acquire their status as *phenomena bene funda* by their grounding upon the actual. They must always be understood as representations or perceptions of the monads of the actual world.

In this context, Arthur (2021) convincingly argues that it would be anachronistic to take the idea of levels seriously as an ontic division. That is, we should not understand the monadic realm as some form of precursor to the Kantian *Ding an sich*. Rather, although the monads are more fundamental than the phenomena resulting from them, a monad 'cannot continue to exist without being in some state, and that state is nothing other than its perception' (*Monadology* §21 (Leibniz, 1998)). According to Arthur's account, 'there is no distinction to be found between monadic and perceptual states'. Rather, 'the picture we have is of a monad passing through a sequence of transitory states or perceptions, each containing smaller ones, and tending by appetition towards future ones in the same series. The appetition or tendency to pass from one perception to another is identified by Leibniz with the primitive force of a substance.' (Arthur, 2021, pp. 11–12).

Leibniz's view of the ontological status of relations is subtle and significant for his view of time. Following the account of Arthur (2014) and Mugani (2012)), as described in Thébault (2021b), we can take Leibniz to believe that the relations of situation and succession supervene on intrinsic modifications of the monads: *changes* in their perception and appetition. Thus, with regard to time, real temporal relations of succession, both between the states of the same monad and the states of different monads, are founded upon the appetitive activity of individual monads. With regard to the monads of the actual world, all we have are changes of the monadic states which are coordinated via the principle of pre-established harmony. Well-founded phenomena then arise as representations (i.e. perceptions) of the actual monads. This means that with regard to time we thus have the temporal relations of succession which are supported by the actual monadic states. Finally, temporal relations of succession supported by possible monadic states as *entia rationis*.

Founded on the solid platform of the relational metaphysical core of the Leibnizian view, it is fairly straightforward to reconstruct two crucial further features of Leibnizian time based on his conception of what it means for time to have 'parts' and the ontological dependence of these parts. Crucially, for Leibniz, the order of the 'parts of time' is parasitic on such an order obtaining among the determinate states of things that individuate them (Arthur, 2021, p. 92). This means that his metaphysics of time does not allow for two universes to differ solely in virtue of the situation 'in time' of the same set of dynamical motions—to be distinct, universes must have different dynamical histories. Furthermore, it also leads naturally into a requirement that two universes that involve different sets of dynamical motions—different determinate states of things—should involve the realization of distinct temporal structures (at least, in the sense of individuated sequences of moments). In the terminology we introduced to discuss the Newtonian view of time this means that we can plausibly read Leibniz as being committed to a view in which time must be realized by a structure which is *invariant* and *dynamically relative* in our terminology.

It is in this context that the Leibnizian metaphysics of time runs into two immediate problems. These are each precursors to the problem of time in classical mechanics. Our focus henceforth will be upon explicating these two problems and then connecting them back to our discussion of Newton and the relevant parts of the Leibniz–Clarke correspondence.[4] The two problems for Leibniz are as follows: first, what determines the 'order of succession' needed to fix the actual time ordering of the monadic states and consequently the phenomenal relations of succession? The *chronordinal problem* is to fix *determinate chronordinal structure* as both invariant and dynamically relative. Second, what determines the duration measure needed to fix the 'quantity of time' between successive actual phenomenal states as needed for a functional system of mechanics? The *chronometric problem* is to fix *determinate chronometric structure* that is invariant and dynamically relative.

3.2   The Chronordinal Problem

The chronordinal problem is subject to an earlier literature; in particular, Rescher (1979) and McGuire (1976) follow Russell (1900) in convicting Leibniz of a vicious circularity that requires a non-relational concept of time at the basic monadic level. In that context, the supposed problem is that monads do not exist in time and cannot bear temporal relations, such as ordering, between themselves. Thus time can only be consistently constituted within Leibniz's system as an ideal relation among phenomena.

Such a view is, however, in tension with Leibniz's repeated reference to the ordering of monadic states. In particular, Arthur (1985, 2021) points to the following passages:

The *succeeding substance* is held to be the same when the same law of the series, or of continuous simple transition, persists; which is what produces our belief that the subject of change, or monad, is the same. That there should be such a persistent law, which involves the *future states* of that which we conceive to be the same, is exactly what I say constitutes it as the same substance.[5]

...the nature of every simple substance, soul, or true monad, is such that its *following state* is a consequence of the *preceding one.*... [6]

On Arthur's view, Leibniz's theory of time is then in essence a theory of *ordered changes in monadic states* which is applicable also to the states of the things aggregated from them. At the level of the single monad such a view seems unavoidable given our discussion of monadic change in the last section. In particular, *intra*-monadic temporal

---

[4] This will inevitably mean some subtleties of interpretation will be neglected. In particular, we will not consider the relationship between Leibniz's theory of time and key philosophical debates relating to determinism, the arrow of time, dynamic versus static theories of time, causal theories or time, or the temporal continuum. For detailed discussion of such topics the reader is directed to the outstanding recent monograph by Arthur (2021).

[5] Letter to De Volder, 1704/1/21: GP II 264; Arthur 1985, 273–4, emphasis added.

[6] Fifth Paper for Clarke §91; GP VII 412; Arthur 1985, 274.

ordering must surely be taken to be a conceptually necessary part of monadic appetition. However, if time ordering is merely a private intra-monadic relation then it will
be difficult for any kind of unified concept of time to emerge from the monadic base
structure. There would be no determinate chronometric structure that would allow us
to organize the states of the things that aggregated from monads, and thus a mechanical theory with determinate chronordinal structure could not emerge. Arguably the
emergence of such structure is not a conceptual necessity. However, Leibniz was, as we
have discussed, a *mechanist* for whom all natural phenomena are explicable via explanations involving the motions of bodies. The recovery of a determinate chronordinal
structure is plausibly explanatory, and thus pragmatically necessary for Leibniz, since
without it there is no means of ordering the motions of bodies.

The central issue is, then, how we are to conceive of the basis for this *inter*-monadic
ordering upon which an explanatory mechanical superstructure can be erected. Since
the monads are causally isolated, what can it mean for the successive states of different
monads to be temporally ordered? In order to resolve this problem, Arthur (1985,
2014, 2021) convincingly argues that we can understand time at the monadic level in
terms of a (non-circular) inter-monadic notion of temporal succession based upon the
relation of *compossibility*, which is in turn based upon the relation of 'involving the
reason for' (see also De Risi 2007, p. 270–7). Of particular significance to supporting
this interpretation is the following passage from Leibniz's 1714 *Initia rerum*:

*If a plurality of states of things is assumed to exist which involve no opposition to each other,
they are said to exist simultaneously.* Thus we deny that what occurred last year and this
year are simultaneous, for they involve incompatible states of the same thing.

If one of two states which are not simultaneous involves a reason for the other, the former
is held to be *prior*, the latter *posterior*. My earlier state involves a reason for the existence
of my later state. And since my prior state, by reason of the connection between all things,
involves the prior state of other things as well, it also involves a reason for the later state of
these other things and is thus prior to them. *Therefore whatever exists is either simultaneous
with other existences or prior or posterior.*[7]

Arthur's key idea is that since states of the world involve different monads existing
at the same time, such states are highly constrained by the fact that each monad
contains within it a representation of all the others. He introduces the idea as follows:

According to Leibniz, "My earlier state involves the reason for the existence of my later
state": that is, each of the states of any series of states *involves the reason for*, or *contains
the ground for*, any subsequent state in the series, so that the states in each series are ordered
by this relation ...[8]

According to Arthur, the 'involving the reason for' relation, then fixes a notion of
compatibility of states, which in turn fixes a compossibility relation. Thus, even though
the monads are not causally interacting with each other, to be co-existing a given
class of monads must be compossible. Compossibility fixes a trans-monadic notion of
simultaneity, which in turn fixes a total temporal ordering of monadic states: 'whatever
exists is either simultaneous with other existences or prior or posterior'.

Following Arthur's account chronordinal structure is then determinate within Leibniz's metaphysical framework, since it is fixed uniquely fixed for any monadic state
by the compossibility relation. Of course, such a view is not fully articulated by Leibniz himself. However, as shown explicitly in Arthur (2021, Appendix A1) a formally

---

[7] Loemker (1969, p. 666), emphasis in the original.
[8] Arthur (2021, Appendix A1.1), italics in original.

consistent definition of simultaneity from compossiblity can be built upon a relational core in terms consistent with Leibniz's philosophy.[9] We thus see that there is a natural avenue for the Leibnizian to deflect the Russellian circularity charge.

In summary, we considered the problem of fixing a determinate chronordinal structure that is dynamically relative together with a proposed Leibnizian solution. This *chronordinal problem* will also be central to our discussion of time within the philosophy of Ernst Mach, and, moreover, forms the first precursor problem of time.

3.3   The Chronometric Problem

The second problem that we isolated for the Leibnizian metaphysics of time is related to duration seen as a quantitive measure of time. If time as a well-founded phenomena consists only in the relation of succession, then how can there be a determine quantity of time between successive phenomenal states. It is here that it will prove highly significant to bring in the comparison with Newton and the correspondence with Clarke.

Recall from the previous section that we interpreted Newtonian time as being partially realizable by a single chronometric structure which is dynamically absolute, in the sense of being independent of variation of the motions of physical bodies, and yet surplus structure, in the sense of being dependent of the situation in time of the same set of dynamical motions. We also took a relatively strong ontological interpretation in which time for Newton is absolute in a metaphysical sense plausibly read along the lines of the modern terminology of substantival or having self-subsistent parts.

Leibniz and Newton shared the belief that time was appropriately 'mathematical' in the sense of allowing us to define a single time parameter with uniform metric structure. As such they would have agreed that time could be realized by a single uniform chronometric structure. Indeed, the need for such structure is one of the key common intellectual threads than links the mechanical philosophy of Descartes, the science of time keeping via clocks and astronomical measurements, and, moreover, the conceptual roots of the calculus. Given this, the core question becomes picking out which aspect of the Newtonian concept we should take Leibniz to be denying. In particular, is he objecting to chronometric structure being dynamically absolute and/or surplus? If so, on what basis? If not, then can we reframe Leibniz's objections to Newtonian time in purely metaphysical terms. That is, as solely an objection to the claim that time is absolute in a metaphysical senses of being substantival or having self-subsistent parts.

Here it will prove worth quoting the full discussion of DiSalle who evidently took Leibniz's focus to be solely upon the substantival claim:

The Leibnizian critique is based on pre-conceptions of the terms that Newton is using: if Newton is claiming that time is absolute, he must be implying that time is a substance, and for Leibniz real substances are, or are composed of, distinct individuals. No difference could be discernible, however, between our Universe and one in which all events were arbitrarily

[9] Plausibly, compossibility is not, on its own, sufficient to ground a directed ordering of temporal states. Such a notion depends on further structure implicit in the monadic appetition. See Arthur (1985, 2014, 2021) and De Risi (2007, p. 270–7). Furthermore, arguably, however, compossibility still underdetermines the actual temporal ordering since it does not give us grounds to distinguish it from merely possible temporal orderings (Cover, 1997; De Risi, 2007; Arthur, 2021).

shifted forward or backward in time; for time is only an 'order of succession', not a collection of moments that possess distinct individual natures. Such a shift would therefore be an empty distinction between things that are truly indiscernible. But Newton's definition does not imply any such distinction: the only distinctions that Newton's concept requires are between simultaneous and non-simultaneous events, and between equal and unequal time intervals.[10]

We already saw in the last section, in light of what Newton says in paragraph 6 of the Scholium and the General Scholium, that the strong dynamical interpretation offered by DiSalle is not entirely textually adequate. In particular, Newton explicitly indicates that his view involves time and space having 'parts' which have 'essences' and thus that there are self-subsistent 'particles of space' and 'moments of duration'. This notwithstanding, it might still be that DiSalle has Leibniz correct: the objection to the absoluteness of Newtonian time does indeed turn purely on the substantival point.

The most relevant passage in the correspondence can be found in Leibniz's third letter where he introduces the 'temporal shift scenario' referred to by DiSalle. Immediately following the appeal to the much discussed 'spatial shift scenario' to argue against absolute space, Leibniz continues:

The case is the same with respect to time. Supposing any one should ask, why God did not create every thing a year sooner; and the same person should infer from thence, that God has done something, concerning which 'tis not possible there should be a reason, why he did it so, and not otherwise: the answer is, that his inference would be right, if time was any thing distinct from things existing in time. For it would be impossible there should be any reason, why things should be applied to such particular instants, rather than to others, their succession continuing the same. But then the same argument proves, that instants, consider'd without the things, are nothing at all; and that they consist only in the successive order of things: which order remaining the same, one of the two states, viz. that of a supposed anticipation, would not at all differ, nor could be discerned from, the other which now is.[11]

That this passage involves a denial of the substantival time thesis is fairly obvious. This is of course unsurprising in the context of the positive view of time as we have presented it. Furthermore, denial of any surplus chronometric structure is fairly clear: surplus chronometric structure of Newtonian time is precisely the reason why God would have no reason for creating the world sooner or later.[12] In our terms then, Leibnizian time is then required to be an invariant chronometric structure on pain of violating the principle of sufficient reason.

With regard to the 'dynamical absoluteness' thesis we will need to dig a little deeper into the correspondence. In his third reply in response to the passage quoted, Clarke notes that:

[I]f time was nothing but the order of succession of created things; it would follow, that if God had created the world millions of ages sooner than he did, yet it would not have been created at all the sooner. Further: space and time are quantities; which situation and order are not.[13]

What is significant here is not the line of argument regarding God's creation of the world sooner or later; for more on that issue see Arthur (1985, pp. 285–290). Rather,

---

[10] DiSalle (2006, p. 22).

[11] Alexander (1998, p. 26-7).

[12] This corresponds to the choice of a temporal origin point or offset within a theory where time is represented as homogenous.

[13] Alexander (1998, p. 32).

it is the final remark that time is a *quantity* but order is not. This line of objection (which Leibniz does not appear to respond to in the fourth letter) is amplified in Clarke's fourth reply:

. . . that time is not merely the order of things succeeding each other, is evident because the quantity of time may be greater or less, and yet that order continue the same. The order of things succeeding each other in time, is not time itself; for they may succeed each other faster or slower in the same order of succession, but not in the same time.[14]

Leibniz, in his fifth and final letter, responds as follows:

The author objects here, that time cannot be an order of successive things, because the quantity of time may become greater or less, and yet the order of successions continue the same. I answer; this is not so. For if the time is greater, there will be more successive and like states interposed; and if it be less, there will be fewer; seeing there is no vacuum, nor condensation, or penetration, (if I may so speak), in times, any more than in places.[15]

Crucial in understanding what Leibniz is getting at in this passage is how we should understand the notion of a 'quantity of time'. Following Arthur (2014, p. 149–150), it seems very plausible to see Leibniz as drawing upon the Scholastic distinction between duration and time. Whereas, duration is an attribute of enduring things, time is an *abstract measure of this duration*. This is the Scholastic sense of what it means for time to be a 'quantity'. Such a distinction is also found in Descartes who notes that 'in order to measure the duration of all things, we compare the duration of the greatest and most regular motions which give rise to years and days, and we call this duration time' (Descartes, 1984, p. 212).[16] In a manuscript from 1680 Leibniz makes almost exactly the same claim regarding the quantity of time as Descartes: 'the basis for measuring the duration of things is the agreement obtained by assuming different uniform motions (like those of different precise clocks)' (quoted in Arthur 2014, p. 206).

In his later work, Leibniz explicitly endorses such a view in the dialogue *Conversation of Philarète and Ariste* that was written roughly around the same time as the correspondence but with the views of Malebranche rather than Newton as its target (Loemker, 1969, pp. 621–2). Similarly, in the passage of *Initia rerum* immediately following that quoted above Leibniz asserts:

Time is the order of existence of those things which are not simultaneous. Thus time is the universal order of changes when we do not take into consideration the particular kinds of change.

Duration is magnitude of time. If the magnitude of time is diminished uniformly and continuously, time disappears into a moment, whose magnitude is zero.[17]

Returning to the fifth letter, and following Arthur (2021), we can present the crucial point of interpretation as taking seriously what Leibniz means in the fifth letter by

---

[14] Alexander (1998, p. 52).

[15] Alexander (1998, p. 89–90).

[16] Whilst the connection between Leibniz's view of time and that of Descartes is a subtle one, it certainly does not seem to be correct to categorize the Cartesian view of time as 'relationalist' in anything like Leibnizian sense. In particular, in his *Principles*, Descartes draws the contrast between what he takes to be the incorrect view of duration, as the persistence of a thing in time, and the correct view, as a mode under which we *conceive* of a thing, so long as it continues to exist (Gaukroger 2002, p. 89, see also Schmaltz 2009). It thus seems plausible to place Descartes (but not Leibniz) in the broadly Aristotelean tradition that takes time to be dependent upon the human mind or the soul. See Thomas (2018) for a wider contextualisation of this tradition both back to Aristotle via the Scholastics and forward to critics of absolute time such as Cavendish, Conway, and Law.

[17] Loemker (1969, p. 666), emphasis in the original.

like states interposed. That is, to understand the 'like states' as an abstraction based upon relevant similarities between *states of objects that appear among the phenomena.* Arthur here suggests the comparison with successive swings of Huygens's pendulum clock. These oscillations would be different in themselves, involving different sequences of states of their constituent monads, but alike in relevant respects (ibid p. 95). Arthur further evidences his reading by drawing upon Leibniz's critical response to Locke (pp. 96–7). In particular, in his *Nouveaux essais*, a chapter-by-chapter rebuttal of Locke's *An Essay Concerning Human Understanding* written in dialogue form, he has the Locke character Philalethes doubt the Aristotelian claim that time is the measure of motion, and then offers a respons that runs:

Indeed we could say that a duration is known by the number of equal periodic motions, each beginning when the preceding one ceases, for instance, by so many revolutions of the earth or the stars.[18]

There is thus good textual evidence on the basis of which we can interpret Leibniz's meaning in his final letter, and construct a robust further reply to Clarke on his behalf. Clearly Leibniz did have at his disposal ample conceptual resources to cash out what he means by the 'quantity of time' and we can anticipate how he would have done this had his untimely death not precipitated the premature termination of the correspondence with Clarke.

In sum, following the interpretation of Arthur, we can plausibly understand Leibniz in his fifth letter as putting forward a *dynamically relative* view of chronometric structure, the 'quantity time' in terms whereby the 'like states interposed' are durations of similar phenomena such as the uniform motions. In adopting this interpretation of Leibniz's positive view, we are clearly also endorsing an interpretation of his criticism of Newtonian absolute time which includes the dynamical absoluteness as well as the substantival sense of absolute. That is, on this view we *should* take Leibniz to be denying Newton's dynamically absolute distinction between equal and unequal time intervals. The dynamically absolute quantity of time involved in two Newtonian intervals can be judged to be equal, irrespective of the motions involved. In contrast, the quantity of time in two Leibnizian intervals covaries with the quantity of 'like states interposed' and so is explicitly dynamically relative.

Arthur's exegesis of the view implicit in the fifth letter goes a long way not only to overturn the largely negative verdict regarding the cogency of Leibniz's initial response on this point but also to open up argumentative space for a convincing reply in an imagined sixth letter.[19] However, it still remains to be shown that Leibniz's dynamically relative conception of quantity of time is as functionally adequate. In particular, in the previous section we saw that one of the key motivations for Newtonian time was its functional role in time measurement, astronomy and mechanics. Here is the crucial connection to chronobservable structure, which is the empirical aspect of temporality.

In this context we should return to the idea of the equation of times that Newton discusses in paragraph 5 of the Scholium. The crucial question is whether or not the dynamically relative chronometric structure, which we can take to be implicit in

---

[18] §22 in Nouveaux essais, quoted in Arthur (2021, p. 96).

[19] In his fifth reply, left unanswered due to Leibniz's death, Clarke explicitly accuses Leibniz of contradicting himself. Similarly, Zwart judges the response to be 'not very satisfactory' (Zwart, 1972, p. 136).

Leibniz's view, meets the standards to be *mathematical* time in the sense of allowing a variety of observed motions, including nonuniform motions, to be aggregated into a single time parameter with uniform metric structure. This point is put extremely well by De Risi:

[P]hysics absolutely needs the possibility for time to be measured, thus the possibility for the quantity of succession to be apprehended, and the possibility for the congruence of temporal intervals to be determined. Actually, what is needed is even more. From the point of view of physical science, time must be *uniform* according to Leibniz's definition, i.e. it must have a determinate metric structure that makes it possible for us to assume time as an independent variable of classical mechanics.[20]

We can thus isolate a second aspect precursor problem of time, what we will call the *chronometric problem*. Like the problem fixing determinate yet dynamically relative chronordinal structure, the chronometric problem is the problem of fixing determinate yet dynamically relative chronometric structure. However, the problem is in fact rather more difficult in this case since the desire for determinate chronometric structure is more than a pragmatic one. It is a requirement for the *empirical adequacy* of a theory of time in mechanical theory. As noted in the our discussion of §2.3.1, awareness of this problem is evident in Newton's discussions in the Scholium, and could perhaps be envisaged as the *coup de grâce* in an imagined sixth reply by Clarke.

Tantalizingly, Leibniz can be read as anticipating something like a solution to the problem. In particular, Arthur (1985, 2014), suggests that in *Initia rerum* Leibniz should be understood as indicating that a means for determining dynamically relative temporal distances should be conceived in terms of 'maximally determined' or 'simplest path' through interposed constituents (see also Vailati 1997, p. 136).

Furthermore, Rescher (1979, p. 66) suggests that based upon the principle of perfection we might expect that 'nomic harmony' be sufficient to establish a natural phenomenal measure of duration as a contingent feature of the actual world. Plausibly such a conception of time would also necessarily have to be invariant since if chronometric structure is 'nomologically determined' in this way, then two universes, which are each consistent with the laws of nature, with the same set of dynamical motions should not chronometrically differ.

Finally, in his recent discussion, Arthur (2021, §3.5) offers a fascinating fuller exegesis of how Leibnizian foundations could be given to an account of uniform rectilinear motions within the context of his views on space and motion. In particular, Arthur takes Leibniz in *Initia rerum* as indicating that '[i]t is on the assumption of the continued existence of a given thing through time that we can compare ratios of the spaces traversed by it with ratios of the times taken, assuming a uniform motion, and thus apply quantity to time. The possibility of motion requires successive congruence, the substitutability for one another of equal intervals of space traversed by a uniform motion' (p. 338). The idea is then that 'just as space can be expressed in terms of groups of transformations of equivalent situations, so motions through space can be expressed in terms of groups of transformations of equivalent changes of situation' (p. 339).

Arthur thus provides a plausible reading in which chronometric structure can be rendered determinate within Leibniz's dynamically relative framework in virtue of the motion of a given system of bodies taken as a reference. Significantly, such motions

---

[20] De Risi (2007, p. 273), emphasis in the original.

need not be inertial nor be something pre-existing. Rather, they are understood as calculable relative to the situation determining the initial conditions. There is thus a precise sense in which Leibniz, on Arthur's interpretation, foreshadows the 'second Mach's principle' to be discussed in the next chapter. These shadows notwithstanding, in Leibniz, as indeed in Mach, we still do not find a satisfactory quantitive response to the problem of quantifying time in a dynamically relative manner. In this regard, Newtonian dynamical absolutism with regard to time was destined to long reign supreme, notwithstanding the cogency and power of the Leibnizian, and later Machian, critique.

3.4   Chapter Summary

The basic desiderata of a Leibnizan account of time are that chronometric and chronordinal structure will be determinate, invariant, and dynamically relative. This is because his metaphysics does not allow for the possibility that two universes differ solely in virtue of the situation 'in time' of the same set of dynamical motions nor two universes that involve the same sets of dynamical motions should involve the realization of distinct temporal structures. The Leibnizian metaphysics of time runs into two immediate problems however. First, what determines the order of succession needed to fix the chronordinal structue, and second, what determines the duration measure needed to fix the 'quantity of time' between successive actual phenomenal states as needed for a functional system of mechanics. The first problem is principally pragmatic in nature, and can be largely resolved via appeal to the idea of compossibility. The second is an empirical problem and threatens the empirical adequacy of approaches to mechanics in which chronometric structure is dynamically relative. There are seeds of a response within Leibniz's work but no solution.

4
On Mach On Time

The mature Machian view on the nature of time can be rationally reconstructed as follows. All temporal structure should be functionally integrated with chronobservable structure. Chronometric and chronordinal structure is locally determinate and dynamically relative given a stable environment. Chronometric and chronordinal structure is spatially globally determinate and dynamically relative if it can be connected to chronobservable structure via the change in entropy of the universe. Approximate dynamically absolute chronometric and chronordinal structure is meaningful to the extent to which sufficiently isolated bodies can be assumed to play the role of inertial clocks.[1]

4.1 Mach's Philosophical Outlook

Ernst Mach was a groundbreaking experimental scientist, historian of science, and natural philosopher. For us to fully understand his views on the nature of space and time, some appreciation of his overall philosophical outlook is necessary.

Mach has often been mischaracterized as a naïve positivist or phenomenalist who only believed in the reality of human sensation, however, recent scholarship has pointed to a more nuanced reading.[2] In particular, drawing upon the accounts of Banks (2003, 2014, 2021) and Pojman (2011),[3] we can isolate the aspects of Mach's philosophy most relevant for understanding his view of time as follows.

The most significant idea is that of 'elements'. Elements are characterized as neutral and monadic: *both* sensations, when understood psychologically, *and* objects, when understood physically. Just as we may think of sensations in terms of their material-neurological correlates, we might conceive of atoms or molecules as 'mental symbol[s] for a relatively stable complex of sensational elements' (Mach, 1914, p. 311). In the

---

[1] This chapter substantially reproduces material first published in Thébault (2021a) with permission of the publisher. There are minor changes for reasons of terminological consistency.

[2] The history behind the misreading of Mach and its connection to the logical empiricists is a complex and fascinating story. See Hintikka 2001; Pojman 2010; Banks 2014; Stadler 2021 for extended discussions. For a highly illuminating analysis of the relationship between Mach's thought and the logical empiricists in the particular context of the principle of least action see Stöltzner (2003). For a situation of Mach's through in the wider scientific and philosophical intellectual context of the early twentieth century see Stöltzner (2021).

[3] As already noted, the view described here is also very much in the same spirit as the interpretation of Mach's *Analysis of Sensations* provided in Preston (2021b). Moreover, in the context of contemporary Mach scholarship the broad understanding of Mach's thought that we will work with here is now very much the established mainstream. Two recents volumes which together provide a comprehensive overview of Mach's thought are Stadler (2019) and Preston (2021a).

context of Mach's wider framework of elements, the following aspects of his views on the foundations of physics are then worth noting.

First, from Mach's perspective, we should think of there being no logical or empirical loss in reconstructing physics purely in terms of psychophysical elements. Second, on Mach's view this approach is heuristically fruitful, since it helps economize science and rid us of *idle* metaphysical conceptions.[4] Third, Mach takes his view to be naturalistically well motivated, since it forces us to situate physics within an evolutionary and anthropological context.[5] Fourth, and most significantly, we should certainly *not* understand Mach's view as involving the claim that all unobserved elements are superfluous to science and should be eliminated from our theories. Rather, Mach insists that to ensure science performs its proper function, unobserved elements must be established as appropriately connected with experience in terms of a being part of a 'continuous fabric of experience-reality consisting of events and causal-functional relations' (Banks, 2014, p. 33). Significantly, in the context of such a framework, the observable/unobservable distinction is not taken to be a fundamental one. Moreover, according to Mach, theoretical abstractions are admissible so long as they are suitably circumscribed and can be connected with experience by a series of approximations, cf. (Feyerabend, 1984, p. 19).

This fourth point is of particular relevance for the rational reconstruction of Mach's conception of time that will be provided in the following sections. In particular, on the Pojman–Banks reading it would be a mistake to assign to Mach a position of straightforward positivism or operationalism about theoretical concepts. In particular, Mach explicitly did *not* think that 'a concept or quantity is metaphysical [and thus eliminable] if it is not operationally defined; hence every notion which refers to hidden entities—i.e. to entities inaccessible to direct observation is metaphysical' (Zahar, 1977, p. 202) or that 'phenomena should constitute the whole domain of discourse of theories' (Zahar, 1981, p. 268). Whilst positivistic and operationalist views of theoretical concepts have indeed been defended in Mach's name, it is not the view that is identifiable based upon anything like a careful and balanced appraisal of his writings.

Rather, what we will call Mach's 'phenomenological thesis' later is the weaker claim that theoretical concepts that we employ relating to time should be *continuously connected to experience via causal-functional relations*. We use this phrase in the sense described by Banks as meaning theoretical concepts should be woven into a 'continuous fabric of experience-reality consisting of events and causal-functional relations' (Banks, 2014, p. 33). On this view phenomena are not the 'whole domain' of discourse in science but rather the ultimate subject matter to which all discourse can ultimately be connected via the relevant casual functional relations, that themselves are ultimately

---

[4] It is important to note here that the principle of economy functions within Mach's framework as a methodological ideal or regulative principle, rather than a (hypocritically) metaphysical precept. See Stöltzner (1999) for discussion of both this issue and other important aspects of Mach's epistemology. Further insightful discussion of Mach's epistemology can be found in Staley (2021).

[5] There is a suggestive analogy between Mach's drive to situate physical concepts and physical reasoning within a historical and developmental context, and Friedrich Nietzsche's contemporaneous project for naturalizing moral concepts and moral reasoning, particularly as embodied in *On the Genealogy of Morality*, published in German in 1887 (Nietzsche, 1998). See Gori (2021) for a hugely insightful study of the connections between the thought of these two thinkers.

reducible to his neutral monadic elements. We will return to the implications of this functionalist form of phenomenalism later in the context of an interesting analogy, due to Hintikka (2001), between Mach's view of concepts and the notion of 'identifiability'.

In this vein, it is also worth remarking that as well as being clearly distinguishable from positivism and operationalism, this viewpoint on theoretical concepts should also not be conflated with an unalloyed commitment to conventionalism. In particular, as we will see, Mach's commitment to the conventionality of choice in motions that are used to define duration (mathematical time) is predicated on assumptions about the relative stability of the environment. Thus, contra the otherwise exemplary analysis of Tal (2016), we should *not* see Mach as a conventionalist who assumed, (supposedly) alongside Poincaré, Carnap, and Reichenbach, that the conditions under which magnitudes of duration are deemed equal to one another is *arbitrary*.[6] Furthermore, it is certainly *not* the case that for Mach the choice of a 'coordinative definition' for uniformity of temporal processes is *independent of experience*. As we shall see below, for Mach, such freedom essentially depends upon the scientist's ability to abstract away the environment, which in turn depends upon that environment being relatively stable.[7]

Finally, with regard to the motivation of his viewpoint, it is important to emphasize the centrality of the *economy of thought* within Mach's framework. Ultimately, Mach takes himself to be engaged within a historically and evolutionary embedded process of the continuous refinement and re-articulation scientific thought (in the broadest sense of the term) in more efficient and effective terms. The economy of thought is a regulative principle that drives this enterprise productively forward and motivates his project of reconstructing physics purely in terms of psychophysical elements. There is thus a fundamentally pragmatic or methodological aspect to Mach's form of empiricism, coming from a rather different route to the type of epistemologically motivated empiricism associated with both the early modern British empiricists and contemporary constructive empiricists.[8]

4.2 The Early Machian View of Time

4.2.1 The Principal Discussions

Bearing Mach's philosophical outlook in mind, let us review Mach's principal discussions of time. The most significant, and much quoted passage can be found at the beginning of the section entitled 'Newton's Views of Time Space and Motion' in Mach's *Science of Mechanics*, first published in German in 1883. Immediately following a full quotation of paragraph 1 of the Scholium and a quotation of the bulk of paragraph 5, Mach continues:

---

[6] This crucial difference is supported by Mach's own remarks: 'Though I esteem Poincaré with his conventions, still something is lacking in that approach in my opinion. Conventions are not at all arbitrary, but are pushed through with gruesome pressure.' *Letter to Dingler, 26th January 1912* (Blackmore, 1992, p. 100). Whether Poincaré himself held such a view is at least contestable, see Gray (2013, pp. 369–72).

[7] There is thus, in fact, an interesting parallel between Mach's actual view of time measurement and that developed by Tal himself in the context of his detailed study of modern atomic clocks.

[8] For more on the economy of thought and the connection to pragmatism see Banks 2014; Klein 2021; Uebel 2021; Patton 2021.

It would appear as though Newton in the remarks here cited still stood under the influence of the medieval philosophy, as though he had grown unfaithful to his resolve to investigate only actual facts. When we say a thing A changes with the time, we mean simply that the conditions that determine a thing A depend on the conditions that determine another thing B. The vibrations of a pendulum take place *in time* when its excursion *depends* on the position of the earth. Since, however, in the observation of the pendulum, we are not under the necessity of taking into account its dependence on the position of the earth, but may compare it with any other thing (the conditions of which of course also depend on the position of the earth), the illusory notion easily arises that *all* the things with which we compare it are unessential. Nay, we may, in attending to the motion of a pendulum, neglect entirely other external things, and find that for every position of it our thoughts and sensations are different. Time, accordingly, appears to be some particular and independent thing, on the progress of which the position of the pendulum depends, while the things that we resort to for comparison and choose at random appear to play a wholly collateral part. But we must not forget that all things in the world are connected with one another and depend on one another, and that we ourselves and all our thoughts are also a part of nature. It is utterly beyond our power to *measure* the changes of things by *time*. Quite the contrary, time is an abstraction, at which we arrive by means of the changes of things; made because we are not restricted to any one *definite* measure, all being interconnected. A motion is termed uniform in which equal increments of space described correspond to equal increments of space described by some motion with which we form a comparison, as the rotation of the earth. A motion may, with respect to another motion, be uniform. But the question whether a motion is *in itself* uniform, is senseless. With just as little justice, also, may we speak of an "absolute time"—*of a time independent of change*. This absolute time can be measured by comparison with no motion; it has therefore neither a practical nor a scientific value; and no one is justified in saying that he knows aught about it. It is an idle metaphysical conception.[9]

Similar sentiments, with a slightly clearer emphasis on the relevance of the environment, can be found in the parallel discussion in his *Economic Nature of Physical Enquiry*, first published in 1882 as part of his *Popular Scientific Lectures*:

Newton speaks of an *absolute* time independent of all phenomena ... For the natural inquirer, determinations of time are merely abbreviated statements of the dependence of one event upon another, and nothing more ... because all are connected and each may be made the measure of the rest, the illusion easily arises that time has significance independently of all ... The condition of science, both in its origin and in its application, is a *What it teaches us is interdependence*. Absolute forecasts, consequently, have no significance in science. With great changes in celestial space we should lose our co-ordinate systems of space and time.[10]

A point that will be of great significance later, but is only implicit in the above, is that on Mach's view, it is not cogent to talk of an overall time for the universe. Consider the following from his *History and Root of the Principle of the Conservation of Energy*, first published in 1872:

The universe is like a machine in which the motion of certain parts is determined by that of others, only nothing is determined about the motion of the whole machine. If we say of a thing in the universe that, after the lapse of a certain time, it undergoes the variation A, we posit it as dependent on another part of the universe which we consider as a clock. But if we assert such a theorem for the universe itself, we have deceived ourselves in that we have nothing over to which we could refer the universe as to a clock. For the universe there is no time. Scientific statements like the one mentioned seem to me worse than the worst philosophical ones.

People usually think that if the state of the whole universe is given at one moment, it is completely determined at the next one; but an illusion has crept in there. This next moment is

---

[9] Mach (1919, pp. 223–4), emphasis in original.
[10] Mach (1895, pp. 204–6), emphasis in original.

given by the advance of the earth. The position of the earth belongs to the circumstances. But we easily commit the error of counting the same circumstance twice. If the earth advances, this and that occur. Only the question as to when it will have advanced has no meaning at all.[11]

Finally, we find a specific, and thankfully more pithy, characterization of what time actually is on Mach's view, within his *Contributions to the Analysis of Sensations*, first published in 1886:

Space and time, closely considered, stand as regards physiology, for special kinds of sensations; but as regards physics, they stand for functional dependancies upon one another of the elements characterised by sensations.[12]

4.2.2 Rational Reconstruction

We have emphasized the dates that these books were first published in German here to highlight that they are all from the 'early Mach'. Let us draw from the quotations a rational reconstruction of the early Mach view of time. We can do this in terms of five interconnected claims:

i) **Phenomenological Thesis**: The only temporal concepts that are meaningful are those that can be continuously connected to experience via causal-functional relations.

ii) **Dynamical Relativity Thesis**: Temporal concepts can be meaningfully abstracted from relative variation since these are continuously connected to experience..

iii) **Interconnection Thesis**: All local abstractions of time are subject to global effects since all motions of bodies are interconnected.

iv) **No Dynamically Absolute Time Thesis**: Dynamically absolute time is meaningless since it cannot be continuously connected to experience since there are no bodies whose motion to which it would correspond.

v) **No True Time Thesis**: The idea of there being one true time is meaningless since change within the universe as a whole is not a concept continuously connected to experience.

The first, second, and third claims constitute Mach's positive view of time and are stable throughout Mach's writings. The fourth and fifth, as we shall see, undergo significant modifications as he revises his views regarding the plausibility of grounding inertial reference frames and the entropy of the universe in experience. The third claim is that most closely related to the work of Barbour on the so-called second Mach's principle. We will consider the connection to Mach in §4.4.3 and provide a detailed examination of the relevant formal ideas in Part III.

In our terminology, the first claim amounts to a requirement that all temporal structure should be *functionally integrated* with chronobservable structure; that is, embedded in a network of causal-mechanical relationships with empirical temporal phenomena such as clocks or other regular processes. The other claims can also, to a degree, be cashed out in our terminology. In doing this it will prove instructive to focus on the contrast with the views of Newton.

---

[11] Mach (1911, p. 62–3).
[12] Mach (1914, p. 348).

In general terms, the early Mach clearly disagrees with Newton regarding the coherence of a notion of temporal structure that is 'absolute' in a number of the relevant senses. In particular, a) he denies that temporal structure is substantival, based upon the phenomenological thesis; b) he denies that temporal structure is dynamically absolute, based upon the dynamical relativity thesis and the no dynamically absolute time thesis; and c) he denies that time is true (i.e. globally well defined), based upon the no true time thesis. This last is to deny the determinate status of chronordinal structure at least at the global level.

Clearly, however, Mach would assert that time can be mathematical, in the sense of there being a uniform temporal metric *given a stable environment*. He is thus *conditionally endorsing* the determinate status of chronometric structure within mechanical theory. In this context it is worth returning to the issues regarding the translation of paragraph 5 of the Scholium that were discussed in §2.3 above. In particular, in the context of Mach's criticisms of Newton, what is significant to consider is the *Latin to German* translation of paragraph 5 that Mach quotes from in full. In the context of our analysis of Mach's view of time, what is significant first and foremost is what Mach thought Newton thought about time. Thus the issue, discussed in §2.3, of whether the strong dynamical reading of Newton is justified or not is largely independent of our analysis of Mach.

By contrast, what is of great significance to our analysis is whether it is at all plausible to take Mach to have taken Newton to have endorsed something like the strong dynamical reading. On the strong dynamical reading, Newtonian time closely parallels Mach's own mature view. Thus, if Mach's reading of Newton could have plausibly taken such a form, then the dialectic between the two would become increasingly fraught.

The significant point here is that, the *Latin to German* translation of paragraph 5 that Mach used *makes implausible the strong dynamical reading*. In particular, the rendering of paragraph 5 (which might reasonably be categorized as a mistranslation) makes reference to a *non-comparative* 'true' time that all but rules out the strong dynamical reading. Mach, following the 1872 German translation due to Wolfers (Mach, 2012*a*, pp. 250–1), has paragraph 5 as:

Die absolute Zeit wird in der Astronomie von der relativen durch die Zeitgleichung unterschieden. Die natürlichen Tage, welche gewöhnlich als Zeitmaasse für gleich gehalten werden, sind nämlich eigentlich ungleich. Diese Ungleichheit verbessern die Astronomen, indem sie die Bewegung der Himmelskörper nach der richtigen Zeit messen. Es ist möglich, dass keine gleichförmige Bewegung existire, durch welche die Zeit genau gemessen werden kann, alle Bewegungen können beschleunigt oder verzögert werden; allein der Verlauf der absoluten Zeit kann nicht geändert werden. Dieselbe Dauer und dasselbe Verharren findet für die Existenz aller Dinge statt; mögen die Bewegungen geschwind, oder langsam oder Null sein. Ferner wird diese Dauer von ihren durch die Sinne wahrnehmbaren Maassen unterschieden und, mittelst der astronomischen Gleichung aus ihnen entnommen. Die Nothwendigkeit dieser Gleichung bei der Bestimmung der Erscheinungen wird aber sowohl durch die Anwendung einer Pendeluhr, als auch durch die Verfinsterungen der Jupiters-Trabanten erwiesen.

Crucially, whereas the original Latin ('ex veriore tempore mesurent') and both English translations ('on the basis of a truer time', 'by a more accurate time') frame absolute time here via a comparative, Wolfers has 'der richtigen Zeit', which is purely positive. Thus, notwithstanding issues of Latin to English translation and the controversy within contemporary Newton scholarship, it is thus evident that on Mach's understanding,

Newton clearly did endorse a view in which time was dynamically absolute, and that is the view to which he was objecting. The crucial issue is therefore well settled for our current purposes.[13]

Mach's objections to substantival time do not, of course, equate to an endorsement of relational notions of time. Asserting that time is fundamentally a relation would be just as inconsistent with the phenomenological thesis as taking it to be a substance.[14] We can, however, plausibly interpret Mach as implicitly endorsing the general idea that temporal structure should be invariant: due to the phenomenological thesis, we should demand that the realization of our temporal structure cannot vary between dynamically indistinguishable universes. We then have that Mach is committed to chronometric structure being dynamically relative *given a stable environment*, since he explicitly talks about time covarying with changes in the motion of bodies.. Plausibly, *given the possibility of an unstable environment*, in our terminology, Mach would hold that chronometric structure was dynamically incomplete since such structure would fail to be well defined under transformations between dynamical universes that were stable and unstable.

4.2.3 Bunge's Criticisms

The Machian view on time has been subjected to a rather scathing critical appraisal by Bunge (1966). Since Bunge's choice of sources is such that he almost entirely focuses upon the 'early' Machian view of time, it will prove highly instructive to consider the objections first, before tracing the subsequent development to a 'middle' or 'mature' Machian view (i.e. 1889–1905).[15]

Bunge makes three general negative claims regarding Mach's view of time and one specific negative claim. Let us consider them one by one. First, Bunge criticizes Mach's supposed desire to eliminate the theoretical conception of time from mechanics on the grounds that 'it would be foolish to eliminate the temporal concept: if we did it we would lose valuable information and we would be unable to build theories of change' (p. 587). This criticism is part of a general attack on Mach's 'narrow and dogmatic empiricism' that, according to Bunge, led Mach to be hostile to theoretical physics altogether. Needless to say, this line of attack is not easy to sustain in the face of what Mach actually thought and said. As noted above in terms of the phenomenological thesis, Mach was not advocating an elimination of all theoretical concepts but rather such concepts should be made appropriately continuous with experience.

As pointed out by Hintikka (2001), it would be to entirely misread Mach to take him to be advocating a radical revisionary programme for mechanics whereby theoretical terms such as mass should be eliminated as primitive terms via explicit definition from an underlying theory. Hintikka takes Mach to be advocating something closer to what econometricians call 'identifiability', wherein 'a parameter is identifiable on the basis

---

[13] Particular thanks are due to Tzu Chien Tho for help with Latin and German translations and Richard Dawid and Radin Dardashti for help with English to German translations in the above analysis.

[14] Banks (2014, p. 55) makes a similar point.

[15] Since our analysis is concerned with the relevance of Mach's views for classical mechanics, we neglect here discussion of the contested historical issues related to Mach's 'late' view on time within his response to the special theory of relativity. See Thébault (2021*a*, §5).

of an economic theory if and only if its value can be determined on the basis of the theory plus possible data' (Section I, accessed online).[16] Although couched in much less vivd language, the spirit, if not the details, of Hintikka's interpretation corresponds well with that of Bank's discussed above: Mach does not seek to eliminate theoretical conceptions but rather make sure that they are woven into the 'continuous fabric of experience-reality'. The view criticized by Bunge is not the view held by Mach, rather it is best identified with the naïve positivist reading of Mach that was, in fairness to Bunge, the mainstream view at the time of his article (Hintikka, 2001; Pojman, 2010).

It is worth noting here that a stronger reading of Mach's sceptical view on spatiotemporal structure might be seen to be implied by the discussion of (Banks, 2021, pp. 270–2). In particular, Banks suggests that Mach's early discussions imply a form of eliminativism: 'Mach indicated his desire to eliminate spatio-temporal representation from physics altogether, not just to reduce space and time to spatio-temporal relations between bodies' (p. 271). This appears in contrast to our earlier discussion, itself motivated by Banks, wherein we suggested Mach's view was best read as a form of phenomenalist functionalism about time: the only temporal concepts that are meaningful are those that can be continuously connected to experience via causal-functional relations. The seeming tension between the two readings is, however, rather less significant than it seems. Plausibly, Mach can be read as a phenomenalist-functionalist with regard to spatiotemporal *theoretical structure* (the aspect we focus on) whilst also being an phenomenalist-eliminiativist with regard to spatio-temporal *representations* (the aspect Banks is focused on in the quotation). The important point is that Mach's view of space and time is focused on a reduction of space and time to causal-functional dependancies between his elements which means that ultimately they are the basis for everything else. However, this does not mean temporal concepts per se become meaningless, just as saying the money can be reduced to functional-social relations is not equivalent to saying that the concept money is meaningless.

The second line of criticism that Bunge offers against the early Machian view of time is more plausible and can be understood independently of the first. Bunge notes that for a functionally adequate physical theory we require a complex concept of time that includes a quantitive concept of duration, and thus a 'mathematical time'. The point is a valid one. However, Bunge's framing of Mach's view on this front is again inadequate. According to Bunge, Mach 'had in mind pairs of simultaneous events and was content with remarking that timing an event consists in pairing it to another event—its match in the standard sequence' (p. 586). It is difficult to follow precisely what Bunge means here, since there does not seem to be any mention of 'simultaneous events' in Mach's various discussions of time, and his uses of 'comparisons' and 'dependence' surely can only be read as quantitive functional relations rather than imply the existence of purely qualitative temporal relations of 'pairing' and 'matching'.[17] This notwithstanding, Bunge is correct that Mach does not provide anything like a

---

[16] It is not entirely clear how closely the notion of 'identifiability' Hintikka's provides in the relevant passage fits with that deployed in contemporary applied statistics. A statistical model is given by a parameterized collection of probability models for the same data. Such a model is called 'identifiable' when each element within it yields a distinct probability distribution for the data.

[17] Furthermore, even if these are read as purely qualitative relations, this does not preclude a relational view of time building quantitive temporal relations from them.

precise, mathematized concept of dynamically relative chronometric structure. We will return to this issue in the context of the Machian version of the chronometric problem shortly.

For the third general line of criticism offered by Bunge it is worth quoting the relevant passage in full:

According to Mach "we select as our measure of time an arbitrarily chosen motion (the angle of the earth's rotation, or [the] path of a free body) which proceeds in almost parallel correspondence with our sensation of time." This is false: (a) time standards are supposed to be regular rather than arbitrary and are corrected or replaced as soon as they are suspected of some irregularity; (b) since the beginning of scientific time reckoning the perception of duration has had little saying in the choice of standards; it has none in the adoption of 'atomic' (actually molecular) clocks; (c) what is decisive in the adoption of a time standard is not our unreliable time sensation, which is rarely parallel to physical duration, but some theory which can justify the assumed regularity of the chosen process.[18]

These criticisms look altogether more damming than the previous two. All three of Bunge's points labelled (a)–(c) are highly plausible and look, at first sight, to directly contradict what is asserted in the quotation from Mach. However, putting the quotation in context makes things look rather different. The excerpt quoted was added to the *Mechanics* from the second 1889 edition onwards (so technically is in our mature Mach period, 1889–1905). The full passage runs as follows:

My views concerning physiological time, the sensation of time, and partly also concerning physical time, I have expressed elsewhere (see *Analysis of the Sensations*, 1886, Chicago, Open Court Pub. Co., 1897, pp. 109–118, 179–181). As in the study of thermal phenomena we take as our measure of temperature an *arbitrarily chosen indicator of volume*, which varies in almost parallel correspondence with our sensation of heat, and which is not liable to the uncontrollable disturbances of our organs of sensation, so, for similar reasons, we select as our measure of time an *arbitrarily chosen motion*, (the angle of the earth's rotation, or path of a free body,) which proceeds in almost parallel correspondence with our sensation of time.[19]

Three points are worthing noting. First, when read in context it is clear that this passage is not principally concerned with presenting Mach's full and detailed view of physical time. Rather, the phrase occurs in the context of a paragraph added to the second edition of the *Mechanics* with the main object of highlighting Mach's view of physiological time that is discussed in the *Analysis of the Sensations*, which had been published in between the two editions of *Mechanics*. The parallel drawn between physical and physiological time is thus serving to a large degree as a book plug. Moreover, the relevant passage is certainly not intended to contradict the earlier view of physical time already fully espoused in the earlier passages of the *Mechanics* quoted extensively above. This is fairly clear already from putting the small excerpt used by Bunge in the context of the relevant paragraph, but is undeniable in the context of a reading of the full text. Furthermore, it is also worth noting that Bunge's removal of Mach's emphasis from the phase 'arbitrarily chosen motion' somewhat modifies the meaning compared to the phrase when read in context.

Second, and most significantly, it is clear from the context of these remarks within the *Mechanics*, and not least the implied meaning of the emphasis, that Mach is not suggesting that time standards are 'arbitrary' in the sense that any one is just as good as another, nor that those we should choose for physical theory must be those that are based upon our unreliable time sensation. As is explicit from what we called

---

[18] Bunge (1966, p. 587).
[19] Mach (1919, p. 541), emphasis in original.

the interconnection thesis above, in fact Mach was greatly concerned with the need for a stable environment with regular processes.[20] Furthermore, Mach, in fact, even himself suggested that the rotation of the earth would not be a stable enough process and, with impressive prescience, actually suggested that eventually something like an atomic standard of chronometry would emerge. Consider the following from *Economic Nature of Physical Enquiry*, first published in 1882:

If it be objected, that in the case of perturbations of the velocity of rotation of the earth, we could be sensible of such perturbations, and being obliged to have some measure of time, we should resort to the period of vibration of the waves of sodium light, all that this would show is that for practical reasons we should select that event which best served us as the *simplest* common measure of the others.[21]

The obvious interpretation of this passage is that the 'simplest common measure' is that which is least sensible to perturbations. The choice of the period of vibration of the waves of sodium light as 'simpler' is on account of our greater *practical* ability to use such a natural process as a stable measure of some other process, not due to a closer correspondence to an absolute time.

It seems fair to forgive Bunge for neglecting these footnoted remarks in one of Mach's more minor texts. What is less excusable is that the contrast between physiological time and physical time is in fact explicitly drawn by Mach in the second of the passages from *Analysis of the Sensations* that is referred to just before the text that is quoted by Bunge. *Analysis of the Sensations*, first published in 1886, is Mach's other major text besides the *Mechanics*. The most relevant text runs as follows:

The time of the physicist does not coincide with the system of time-sensations. When the physicist wishes to determine a period of time, he applies, as his standards of measurement, identical processes or processes assumed to be identical, such as vibrations of a pendulum, the rotations of the earth, etc. [22]

That Mach specifically directs the reader to this text in the full quotation given above is the clearest indication that the views of physical and physiological time should be expected to be read together consistently. On Mach's view, physical and physiological time do not *coincide quantitively* since physical time is constructed via the identification of highly regular physical processes but physiological time is liable to the uncontrollable disturbances of our organs of sensation.[23] However, the two do *proceed in almost parallel correspondence* since physiological time is still ultimately also connected to regular physical processes, albeit with less precision, such as the connection between the rotation of the earth and the physiological daily cycle. Although it is not made explicit here, the obvious further implication of the comparison with thermodynamics in the full quotation is the connection between what we would now call the physiological and physical 'arrows of time'. We will return to the question of the arrow of time in the context of Mach's mature view and the idea of the entropy

---

[20] Admittedly, this then brings with it the problem of defining a suitable notion of 'stable' in purely relative terms. While this is without doubt a difficult problem, it is a tractable one in the context of modern relational particle mechanics (Barbour and Bertotti, 1982; Barbour, 2009*b*; Barbour, 2003; Barbour, Koslowski and Mercati, 2014). Moreover, the need for a relative notion of stability is certainly not the problem that Bunge is pointing to here.

[21] Mach (1895, footnote p. 205), emphasis in original.

[22] Mach (1914, p. 349).

[23] Banks (2003, p. 246) notes in a similar vein that 'Mach said he did not believe physiological conceptions of space, time, and matter should be mirrored in the concepts of physical theory'.

of the universe in the following section.

The important point is that a full and systematic reading of Mach's extensive writings on time in no way supports the interpretation given in the short excerpt given. In fact, Mach would agree with at least the first two of Bunge's three statements; that is, it is entirely consistent with the early Machian view of time to say that: a) time standards are supposed to be regular and are corrected or replaced as soon as they are suspected of some irregularity; and b) the choice of standards in scientific time reckoning has little to do with the perception of duration. The third statement was c) what is decisive in the adoption of a time standard is some theory which can justify the assumed regularity of the chosen process. Whilst there is no direct contradiction between this statement and Mach's, it is perhaps not in the spirit of his approach to theoretical reasoning. However, this is certainly not because of a lack of reference to our unreliable time sensation. Rather, for Mach, what is significant is not merely the regularity of a particular process and the theory of that process, but also the stability of the environment in which the process, and the scientists applying the theory, are embedded. This crucial element of Mach's thought will be refined and further articulated in the mature Machian view of time to be discussed shortly.

In this context, it makes sense to return to the idea of the chronometric problem that we have encountered already in our discussions of Newton and Leibniz and which was mentioned briefly above.. Recall that the chronometric problem is the problem of fixing determinate yet dynamically relative chronometric structure and that solution to the problem is a requirement for the empirical adequacy of a theory of time in mechanical theory. To solve the problem one must show how one can save the appearances of chronobservable structure in mechanical theory within a framework where chronometric structure is dynamically relative. One can frame the pertinent aspect of Bung's criticism of Mach in precisely these terms. Chronobservable structure is given by the time standard of a theory as encoded in regularity of a given set of phenomena. The challenge is to find a dynamically relative means of fixing chronometric structure such that it has the appropriate functional relationship with such structure. The time metric of stable physical process and the time metric of the mechanical theory in question are required to march in step for the theory to be empirically adequate. Clearly this is a substantive challenge for the Machian view point, and the steps towards its at least partial resolution will occupy us in much of what follow.

The final of Bunge's criticisms is more specific and relates to the no true time thesis. Referring to the passage from *Conservation of Energy* quoted above, Bunge notes that, on Mach's view, there 'is no cosmic time' and '[c]onsequently certain questions, such as the age of the universe, are meaningless (p. 586). Now, it certainly does appear consistent with the view of the early Mach to deny that the universe could have an age. And at first sight, this seems deeply problematic for the adequacy of Mach's view. However, it is important to remember the relevant historical context of steady-state cosmology. Whilst there is a fairly straightforward (but not entirely unproblematic) meaningfulness to questions regarding the age of the universe when asked in the context of 'big bang' cosmology, in the context of a steady-state cosmology, where there is no expanding scale factor to play the role of a clock, Mach's view seems a more defensible one. This notwithstanding, Bunge does appear to have a point: if Mach is asserting that it is conceptually inconsistent to talk about a single time for the universe, his theory of time looks rather rigid and conceptually sparse in the light of modern cosmology. However, as we shall see, there is an important development in Mach's thought regarding the true time thesis during what we are calling his mature period. Before then, it will prove useful to consider the development of Mach's view on time spurred by the work of his contemporaries on clocks defined with reference to inertial frames of reference.

4.3 Inertial Clocks

Mach was not the only person thinking deeply about the conceptual foundations of Newtonian mechanics in the late-nineteenth century. Between the publication of the first edition of the *Mechanics*, in 1883, and that of the second, in 1889, he became aware of work on the definition of inertial systems that forced him to change, or at least adapt, his view. Of most significance is a proposal due to Lange (1885) which built on earlier ideas due to Neumann (1870).[24] It will prove well worth setting out both proposals briefly, drawing upon the discussions of DiSalle (1990, 2002, 2020) and Barbour (2001, 2009a). See also Torretti 1996, §1.5, Friedman 2001, p.76, Brown 2005, p.19, Pfister 2014 and Pfister and King 2015, §1.2.

Neumann's idea was to transform reference to absolute space and time in Newton's mechanics to material bodies moving such that they can play equivalent functional roles. With respect to absolute space, this is his hypothetical body—'body alpha'—with respect to which the motion of a free particle is rectilinear. With respect to absolute time, this is the time scale given by two freely moving particles. The key idea behind this 'inertial clock' is that equal intervals of time will be marked out by any two such particles via the period in which they move mutually proportional distances. The idea is then that body alpha and an inertial clock can be combined to give an inertial reference system with respect to which the first law of Newtonian mechanics can be said to hold.

As a solution to the problem of inertial motion, Neumann's introduction of the mysterious body alpha seems to have all the benefits of larceny over labour. Transmuting absolute space into a mysterious material body does nothing to remove the philosophical objections to the concept. That said, the idea of an inertial clock looks very plausible, and the definition offered by Neumann leads naturally into the proposal of Lange. The idea is to use the mutually uniform and rectilinear motion of three free bodies to define a standard for inertial motion of a fourth. Three bodies moving inertially can be used to define a coordinate system such that any fourth body can be judged to be moving inertially precisely when its motion relative to the coordinate system is uniform and rectilinear. Crucially, defining such a reference frame will also

---

[24] Mach had in fact discussed Neumann's proposal already in his 1872 *History and Root of the Principle of Energy* (Mach, 1911, p. 75–80). We will consider the relevance of these remarks, which do not feature in the 1883 1st edition of the *Mechanics*, shortly. Related proposals are those of Euler 1748, Kelvin and Tait 1867, Streintz 1883, Thomson 1884, Tait 1884. Mach makes explicit reference to the work of Euler, Neumann, Lange, and Streintz in the appendix. For our purposes it will be adequate to focus on his reaction to Lange. For discussion of Tait's proposal see Barbour 2001, pp. 101–4.

provide means to define correlates of absolute rotation and acceleration. Lange appears to have provided a means by which Newtonian spatiotemporal absolutism can be rendered entirely innocuous.

The significance of Lange's ideas for Mach's various critiques of Newtonian mechanics is not a matter of broad scholarly agreement. This is partially a function of the differing interpretations of what Mach was actually proposing. On the one hand, DiSalle (1990, 2002, 2020) and Norton (1995) take Mach to be proposing a *redescription* of Newtonian mechanics without absolute concepts; on the other hand, Barbour (1995) takes Mach to be proposing a *new theory of inertia*. Settling this debate would involve a lengthly investigation of the so-called Mach's principle and the famous bucket thought experiment and will not be attempted here.[25]

In our discussion, we will train our focus upon Lange's work only so far as it relates to dynamically absolute time (which is of course to focus on the aspect that he took from Neumann). In that context, the two most relevant questions are: a) does the substitution of Newtonian time by an inertial clock render Mach's criticisms of dynamically absolute time obsolete? That is, does it contradict the no dynamically absolute time thesis?; b) is the suitably reformulated version of Newtonian time consistent with the other aspects of Mach's positive view of time? That is, does it contradict the phenomenological, dynamical relativity thesis, interconnection thesis, and no true time theses?

In a certain limited sense, the answer to the first question is obviously yes. If we take the motion of two freely moving particles to be continuously connected to experience, then absolute time becomes meaningful via the definition of an inertial clock. And in this sense, Mach was clearly a little too quick to dismiss absolute time as meaningless. In this light, it is strange to note that Mach's reading of Lange spurred him merely to add an appendix to the second edition of the *Mechanics* (later expanded and then included into the main text) but not to change the critical passages quoted above. In the appendix he not only gives Lange's work glowing praise but also seems to endorse the proposal:

Lange's treatise is, in my opinion, one of the best that [has] been written on this subject. Its methodical movement wins at once the reader's sympathy. Its careful analysis and study, from historical and critical points of view, of the concept of motion, have produced, it seems to me, results of permanent value ... A system of coordinates with respect to which three material points move in a straight line is, according to Lange, under the assumed limitations, a simple *convention*. That with respect to such a system also a fourth or other free material point will move in a straight line, and that the paths of the different points will all be proportional to one another, are *results of inquiry* ... we shall not dispute the fact that the law of inertia can be referred to such a system of time and space coordinates and expressed in this form.[26]

In the context of this passage and the presumed failure of the no absolute time thesis, various scholars have taken Mach to be either guilty of a degree of stubbornness in not explicitly correcting his earlier 'misunderstandings' (Borzeszkowski and Wahsner, 1995) or of revising his view such that Newtonian space and time are not that conceptually objectionable after all, although they might be still methodologically

---

[25] From a textual perspective it is surely of significance here to follow DiSalle (2002, 2020) in placing emphasis upon the additional discussions of Newton's Corollary 5 that were added to later editions to the *Mechanics*. Unfortunately a detailed comparative discussion of the early and mature Machian views on inertial frames is beyond to scope of our current project.

[26] Mach (1919, p. 544–6).

problematic (DiSalle, 1990). This latter view is well supported by the change in tenor of Mach's later remarks in his *Knowledge and Error*, first published in 1905:

[A] pair of precisely defined physical processes that coincide in time at both ends and are thus temporally congruent retain this property at all times. Such a precisely defined process can now be used as a time scale, and that is the basis of chronometry . . . in each measurement we use bodies, so that in setting up geometrical concepts we must start from bodies . . . This seems to point to the fact that bodies are rigid and impenetrable, which manifests itself when they touch each other, which is the basis of all measurement. However, things have moved on since the beginning of the 19th century. We still need rigid bodies to build our equipment, but we can use light interference to mark points and measure stretches by wave length in seemingly empty space much more accurately than would be possible by means of rigid bodies that abut and touch each other. It is even likely that light waves in vacuo will furnish future physical standards of length and time, in terms of wave length and period of oscillation respectively and that these basic standards will be more appropriate and generally comparable than any others. Through such changes space and time increasingly lose their hyperphysical character [27]

Given such quotations, it appears that Mach has conceded the central point, and accepts that the process of rendering absolute time perfectly respectable (i.e. not hyperphysical) is merely one of finding appropriate practical means by which to construct clocks (i.e. 'light' clocks rather than inertial clocks).

It is without doubt true that Mach must have rethought or at least refined his view about absolute time in response to the work of Lange and others. That said, it would be a mistake to read these changes too strongly. The transmutation of absolute time into an inertial clock is only respectably Machian to the extent to which force-free bodies can in fact be identified empirically, and thus be functionally integrated with chronobservable structure in the relevant sense. Then, due to the interconnection thesis, this is not strictly speaking possible since no such clocks in fact exist.

This point was explicitly noted by Russell (1903) in the context of the proposal of (Streintz, 1883) that parallels that of Lange. See also Saunders 2013, §3 and Barbour 2001, §12.3. It is also indicated in Mach's own discussion of Lange's proposal. In particular, following the more positive discussion of Lange's work in the later editions of the *Mechanics* just quoted, Mach continues:[28]

[A] number of years ago I was engaged with similar attempts . . . I abandoned these attempts, because I was convinced that we only *apparently* evade by such expressions references to the fixed stars and the angular rotation of the earth. This, in my opinion, is also true of the forms in which Streintz and Lange express the law.

In point of fact, it was precisely by the consideration of the fixed stars and the rotation of the earth that we arrived at a knowledge of the law of inertia as it at present stands, and *without these foundations* we should never have thought of the explanations here discussed. The consideration of a small number of isolated points, to the exclusion of the rest of the world, is in my judgment inadmissible.

It is quite questionable, whether a *fourth* material point, left to itself, would, with respect to Lange's "inertial system," uniformly describe a straight line, if the fixed stars were absent, or not invariable, or could not be regarded with sufficient approximation as invariable.

The most natural point of view for the candid inquirer must still be, to regard the law of inertia primarily as a tolerably accurate approximation, to refer it, with respect to space, to the fixed stars, and, with respect to time, to the rotation of the earth, and to await the

---

[27] Mach (1976, pp. 337–348).

[28] Here Mach's references to discussions earlier in the *Mechanics* have been removed.

correction, or more precise definition, of our knowledge from future experience . . . [29]

We must, therefore, surely understand the mature Machian view of time as involving a partial revision, rather than a retraction, of the no dynamically absolute time thesis. This reading is fully consistent with that of Barbour (1995), but in contrast, to DiSalle (1990, 2002), who sees a more significant shift in Mach's view on the cogency of the concept of absolute time and takes the quotation just given to indicate a persistent *methodological worry*.[30] Ultimately, the key to assessing the strength of this shift in viewpoint is the extent to which the early Mach's concerns with regard to absolute space and time were *already* expressed in methodological terms. If we understand Mach's overall empiricist outlook as itself being a principally methodological one, then the early Mach's criticisms of Newtonian absolute space and time are then largely in parallel to mature Mach's criticisms of Lange's inertial system.

Furthermore, we can find already in the early Mach's discussion of Neumann's proposal, sceptical remarks that already express the key *methodological* worry regarding inertial reference systems. In particular, as noted by Staley (2021, p. 39), in Mach's 1872 *History and Root of the Principle of Energy* we find Mach pointing out the limitations of inertial reference systems in terms of the requirement for the assumption of isolation of the reference body from the distant stars (Mach, 1911, p. 75–80). Clearly, once more, Mach is in essence providing a methodological critique: 'such a system of co-ordinates has a value only if it can be determined by means of bodies' (p. 77). Thus, while Lange's more powerful system evidently caused a degree of change in Mach's thinking in that he appears to have taken it as a spur to thinking more precisely and carefully about the implications of such proposals, there is a high degree of consistency across the various critical remarks directed at both Neumann and Lange but also at Newton.

In sum, within the 1872 discussion of Neumann's inertial reference system, the 1883 discussion of Newtonian on space and time, and the 1889 discussion of Lange inertial reference system, Mach's core worry is a methodological concern regarding the general validity the assumption of relative stability of our environment. Thus while we *can* plausibly see a degree of revision within the discussion of Lange, there are good reasons, not to take this revision too strongly. We will return to the issue of formulating the revised no dynamically absolute time thesis within our rational reconstruction project at the end of this section.

Other commentators appear at pains to underplay the significance of the critical discussion of Lange in the second edition of the *Mechanics*. In particular, in a note added at the end of the paper by Borzeszkowski and Wahsner (1995) in response to the discussion of Barbour (1995) in the same volume, the authors appear to beg precisely the relevant question:

Mach's criticism of Lange that one finds in some editions of his *Mechanik* shows that he did not mention that, accepting—as he did—Lange's construction of an inertial system, for reasons of logical self-consistence, a fourth force-free material point *must* follow with respect to one of Lange's inertial systems a straight line (uniformly). The passage here under consideration

---

[29] Mach (1919, p. 546–7), emphasis in original.

[30] It is worth nothing that DiSalle's methodological worry interpretation of the change in Mach's thought *with regard to space at least* is supported by the additional discussions of Newton's Corollary 5 which were added by Mach to later editions of the *Mechanics*. See DiSalle (2002, 2020).

shows again Mach's initial misunderstanding, not only of Newton but also of Lange. This led him to a dim formulation.[31]

Putting to one side the unnecessary condescending tone, this line of response does not seem very plausible. Surely the internal logical consequences of Lange's systems are not something that Mach has simply 'misunderstood' in this passage. Rather, it seems much more plausible to take Mach to be pointing to the problems inherent in de-idealizaing Lange's system within a real physical setting. That reading, following Barbour, seems the only way of making sense of the phrase 'with sufficient approximation'. Thus, the 'misunderstanding' reading of the passage advanced by Borzeszkowski and Wahsner (1995) is simply not adequate.

More interestingly, Borzeszkowski and Wahsner (1995) point out that 'Mach himself dropped this passage later. In later editions, in particular in the last edition supervised by the author, Mach agrees with Lange. There his point then was to state that Lange's point of view need not be the last word.' (p. 66). This claim is worth investigating, especially since the later editions are not available in translation. First, we find that the relevant passage is still present verbatim in the sixth edition (see p. 256) which was published in 1908. Thus the addition is present in unmodified form in the second, third, fourth, fifth, and sixth editions, and thus consistent within the mature Mach period.

In the seventh edition published in 1912, the last with which Mach was involved in the publication of and that which the authors in question are presumably referring to, the evidence to substantiate the claims of Borzeszkowski and Wahsner (1995) is at best slim. It is true that Mach's explicit doubts about the fourth material point are indeed no longer part of the relevant section. And thus, we might think the change marks the emergence of a new *late* Mach position with regard to inertial clocks. However, what we find in place of Mach's own critical remarks, quoted above, are in fact references to recent critiques of Lange's system, in particular due to Petzoldt, see (Mach, 2012*a*, p. 267–9). These worries, according to Mach, are also of concern to others, and 'cannot be eliminated so quickly' (*nicht so rasch zu beseitigen sind*). Mach concludes that, 'until the fog clears' (*bis sich die Nebel verziehen*) the relevant discussion should be 'temporarily suspended' (*vorläufig abbrechen*) (author's translations). Thus, Mach's position of positivity tempered by scepticism with regard to Lange's system remains unchanged.[32]

Furthermore, it should be noted in this context, following Pfister (2014), that the criticism of Lange's system on the grounds that it provides no clear definition of 'free particles' nor means to identify such objects, were also made by Gottlob Frege in an 1891 paper, and in fact accepted by Lange himself in an article of 1902. Thus, Borzeszkowski and Wahsner (1995) are simply incorrect both with regard to the plausibility of Mach's own specific criticism of Lange in the second to sixth editions, since thoughts along similar lines were commonly shared, and, moreover, with regard to his supposed 'agreement' with Lange in the seventh edition.

---

[31] Borzeszkowski and Wahsner (1995, p. 66) emphasis in original.

[32] This surely also holds also for the ideas of Neumann also, notwithstanding the remarks relating to Newton's Corollary 5 mentioned by DiSalle (2002, p. 177).

To return to the point in hand, although the mature Mach never explicitly makes the point himself, the obvious idealization upon which the Lange systems stands is that the reference bodies in question can be assumed to be moving without the influence of the gravitational interaction of distant bodies. It is *only* given such an assumption that one can treat them as force-free bodies. Thus, while it does seem correct to say that the mature Mach is wrong to focus upon effects of the *motion* of the fixed stars rather than the effects of their *gravitational interaction*, his essential point still stands.

Identification of a Neumann–Lange inertial clock, and thus an empirical stand-in for absolute time, rests upon the assumption that effects of distant bodies can be neglected. As such, these constructions are, due to the interconnection thesis, strictly speaking not fully admissible within the Machian framework of time. In the context of the reconstructed mature Machian view that we have periodized up to 1905, we can thus reformulate the no dynamically absolute time thesis as follows:

iv*) **Modified No Dynamically Absolute Time Thesis**: Dynamically absolute time is not meaningful since due to the interconnection thesis no bodies can be found to play the role of genuine inertial clocks. However, an approximate absolute time is meaningful to the extent to which sufficiently isolated bodies can be assumed to play the role of inertial clocks.

Put this way, the modification to Mach's position does not appear to be an extreme one and, as Mach himself notes, the essential points made in the original text of the *Mechanics*, are still valid despite the new insights from Lange's work. To return to our two questions, we thus have that: a) the substitution of dynamically absolute time by an inertial clock *does not* render Mach's criticisms obsolete, due to the interconnection thesis; and b) thus the other aspects of Mach's positive view of time together mean that Lange's version of absolute time is still *inconsistent* with Mach's view. [33]

4.4 Machian Tensions

4.4.1 The Interconnection of Things

We ended the previous section considering Mach's remarks towards the inherent limitations of the idea of an inertial system due to the interconnection of things. With some irony, in advancing such observations Mach was opening up fissures that would serve to partially undermine his claim that there is no global time. The chain of reasoning behind this inference is due to Barbour (2001, §12.3). Let us consider a potential response to the Machian objection to the definition of absolute time via inertial clocks. The response runs as follows. It is true that we do not in fact have access to force-free bodies since we cannot rule out the influence of the distant stars on local inertial motions. However, if we have access to the causes of the forces exerted by the stars on the local bodies in question, then we can simply take these into account, extrapolate the motion of force-free bodies, and in so doing construct a genuine, rather than merely

---

[33] There is an interesting connection between Mach's revised view and the contemporary model-based account of the epistemology of time measurement via the international network of atomic clocks due to Tal (2016). A further interesting connection, which we unfortunately do not have space to explore, is to the analysis of time in the physical relativity view developed by Harvey Brown. See in particular Brown (2005, p. §6.2).

approximate, inertial reference system. Thus the force of the Machian objection is blunted.

We can expand upon this line of response in the particular context of time as follows. Consider a pair of reference bodies identified as 'force-free' and thus moving inertially. We can take the mutually proportional motion of these bodies to provide an inertial clock that plays the role of absolute time. The Machian objection to this arrangement is that on account of the 'interconnection of things', these bodies will not in fact be free, rather their motion will always be subject to some deflection from pure inertial motion due to interactions with other bodies. However, as pointed out by Barbour (2001, pp. 660–1), we can examine the sources of these deflections in terms of *identifiable causes*. Once these causes are identified we can in principle calculate what the motion of the reference bodies would have been if they were in fact force-free. This then allows us to extrapolate the time that would be given by an inertial clock even although we do not in fact have access to any genuinely force-free bodies.

Here Barbour suggests two lines of response, both in the spirit of Mach. First, is a general epistemological worry about whether the causes of the deflection need always be identifiable. In principle there is no reason why the potential causes of deflection from inertial motion need be exhausted by the objects that can be identified in our neighbourhood. For example, Mach talked about the distant stars, whose precise constitution and motion was not determined at the time or, even more explicitly the potential for interconnections, possibly mediated via fields, to produce unknown causes:

The world remains a whole so long as no element is isolated, but all parts are connected, if not immediately then at least mediately through others. The concordant behaviour of members not immediately connected (the unity of space and time) then arises only apparently by failure to notice the mediating links. The goal of the cosmic motion remains unknown only because the segment that we can look at has narrow boundaries beyond which enquiry does not reach.[34]

In modern terms we might consider the effects of dark matter or dark energy. The general point is that we have good physical reasons to expect there to be *non-identified* (or perhaps even non-identifiable) causes of a particular local motion. This then indicates that our extrapolation process towards an inertial clock may not be an entirely reliable one.

The second line of response that Barbour marshals on Mach's behalf runs as follows. Let us neglect the first principally epistemological worry and assume that all causes of a particular local motion can be identified. It is still the case that since Newton's theory of gravitation is universal, a full specification of the causes of deflection from inertial motion of any particular body will include all other bodies in the universe. As noted by Barbour, Russell (1903) makes a related point when discussing a proposal by Streintz that is similar to that of Lange. Furthermore, along similar lines, it has been noted by Guzzardi (2014) in the context of a discussion of Mach's view on the causation, determinism and the conservation of energy that '[on Mach's view] each modification of a part of a system *necessarily* reverberates throughout the entire system' (p. 1286) (emphasis in the original). Thus, the Russell-Barbour line of argument can be reasonably taken to be in the Machian spirit, even if it was never explicitly

[34] Mach (1976, p. 346–7).

put forward by Mach himself. By appeal to this counterargument such a Machian can point out that to arrive at a genuine inertial clock, our extrapolation procedure would need to take into account every body in the universe. The universe would have become its own clock! It is at this point that we should return to Mach's writings, in particular his discussions of the idea of universal entropy, that will ultimately link back to this ideas of Barbour.

4.4.2   The Chronordinal Problem and Universal Entropy

Recall that in his early view Mach is explicit in his denial of the possibility of a global time. This denial is most explicit in his 1871 *Conservation of Energy*. What is significant for our understanding of the progression in his thought is that in the text the denial of a global time is explicitly connected to the proposal by W. Thomson and Clausius that the second law of thermodynamics can be applied to the universe as a whole (see p. 62). This strong and persistent connection between the idea of the entropy of the universe and the idea of universal time runs throughout Mach's work. However, as discussed by Brush (1968) and Banks (2003), Mach's views on thermodynamics changed during the mid-1890s, not least spurred by the reaction to the ideas of Boltzmann on irreversibility. What is important for our current purposes is that his view of the plausibility of talking about the entropy of the universe seems to have undergone a shift from an early phase in which the concept of universal entropy is taken to be meaningless to the mature phase in which universal entropy becomes something that could possibly be determined.[35] This, in turn, has an implication for his view on the plausibility of a global time.

Consider the following four quotations. First in the *Popular Scientific Lectures*, published in 1894:

It is to be remarked further, that the expressions "energy of the world" and "entropy of the world" are slightly permeated with scholasticism. Energy and entropy are metrical notions. What meaning can there be in applying these notions to a case in which they are not applicable, in which their values are not determinable? If we could really determine the entropy of the world it would represent a true, absolute measure of time. In this way is best seen the utter tautology of a statement that the entropy of the world increases with the time. Time, and the fact that certain changes take place only in a definite sense, are one and the same thing.[36]

Then, in *Principles of Heat*, first published in 1896:

Hasty readers of my *Conservation of Energy* have supposed that I there (p. 62) denied the existence of any irreversible processes. But there is no passage which could be so understood. What I said about the expected "death of heat" of the universe I still maintain, not because I suppose all processes to be reversible, but because phrases about "the energy of the universe", "the entropy of the universe", and so on, have no meaning. For such phrases contain applications of metrical concepts to an object which cannot be measured. If we could actually determine the "entropy of the universe", this would be the best absolute measure of time, and the tautology which lies in the phrase about heat-death would be quite cleared up.[37]

---

[35] With more than a little irony, given the his low opinion of Mach's thought (Earman, 1989), there is thus a clear connection between the early Mach's denial of the coherence of universal entropy, and Earman's devastating 'not even false' criticisms of the notion of the universal entropy in the context of the so-called past hypothesis (Earman, 2006).

[36] Mach (1895, p. 178).

[37] Mach (2012b, p.439).

Then, in the extra material added to the appendix of the third addition of the *Mechanics* in 1897:

I have endeavoured also (*Principles of Heat*, German edition, page 51) to point out the reason for the natural tendency of man to hypostatise the concepts which have great value for him, particularly those at which he arrives instinctively, without a knowledge of their development. The considerations which I there adduced for the concept of temperature may be easily applied to the concept of time, and render the origin of Newton's concept of "absolute" time intelligible. Mention is also made there (page 338) of the connection obtaining between the concept of energy and the irreversibility of time, and the view is advanced that the entropy of the universe, if it could ever possibly be determined, would actually represent a species of absolute measure of time.[38]

Finally, in a later letter to Adler, quoted by Banks (2003, p.220–1), he remarks:

For me Time-Space questions are essentially physical questions. The riddle of time will be solved, I believe, through the correct conception of the second law.

There is, therefore, a discernible shift from an earlier position that the change in entropy of the universe, and therefore a global time, is 'meaningless', to a weaker position that such an entropy change is a thing that could possibly be determined.[39] Certainly, it would be over-reading to take Mach to be positively asserting that a true time exists; however, it is also clear that he has moved past his earlier complete denial of the conceptual plausibility of a 'time for the universe'. We can suitably reformulate the fifth claim regarding time as a conditional claim:

v⋆) **Conditional True Time Thesis**: True time is meaningful if universal change can be continuously connected to experience via the determination of the change in universal entropy.

Modifying this thesis has great significance for the overall structure of the Machian view. In particular, rather than, as with the early Mach, denying that true time is a cogent notion, it is clear that the mature Mach would assert that if the change in the entropy of the universe were determinable, then a spatially global 'true time' would be induced.

We have just seen that an appeal to universal entropy allows the Machian to establish a dynamically relative spatially global mathematical time given by a measurable physical quantity. Crucially, however, as Mach would probably have been aware, there is good reason to expect that universal entropy, if it could be determined, would not

---

[38] (Mach, 1919, pp. 542–3).

[39] There is also a natural connection here to Bank's more phenomenalist-eliminiativist interpretation of Mach on space and time. In particular, as detailed in Banks (2003, §14) there is a discernible shift in the mature Mach period towards an attempt to interpret the physical meaning of space and time via statistical and thermodynamical concepts. Ultimately, there need not be a direct tension between Bank's view and the phenomenalist-functionalist reading we have put forward since Mach might plausibly be understood to be an eliminiativist about spatiotemporal *representations* and yet a functionalist about spatiotemporal *concepts*. Moreover, as Bank's himself admits, the remarks his interpretation is built upon are rather 'dark', and his interpretation is suggested as a 'tempting' rather than a necessary one. The key point, contra Banks, is that it is not at all clear that thermodynamic concepts could ever provide a foundation for a concept of time that is functionally adequate for physical theory. In particular, it is not at all clear whether an entropic time is sufficient to provide a suitable metric structure. This is in direct contrast to the Lange-type inertial systems idea, which as Mach was well aware provides a formally rigorous means of reconstructing the metric structure of space and time via the motion of bodies.

be a monotonically increasing function. The chronordinal structure it induces would thus not be *temporally* global since we cannot preclude the possibility of recurrence, and thus a reversal in the sign of universal entropy change, leading to a temporal re-ordering.[40] Thus, the 'true' time that forms part of the mature Machian view of time only partially fulfils the Newtonian idea: it is spatially but not temporally global.

This mature Machian approach to global time ordering constitutes a response to the chronordinal problem that we discussed in the context of Leibniz earlier. Recall that the problem was the problem of fixing determinate chronordinal structure that is dynamically relative. We can understand the Machian solution as one based upon appeal to a particular contingent feature of the universe in order to fix of determinate chronordinal structure. The viability of this solution clearly depends upon our ability to single out a specific relevant feature. Moreover, as noted, due to the failure of monotonicity, the solution is only a partial one, since the chronordinal structure in questions is only determinate locally in time. This issue will be of particular significance in the context of the formal problem of time discussed in Part III.

4.4.3   The Chronometric Problem and the Second Mach's Principle

To complete our discussion let us return to the second problem for theories in which time is dynamically relative. The chronometric problem regarding fixing a determinate temporal metric, a mathematical time. We saw this problem in the early Machian view of time as pointed out by Bunge. Furthermore, the mature Mach does not, himself, provide any clear and explicit response to this problem. That said, the remarks about 'interconnection of all things' and the connection with universal entropy suggest a new and powerful line of reasoning. This has been developed by Mittelstaedt (1980) and Barbour (1981), and called by them, in honour of Mach, the 'second Mach's principle'. Barbour's most explicit discussion (p. 46) of the second Mach's principle runs as follows:[41]

Newton had justified the concept of absolute space by means of the undoubted fact of inertial motion; this is reflected in the fact that there exist distinguished frames of reference in which the Newtonian laws of motion take a canonical and especially simple form. To counter this argument of Newton's for the existence of absolute space, Mach had suggested that inertial motion is somehow governed causally by the distant matter in the universe. This could then explain why some frames of reference are clearly preferred to others without there being any apparent reason for this preference. Following a suggestion of Einstein, this is now called Mach's Principle. In the case of time too, we find a similar curious preference by nature for a particular time measure. Prima facie, time is just a one-dimensional sequence, a topological labelling of successive states. It is then striking fact that there exists a unique (up to shift of origin and change of scale) time metric for which the laws of motion take a particularly simple form. What is the origin of this distinguished time metric?

---

[40] This is an implication of what was shown by Boltzmann by reference to Poincaré's recurrence theorem in his 1897 reply to Zermelo's criticism of the H theorem. See Brush (1966) for an English translation of the original papers and Brush (1968, p. 204) for a short discussion.

[41] We reproduce this lengthly quotation in full here since the original text is only available within a relatively rare collection, with all the other essays in German. Furthermore, Mittelstaedt's original remarks, translated here by Barbour, are in an untranslated German monograph. Since, on the view of this author, the second Mach's principle is of considerable physical and philosophical significance, the genesis of its formulation is of more than incidental importance.

Now we have already seen that Mach was greatly impressed by the "peculiar and profound connection of things". Mittelstaedt poses the question: Did Mach envisage—even less concretely than in the case of inertia—that one might eventually be able to discern, in the connection of things, a causal explanation for the existence of a distinguished time metric? Could, for example "...the cosmic entropy causally influence the rate of all other clocks in some as yet unknown manner" (Mittelstaedt, 1980, p. 25)? Although Mach said nothing explicit to the effect, Mittelstaedt, correctly in my view, thinks it is a surmise sufficiently well justified to be honoured with the name Second Mach's Principle. Its verification would require establishing that "...certain cosmic changes of state (of the entropy or radius of the universe) causally influence the rate of local changes of state. Further, it would he necessary to establish which of the known interactions brings about this influence" (Mittelstaedt, 1980, p. 25).

This second Mach's principle was never explicitly endorsed by Mach himself. However, within it we find the kernel of the modern form of Machianism about time that has proved sufficient to fully resolve the chronometric problem. Recall the idea of a perfected measure of duration based upon an extrapolated inertial clock with *all* disturbing interactions considered. In deriving such a determinate chronometric structure, we would have had to take into account the motion of all bodies in the universe. This would be precisely to implement the second Mach's principle at the universal scale. The beauty of the second Mach's principle is that there is no need, however, to posit such an inertial body in the first place. Rather, what is required is a methodology for deriving a privileged temporal metric from the causal-mechanical interactions of the entire universe. This need not make reference to inertial bodies or universal entropy, but rather solely dynamical gravitational interactions encoded in cosmic changes of state.

In general terms, the proposal is that determinate, global chronometric structure can be constructed in Newtonian physics following the second Mach's principle based upon a suitable aggregation methodology based. Such structure would by its nature be dynamically relative since it will provide a measure of duration that will covary with any dynamically change in the motion of any material body. As Barbour notes, this idea of a time derived from the motion of all bodies is no philosopher's fantasy. It is closely connected to the astronomical notion of *ephemeris time* and can be constructed explicitly in analytical formulations of Newtonian mechanics (**?**). We will consider such reconstructions explicitly in Part III.

4.5 Chapter Summary

The mature Machian view on the nature of time can be rationally reconstructed as follows. All temporal structure should be functionally integrated with chronobservable structure. Chronometric and chrondodinal structure is locally determinate and dynamically relative given a stable environment. Chronometric and chronordinal structure is spatially globally determinate and dynamically relative if it can be connected to chronobservable structure via the change in entropy of the universe. Approximate dynamically absolute chronometric and chronordinal structure is meaningful to the extent to which sufficiently isolated bodies can be assumed to play the role of inertial clocks.

Part II

Symmetry and Structure

In the following five chapters we will introduce a novel framework for the analysis of physical theories. This will involve the introduction of a considerable amount of formal machinery, and it will prove worthwhile to provide some remarks of a more general motivational and methodological character before asking the reader to continue with us to the formal stage of our analysis.

Our primary motivation in constructing our framework, and in particular in not relying upon the 'standard formalism' for representing the structure of the model space of a physical theory, is a simultaneous requirement for greater generality and greater specificity. The standard formalism we have in mind here is usually defined as follows. First pick out a set of *kinematically possible models* (KPMs) via a triple $\{\mathcal{M}, F, H\}$, where $\mathcal{M}$ is a differentiable manifold, F stands for the solution-independent fixed fields common to all KPMs, and H for the 'dynamical' (i.e. non-fixed) fields *not* common between all KPMs. Second, consider the transformation behaviour induced by the pullback of the fields by (typically a subset of) the diffeomorphisms of $\mathcal{M}$. Finally, draw interpretational implications from here based on the transformation behaviour supplemented by relevant dynamical, metaphysical, or epistemic interpretative criteria. Our framework can be understood as a generalization of the 'standard formalism' according to the specific needs of this book. For classic applications of the standard framework see Friedman 1983; Earman 1989; Malament 2012; Pooley 2013; Pooley 2017; Roberts 2022.

The most prominent aspect of generalization is that in the standard formalism the transformations considered are invariably the pullbacks of tensor fields by smooth invertible coordinate transformations on the underlying spacetime manifold; that is, diffeomorphisms of the event space. In contrast, the focus of our analysis will be the more general class of transformations of mathematical objects, which need not be tensor fields, induced by the (analogue of) the pullback of those objects by some underlying transformation on the *the space of dynamically possible models*; that is, the dynamical sector of the model space itself. Significantly, in some cases, the transformations we will consider will have an action that does not correspond to a diffeomorphisms on the underlying spacetime manifold, hence the need for greater generality. We will set out this important formal difference explicitly in §9.1.

This desire to focus on transformations on the space of dynamically possible models is also the reason that greater specificity is needed. In particular, our framework will make essential use of a tripartite distinction between types of structure: *constitutive* structure, *nomic* structure, and *spatiotemporal* structure.[1] The first chapter will focus on outlining what we mean by the first two of these and explaining the connection to the key ideas of symmetry and dynamical equivalence in general terms. Our analysis will then become more specific and detailed in the following chapters, building up powerful machinery for the analysis of spatiotemporal structure and symmetry based on what we will call the *AIR* schema.

---

[1] Considerable inspiration has been drawn here from Friedman's iconic *Dynamics of Reason* (Friedman, 2001). Although the three types of structures in our account do not neatly align with those picked out by Friedman, we take our use of 'constitutive' to be broadly consistent with his.

5
Structure and Possibility

The structure of a theory is specified in terms of three types of structure: constitutive structure, nomic structure, and spatiotemporal structure. Constitutive structure is the most basic structure of a theory. This is the structure that one must assume in order to build the space of kinematically possible models. Nomic structure is the structure that represents the dynamical laws of a particular theory. Nomic structure serves to pick out a partition in the space of kinematically possible models and provides an equivalence relation between dynamically possible models. Regular nomic structure is nomic structure that enforces a partition but not a projection. Irregular nomic structure is nomic structure that enforces a partition and a projection.

5.1   Constitutive Structure and Kinematical Possibility

Constitutive structure is the most basic structure of a theory. This is the structure that one must assume in order to build the space of kinematically possible models—that is, the basic 'pre-nomic' set of possibilities. Each of these models can be thought of as something like a bare universe, stripped of dynamical structure. For the most part we will assume that the constitutive structures consist of: i) a manifold structure used to characterize physical events, and represented by a differential manifold; ii) geometric structures, used to characterize relations of ordering, distance and orientation between the events; and iii) matter structures, used to characterize the non-geometric material content. We will say more about the precise formal details of each of these shortly. For the time being the idea to keep in mind is that together the constitutive structures are necessary and sufficient to represent a kinematical universe of *structured events*.

It is worth mentioning here a partial inspiration drawn from Russell's *Analysis of Matter*. According to Russell, '[e]lectrons and protons ... are not the stuff of the physical world: they are elaborate structures composed of events, and ultimately of particulars' (Russell, 1927, p. 386). We are assuming a similar form of 'event ontology' as being implemented via the constitutive structures. That said, we are certainly not assuming that talk about objects *participating* in events is coherent in a physical context. And, moreover, we will also not assume, like Russell, that events can contain each other.[1]

In order to define the constitute structure more formally we can proceed as follows. First, let us consider a theory, $\mathcal{T}$, as a collection of structures:

$$\mathcal{T} = \{\mathcal{S}_I\}\,\forall I, \tag{5.1}$$

[1] See Pashby (2015, p. 15) and Meyer (2013, p. 17).

where I ranges over the (possibly infinite) collection of different types of structures that make up the theory.

We will not always strictly distinguish notationally between types of structure and their tokens. Unless otherwise stated, the symbol $\mathcal{S}$ will stand for some token of the Ith structure of $\mathcal{T}$. We will use an upper-case Latin superscript for labelling types of structures. When we need to distinguish between different tokens of the same structure we will use a lower-case Latin subscript from the start of the alphabet. Thus two tokens of the same structure would be written $\mathcal{S}^a, \mathcal{S}^b$, and the set of all tokens of a particular structure will be written $\{\mathcal{S}^a\}$. We will not typically need to consider tokens of two different types, but when we do this can be written as $\mathcal{S}_I^a, \mathcal{S}_J^a$.

A particular set of structures may be either overly specific, that is, pick out what would typically be understood as a formulation of a theory, or underly specific, that is, the set of structures do not uniquely distinguish a particular theory. We do not take ourselves to here to be attempting to provide a formal definition of theoretical equivalence but rather giving a method for specifying a theory via its structures that is applicable to the particular examples of theories that we want to investigate. That said, equivalence between the structures of two theories, specified in our terms, might be taken as a sufficient but non-necessary criterion for theoretical equivalence.[2]

The constitutive structures are those required to formulate the space, of kinematically possible models (KPMs) of the theory $\mathcal{T}$. A useful way to divide these structures, and one that will be sufficient for all the classical theories we consider, is in terms of *manifold*, *geometric*, and *matter* structures. We can define these as follows.

Definition 5.1 *Manifold Structure*, $\mathcal{M}$, *is the bare structure of the event space. At the most basic level we can assume a set of events, $\{e_1, e_2, ...e_n\}, e_i \in \mathcal{M}$, together with the topological structure needed to define continuity and convergence properties of functions from the event space to itself or the real line.*

Formally, a *topology*, Ω, on a set, X, is a collection of subsets of X that are open sets of X. The ordered pair, (X, Ω), is then a *topological space*. Given such a topological space, the *neighbourhood* of a point, $x \in X$, is a set, U, which contains an open set, V, containing x. So long as the domain and codomain of a function are topological spaces, we can make statements regarding whether or not the function is continuous. In particular, if X and Y are topological spaces, then the function $f : X \to Y$ is *continuous* at the point x_0 iff for each neighbourhood V of $f(x_0)$ in Y, there is a neighbourhood U of $x_0 \in X$ s.t. $f(U) \subset V$. See Willard (2004) and Nakahara (2003). Continuity of a function is thus only definable given a topology on both the domain and codomain. A similar argument holds for convergence.

For our purposes, it will typically be appropriate to assume that the event space is a *differential manifold*. Informally we can think of this as the set of events together with topological structure and appropriately smooth local coordinate structure. More formally, a *differential manifold* is a topological space equipped with differentiable structure. In particular, the differentiable structure of an m-differentiable manifold,

---

[2] For discussions of theoretical equivalence see Glymour 1970; Quine 1975; Glymour 1980; Glymour 2013; Halvorson 2012; Barrett and Halvorson 2016; Lutz 2017; Nguyen 2017; Weatherall 2019*a*; Weatherall 2019*b*.

$\mathcal{M}$, assumed to be topological space, can be characterized in terms of a family of pairs, $\{(U_i, \varphi_o)\}$, where $\{U_i\}$ is a family of open sets which covers $\mathcal{M}$, and φ_i is a homeomorphism from U_i onto an open subset of $\mathbb{R}^m$. Crucially, given U_i and U_j such that $U_i \cap U_j = \emptyset$, the map $\psi_{ij} = \varphi_i \cdot \varphi_j^{-1}$ from $\varphi_j(U_i \cap U_j)$ to $\varphi_i(U_i \cap U_j)$ is infinitely differentiable (Nakahara, 2003).[3] The generalization to more general kinds of events spaces including discrete spaces and vector spaces will be straightforward in most cases.

Definition 5.2 *Geometric Structure, G_i is structure placed upon $\mathcal{M}$ in order to characterize relations between events such as ordering, distance, and orientation. For our purposes, geometric structure will be represented by the metric or affine (or derivative operator) structures of differential geometry. We will denote the collection of all such structures by the set $\{G_i\} \forall i$, where the G_i are the ith representation of a geometric field and the index i runs over all the geometric structures of the theory.*

For more on the formal definitions used above an accessible introduction to differential geometry written for physicists is Nakahara (2003). An extremely concise and formally rigorous overview of all relevant topics can be found in Malament (2012, §1).

Definition 5.3 *Matter Structure, T_α. This is additional structure used to characterize the non-geometric matter content of the theory $\mathcal{T}$. For our purposes, they will be represented by fields (scalar, tensor, spinor, mixed) over $\mathcal{M}$. We will denote the collection of all such structures by the set $\{T_\alpha\} \forall \alpha$, where T_α is the αth matter field and the index α ranges over all matter structures of $\mathcal{T}$.*

In terms of these structures, the total space of kinematical structures, $\mathcal{K}$, is then the exhaustive set of all combinations of manifold, geometric, and matter structures that are defined in the theory:

$$\mathcal{K} = \{\mathcal{M}, G_i, T_\alpha\} \forall i, \alpha. \tag{5.2}$$

We can also define the space of KPMs which is given by the space of all models that can be defined in terms of the kinematical structures. We will denote this space as K.

Constitutive structure is structure that is *necessary and sufficient* to represent a kinematical universe of structured events within the theory. However, our particular formulation of the constitutive structures is not the only possible choice, thus the specification of $\mathcal{K}$ we will work with will in general be sufficient but non-necessary, formally speaking. This is just to say that there are other possible choices as to the mathematical structures that play the role of the bare events space. For example, rather than the starting with the combination of topological and differentiable structure one might choose ordinal structure as the constitutive structure. Carnap in his discussion of formal space in *Der Raum* proceeds in precisely this direction, using (continuous) ordering structure to build up topological structure.[4] Our choice in constitutive structure is, however, that which is best motivated by physical practice and its assumption will not prove overly restraining in the context of the set of physical and philosophical problems that is the focus of our analysis.

---

[3] See Malament (2012, §1.1) for discussion of alternative definitions of a differentiable manifold.

[4] See Carnap (2019, p. 39–41), Friedman (2019, pp. 174–6), and also Haddock (2016, p. 6).

The final key aspect of our analysis of constitute structure that we will set relates to connection between the toke-type distinction and the space of KPMs. As indicated earlier, we will standardly be implicitly referring to an arbitrary token of a structure in our specifications. In the context of constitutive structures this means that the triple $< \mathcal{M}, G_i, T_\alpha > \forall i, \alpha$ picks out a set of arbitrary *tokens* of the relevant manifold, geometric, and matter structures. We can write this $< \mathcal{M}, G_i, T_\alpha >^a \forall i, \alpha$ using the upper Latin index to indicate a token as per above. The space of KPMs, K is then a space of models which constitute kinematical universe of structured events built with the relevant token of constitutive structures. We can write this K^a.

This seemingly rather arcane token-type distinction will in fact prove of considerable analytical use. In particular, we can consider distinct tokens of the same types of constitutive structure, say $< \mathcal{M}, G_i, T_\alpha >^a$ and $< \mathcal{M}, G_i, T_\alpha >^b$, as picking out spaces of kinematically possible models, K^a and K^b, which differ with regard to key physical features such as the number of matter and geometric degrees of freedom. Models with different tokens of the same type of constitutive matter and geometric structure are typically constitutively distinct kinematically possible models of the same theory despite the fact the share the same constitute structure. Thus, for example, two- and three- body particle models can be understood as having the distinct tokens of a common Newtonian constitutive structure.

These distinctions will be of increasing importance as we enrich our analysis to consider nomic and spatiotemporal structure. In particular, the toke-type distinction will allow us to distinguish between the following to cases. First we have structure common between all kinematically possible models which share the same type structure. Such structure is *constitutively fixed*. Second we have structure common between all kinematically possible models which share the same token of matter structure but which varies between at least two distinct tokens of the same type constitutive structure. Such structure is *contingently fixed*.

On their own, neither the tokens nor the types of KPMs that we can pick out are going to correspond to a very interesting set of possible models. Without further structure, theories are next to useless for physical purposes. The main use of constitutive structures and the token-type distinction will be in a context in which we have included 'nomic' structures and thus given ourselves the resources to talk about a space of dynamically possibly models.

5.2 Nomic Structure and Dynamical Equivalence

Nomic structure or law-like structure is the structure that represents the laws of a particular theory. In our framework nomic structure will have two important and distinct functions. It could thus be implemented by two different structures working in concert—although in some cases we will be able to isolate a single structure that is suited to play both roles. First, the nomic structure picks out a *partition* in the space of kinematically possible models: it tells us which models are dynamically possible and which are not dynamically possible. Second, nomic structure is required to provide us with an equivalence relation between dynamically possible models: it tells us which dynamically possible models are 'dynamically distinct' from each other and which are 'dynamically identical'.

In the context of the space of KPMs these two functions can be described as follows. The first function is the *partitioning* the space of KPMs into a sector of dynamically possible models (DPMs) and a sector of non-dynamically possible models (non-DPMs). The second function is the provision of an *equivalence relation* between dynamically possible models: a methodology for determining which DPMs are dynamically distinct and which are dynamically identical. That the nomic structure provides such an equivalence may sound controversial to some readers. Such a function is highly non-trivial and is not guaranteed. Rather, provision of an unambiguous and physically salient equivalence relation is the *basic desideratum* of nomic structure. Much of the following discussion will be focused on the formal and conceptual conditions for such a nomic structure to play such a function.

The relevant notion of distinctness and identity is here a nomic one rather than an ontic one; that is, a distinction based upon a difference that the laws pick out between two models and not a distinction that is necessarily equivalent to a strong metaphysical notion of distinctness and identity. Most significantly, it is perfectly possible for two models to be dynamically identical and yet ontologically distinct. Ontic distinctness and identity operate at a metaphysical level of interpretation that we will largely leave open: almost everything we say will be compatible with multiple interpretative stances regarding criteria for two models to be ontologically distinct. However, it is plausible to assume that any two models which are classified as dynamically distinct must be taken to be ontologically distinct.

Our notion of nomic distinctness thus *supervenes* on ontic distinctness. Supervenience is a useful philosophical concept that can be defined as follows (McLaughlin and Bennett, 2018):

Definition 5.4 *Supervenience*. *A set of properties A supervenes upon another set B just in case no two things can differ with respect to A-properties without also differing with respect to their B-properties.*

So for our case: nomic distinctness supervenes upon ontic distinctness since no two models can differ nomically without also differing ontically. The converse relation does not hold since we can easily provide examples of states which differ ontically but not nomically. Indeed, such states are one of the principal subjects of analysis in the philosophical foundations of physics. The choice of criteria for dynamical distinctness is thus not entirely metaphysically neutral. We will return to these issues several times in what follows.

As already noted, nomic structure in a particular theory may be implemented by two (or more) different structures working in concert. We leave this possibility open by considering at this stage an abstract representation of nomic structure. We will consider particular instances of nomic structure and the attendant complications in the context of concrete examples in what follows. The important thing for us here is that for a given theory we will make the working assumption that one has means to identify nomic structure sufficient to play the functional roles of: i) differentiating DPMs from non-DPMS; and ii) telling us whether any two given DPMs are dynamically identical or not. Worries about the equivocal status of symmetry principles in providing a formal foundation for such an identification will be considered in due course, drawing upon the influential analysis of Belot (2013). In particular, we will devote considerable time to specifying and exemplifying our notions of dynamically distinct and dynamically identical such that they can be made precise and unambiguous in the *use* of physical theory in a dynamical context, even when this is not possible in entirely general and abstract terms. We leave fleshing out the meaning of dynamical identity to the reader's intuition for the time being.

At an abstract level we can represent the nomic structures that play the relevant two functional roles in terms of a *partitioning map* and a *projection map*.

Definition 5.5 *Partitioning Map, n, is nomic structure that partitions the space, K, of KPMs into the proper subspace, $D \subset K$, of DPMs and the subspace, $\neg D$, of non-DPMs such that $K = D \cup \neg D$. In general, such structure can be represented as a non-injective map*

$$n : K \to D. \tag{5.3}$$

The assumption that the map is non-injective is necessary since unless the laws impose no constraints on the kinematical possibilities, there will be at least one KPM that is not a DPM.

Definition 5.6 *Projection Map, π_N is nomic structure that serves to 'project out' distinctions between models according to some dynamical equivalence principle.*

We can represent this structure in four steps:

  i. *First, consider the function $e : D \to \mathbb{R}$ such that two elements $x, y \in D$ are dynamically equivalent, $x \sim y$, iff $e(x) = e(y)$.*
 ii. *Second, use this equivalence relation to define the* dynamical equivalence classes *$[x] = \{y \mid y \sim x, \forall y \in D\}$.*
iii. *Third, designate the space of equivalence classes $\tilde{D}$ of D the space of* Distinct Dynamically Possible Models (DDPMs).
 iv. *Finally we can consider the projection from the space of DPMs to the space of DDPMs:*

$$\pi_N : D \to \tilde{D}. \tag{5.4}$$

The projection map will necessarily be a surjective function since for every element of $\tilde{D}$ we will have that $\pi_N(D)$ is well-defined. When it is non-trivial it will also be non-injective since it will categorize at least two models as dynamically equivalent. An important special case is when the members of the equivalence class produced by $\sim$ are related by a group action. Here, e gives D a principal fibre bundle structure for the corresponding group, where π_N is the bundle projection, $[x]$ are the bundle's fibres, and $\tilde{D}$ is the base space. This case is of particular significance in standard gauge theories such electromagnetism and Yang–Mills theory. We will illustrate below precisely how the nomic structure of such theories encodes such an equivalence relation.

Returning to our abstract considerations of nomic structure, the combined implication of the two maps that make up the nomic structure is to take us from a representation of the possibility space of a theory in terms of the space of kinematically possible models, K, to a representation in terms of the space of distinct dynamically possible models $\tilde{D}$. We represent this combined action schematically in Fig. 5.1.

More formally, given the above definitions, the overall nomic structure can be represented by a map that is the composition of n and π_e:

$$N = \pi_N \circ n : K \to \tilde{D}, \tag{5.5}$$

which maps the space of KPMs to the space of DDPMs. The nomic structure thus leads us to represent the dynamically distinct possibilities licensed by the theory in terms of a single reduced space where each point corresponds to a dynamically unique model.

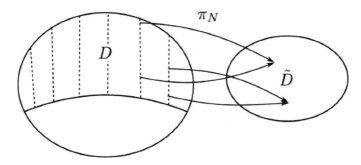

Fig. 5.1: Schematic representation of the nomic structure. D is the partition of dynamically possible models. The dotted lines represent dynamical equivalent models which the projection map, π_N, maps into single points in the space of distinct dynamically possible models, $\tilde{D}$.

It is important to note, once more, that within this framework there is no need for one to require that there is a one-to-one correspondence between the dynamically unique models defined in $\tilde{D}$ and distinct ontic possibilities. This is an additional *interpretational move* corresponding to a particular choice of a model-world representation relation or standard of representational equivalence between models. It is only given a choice of a particular interpretative attitude towards representation that our formal notion of dynamically distinct models can be taken to fix which models should be interpreted as representing ontologically distinct possibilities.[5]

---

[5] Questions of representation and ontological equivalence have been the subject to a recent blooming literature in the particular context of the 'hole argument', which putatively undermines the substantivalist interpretation of spacetime in general relativity. For the original debate see Norton (2019) are references therein. For the most notable contributions to the new representation orientated turn of the hole argument, together with the wider debate regarding representation and equivalence see Roberts 2014a; Weatherall 2018; Belot 2018a; Fletcher 2018; Dewar 2019; Bradley and Weatherall 2019; Ladyman and Presnell 2019; Roberts 2020; Pooley and Read 2021; Martens and Read 2021. It is of particular significance, given our broadly quietist attitude to such matters, that almost all parties to this increasingly labyrinthine debate appear to agree with regard to the relevant standard of dynamical equivalence even when they seem to be at pains to disagree about almost everything else. From our perspective, this focus on representation and ontological equivalence rather misses the most physically salient problem, which is that of dynamical equivalence. Once such a dynamical

5.3 Symmetry and Equivalence

5.3.1 Broad and Narrow Symmetries

The terminology we have introduced thus far already affords us resources to demarcate a number of concepts of symmetry and equivalence. In particular, we are now in a position to disambiguate the idea of a symmetry as a transformation which 'maps solutions into solutions' via of the following three-fold distinction:

Definition 5.7 *Non-Symmetry Maps are maps between dynamically possible models of a theory which are constitutively distinct at the token level; that is, maps between different tokens of the same constitutive structure. These maps take solutions into solution but are not symmetries since they do not preserve constitutive structure at the token level.*

Examples of such maps are mapping between two-body and three-body solutions in Newtonian Mechanics or mapping between Schwarzschild and de Sitter solutions in general relativity.

Definition 5.8 *Broad Symmetry Maps are maps which transform between dynamically possible models that are constitutively equivalent at the token level. A broad symmetry map transforms between dynamically possible models with the same token of the constitutive structure. These are endomorphisms on the space D.*

Definition 5.9 *Narrow Symmetry Maps are the subset of the broad symmetry maps which correspond to maps between dynamically possible models which are constitutively equivalent and dynamically equivalent.*

The difference between non-symmetry maps and broad symmetry maps is precisely that between maps that preserve the constitutive structure at a type level, so DPMs which are such that the constitutive structures are distinct; that is, such that $< \mathcal{M}, G_i, T_\alpha >^a \rightarrow < \mathcal{M}, G_i, T_\alpha >^b$; and maps that preserve the constitutive structure at the token level; that is, such that $< \mathcal{M}, G_i, T_\alpha >^a \rightarrow < \mathcal{M}, G_i, T_\alpha >^a$. The key feature of symmetries in the broad sense is preservation of constitutive structure at the token rather than the type level, thus the 'maps solutions to solutions' idea should always be understood with this qualification in mind.

Constitutive equivalence is then a purely formal relation that is implied by the definition of the constitutive structure itself as a type of mathematical object. There are no existent examples, to our knowledge, of controversy regarding interpretation of constitutively distinct models as equivalent *in any relevant sense*. The second distinction is similarly straightforward since constitutively equivalent models are, by definition, equivalent in the precise sense that they are built from the identical tokens of the same type of constitutive structure. However, any further designation of equivalence will, on its own, quickly collapse into triviality.

equivalence criterion is settled it is not clear to us what fruit further debate can bear. This view is in line with the approach gestured to in Gryb and Thébault (2016*b*), and explicitly defended by Curiel (2018). .

In particular, as noted by Belot (2013), if we designate any two DPMs that are related by a broad symmetry as dynamically equivalent, then dynamical equivalence immediately collapses into constitutively equivalence and ceases to play any functional role within the framework. While merely transforming solutions into solutions is plausible as a necessary condition for a transformation to be between dynamically equivalent models, it is obviously too weak as a sufficient condition. We will return to these issues in the context of an extended discussion of Belot's analysis shortly.

5.3.2 Dynamical Equivalence

This takes us to the final and most subtle distinction, that we have characterized in terms of *dynamical* equivalence. We take to be uncontroversial that *something more* than the idea of constitutive equivalence between DPMs, and thus the broad notion of symmetry, is needed to understand the physical significance of 'symmetry' in the context of the DPMs of a physical theory. What this 'something more' should be is the object of intense debate within the philosophy of symmetry. In the context of our framework we can articulate the controversy as regards to how, if at all, a notion of narrow symmetry, and thus of dynamical equivalence, can be tied to the nomic structure of a physical theory. Our approach thus takes something like the 'dynamical approach' to symmetries as a starting point, but certainly not an ending point.[6]

The most decisive relevant intervention within the recent literature on the dynamical approach to symmetries is that due to Belot (2013). It will prove well worthwhile to conduct a brief detour into this discussion before we lay out our own account in the following chapters. Belot's discussion focuses on difficulties in finding a formal notion of symmetry that can serve as a basis for what he calls 'physical equivalence'. This is in essence the same as our notion of narrow symmetry. Belot considers the standard formal approaches to the definition of symmetry and argues that that are plausible reasons, based upon a stock of physical examples, to believe that in all cases the relevant formal definition of symmetry is simultaneously too weak and too strong to provide a general and unequivocal principle on its own.

Most relevant for our purposes, Belot argues that each of the notions of symmetry coming from a variational principle (Variational Symmetries, Divergence Symmetries) and Hamiltonian formalism (Hamiltonian Symmetries) falls foul of the simultaneously too weak and too strong problem. These can be defined as follows.

Consider a local Lie group, G, with an action on the independent variables, x, and the dependent variables, u and their nth-order derivatives, $u^{(n)}$, and a variational integral:

---

[6] A complementary approach, with somewhat sharper interpretational teeth, is the PESA approach described in Gryb and Sloan (2021). The particular virtue of this approach is that it side-steps tensions between dynamic and epistemic definitions of symmetry discussed at length in the literature (Ismael and Van Fraassen, 2003; Roberts, 2008; Healey, 2009; Belot, 2013; Dasgupta, 2016; Wallace, 2022; ?). Under the PESA proposal, we derive formal properties of a symmetry from contingent facts about the world via precise dynamical criteria that can be used to distinguish observable and surplus structure by considering the number of independent input data required to solve the evolution equations. Such work offers a natural interpretative extension of the intepertationally minimal approach describe here. It is also a distinct alternative to category-theoretic (Weatherall, 2016; Nguyen, Teh and Wells, 2020; Bradley and Weatherall, 2020) definitions of surplus structure or other recent proposals for analysing symmetry (Caulton, 2015; Dewar, 2019; Martens and Read, 2021).

$$\mathcal{L}[u] = \int_{\Omega} L(x, u^{(n)}) dx \tag{5.6}$$

where Ω is an open connected subset of the space of independent variables with boundary $\partial\Omega$. Given this we have the following three notions of symmetry (Olver, 1991):

Definition 5.10 *A variational symmetry is where G leaves $\mathcal{L}[u]$ unchanged. In general, this is equivalent to the condition that the Lagrangian form $L(x, u^{(n)}) dx$ is a contact-invariant two-form, and thus that:*

$$(g^n)^{\star}[L(\bar{x}, u^{(n)}) d\bar{x}] = L(x, u^{(n)}) dx + \Theta, g \in G \tag{5.7}$$

for some contact-form $\Theta = \Theta_g$ possibly depending upon the group element g.

The contact form Θ is required for Lagrangians that have an explicit dependence on the independent variable x (Olver, 1991). For most applications in this book, we will consider Lagrangians that do not have an explicit dependence of x so that Θ is not necessary.

Definition 5.11 *A divergence symmetry is where G leaves the Lagrangian form $L(x, u^{(n)}) dx$ unchanged up to a boundary term. In general this means that we have:*

$$(g^n)^{\star}[L(\bar{x}, u^{(n)}) d\bar{x}] = L(x, u^{(n)}) dx + \Theta + d\Xi, g \in G. \tag{5.8}$$

for some exact two-form $d\Xi$.

Definition 5.12 *A Hamiltonian symmetry is where G leaves the Hamiltonian function H unchanged.*

For detailed and comprehensive treatments of each of these notions of symmetry the magisterial analysis of Olver (1991) is strongly recommend. We will return to the formal analysis of these notions of symmetry in the context of physical examples in the following chapters.

In our terms, the problem is that standard approaches to the analysis of nomic structure in terms of variational symmetries, divergence symmetries, and Hamiltonian symmetries will not furnish a reliable notion of narrow symmetry since there are cases where there are at least some DPMs which we want to think of as dynamically distinct which they designate as equivalent, and at at least some DPMs which which we want to think of as dynamically equivalent which they designate as distinct.

Three points are worth bearing in mind in the context of Belot's analysis. First, and least significant, it should be noted that the two examples that Belot draws upon repeatedly as indicative of the *too strong* aspect of the problem are formally and physically extremely special. These are the Coulomb–Kepler system of motion in a central potential and (isotropic) simple harmonic motion. In each of these systems there is a special class of 'hidden symmetries' that are formal symmetries of both the action principle (variational and divergence) and the Hamiltonian in the relevant sense and yet relate models that we intuitively take to be dynamically distinct.

Most vividly, in the Coulomb–Kepler model, transformations that preserve the so-called Runge–Lenz vector are formal symmetries in all the relevant sense (i.e.

variational, divergence, Hamiltonian), yet correspond to mapping between solutions that can be characterized as ellipses with different eccentricities but the same energy, as determined by the length of the major axis. Explicitly the vector is given by $\mathbf{A} = \mathbf{p} \times \mathbf{L} = mk\frac{\mathbf{r}}{r}$ where $\mathbf{L}$ is angular momentum and $k = GmM$ and the other symbols have their standard meaning.[7]

Prima facie such examples appear highly problematic for the programme of building a notion of dynamical equivalence from the analysis of nomic structure. However, it is important to consider the question of how general this problem of hidden symmetries actually is. A little investigation into the mathematical literature reveals strong formal evidence for hidden symmetries to be peculiar to a very small number of systems. In particular, not only do the Coulomb–Kepler problem and the harmonic oscillator exemplify the class of *super-integrable* systems but they can in fact be proved to be the *only two examples* of such systems for spherically symmetric potentials.

This can be seen as follows: Bertrand's theorem states that the only spherically symmetric potentials $V(r)$ for which all bounded trajectories are closed are the Coulomb–Kepler system and the harmonic oscillator, and this can be shown to imply that no other super-integrable systems are spherically symmetric (Miller, Post and Winternitz, 2013). A straightforward physical explanation of what is going on here is that the extra constants of motion in the Coulomb–Kepler system and the harmonic oscillator are associated with parameters of the relevant *closed* dynamical orbits. Their existing additional constants of motion depends in each case upon the orbits being closed (Goldstein, Poole and Safko, 2014, p.418). Since Bertrand's theorem implies that for spherically symmetric potentials only these two systems have closed orbits, it is no surprise that only these two systems have the extra integrals of motion needed for super-integrability. Finally, and more generally, since super-integrability corresponds to the existence of more integrals of motion than there are independent degrees of freedom, there are good *prima facie* reasons to expect that questions of dynamical equivalence will be especially subtle for such systems.

Where does this leave us? Evidently, the examples of the Coulomb–Kepler system and the harmonic oscillator demonstrate that Belot is correct that formal symmetry principles are too strong to licence completely general claims about dynamical equivalence. However, it remains to be seen whether the too strong side of the problem in fact generalises to examples outside the special class of super-integrable systems, which clearly have special physical features which indicate a more nuanced treatment of dynamical equivalence is required.

This need for care in the context of super-integrability is well illustrated by the relevance of hidden symmetries to quantization. In particular, the treatment of the Runge–Lenz vector symmetries and spacetime symmetries quantum theory is very similar and yet subtly different. The quantum correlate of the Coulomb–Kepler system is of course the basic non-relativistic model of the hydrogen atom. The Runge–Lenz vector symmetries and spacetime symmetries each lead to degenerate energy levels and together they constitute the symmetries of the relevant model. Interestingly, we then find that although an external potential can split the relevant degenerate energy levels

---

[7] See Goldstein, Poole and Safko (2014, §3.9) for an introductory treatment and Prince and Eliezer (1981) for a more detailed analysis.

in both cases, the 'accidental' degeneracy associated with Runge–Lenz symmetry (first discovered by Pauli) corresponds to *reducible* representations but standard degeneracy associated with the spacetime symmetries correspond to *irreducible* representations (Alliluev, 1958). Thus although the case of the Runge–Lenz symmetry is certainly a subtle one, spacetime symmetries and hidden symmetries are formally distinct in the context of the quantum theory.

Ultimately, the principal moral which we take Belot's analysis of symmetry in super-integrable systems to illustrate is not that dynamical equivalence criteria based upon formal symmetry principles are inevitably always going to be too strong, but rather that the physical context of use of a theory or set of dynamical models always needs to be taken into account. This is particularly relevant in the context of system-subsystem relations. We will return to this point when it becomes relevant in the context of our analysis of physical examples below. For a more programmatic analysis that we take to be along these lines see Wallace (2022).

The second point, which will prove more significant to our analysis, relates to the two key examples that Belot uses to illustrate the *too weak* side of his argument. Most familiarly, there are the Galilean boosts which relate DPMs of Newtonian Mechanics which differ by the addition of a uniform velocity. These are *not* variational or Hamiltonian symmetries and thus we have a very straightforward example in which a narrow symmetry principle that relies upon either of these formal notions of symmetry will be too weak. This might seem to motivate us to move directly to the only one of our three candidate formal symmetry principles that does include Galilean boosts, that is divergence symmetries. However, in that context, we find examples of what looks like a *prima facie* narrow symmetry transformation and yet is not divergence symmetry: scaling symmetries or conformal transformations. Most vividly, spacetime rescaling symmetry of the Einstein field equations does not preserve the Einstein–Hilbert Lagrangian even up to a divergence (Anderson and Torre, 1996, 2B) and is thus not a divergence symmetry. Such examples do indeed indicate the need to move beyond the standard symmetry formalism. In particular, in the context of the canonical analysis that is our main focus in this book, the requirement for formal tools to analyse DPMs related by recalling is one of the two key reasons to move beyond the symplectic analysis and introduce the machinery of contact geometry.

We will return to the additional insights that such a formalism affords us in the next chapter. What is important for our current discussion is the implication of finding that a formal symmetry principle is not strong enough to include all relevant examples of what we intuitively would like to call narrow symmetries. Rather than being a devastating problem, we take this issue to be a good motivation for our flexible and rather pragmatic approach to nomic structure and the partitioning map. This is because our approach allows for the combination for a differential analysis for a piece-meal approach to formal symmetry principles in which we identify narrow symmetries via a partitioning map that is built up in stages, drawing upon different aspects of the nomic structure.

5.3.3 The Preliminary Stance on Narrow Symmetries

Our methodology will be to proceed in stages with the first stage the formulation of a deliberately too weak definition of narrow symmetry built upon an appropriate refinement of one of the standard formal symmetry definitions. The second step will be then to modify the nomic structure such that the further physically well-motivated narrow symmetries can be stipulated. We will provide more details regarding each of these steps in what follows. For the time being we can specify our weak notion of narrow symmetry via a sufficient but not necessary condition as follows:

Definition 5.13 *Preliminary Stance on Narrow Symmetries is that all broad symmetries which are variational symmetries of the full system under study, are narrow symmetries.*

Here we are using the word 'stance' following a partial inspiration from van Fraasen's epistemic voluntarism in the context of physical reasoning regarding symmetries, laws, and theoretical models. In particular, we take there to be significant room for 'operation of the will' when it comes to our choice of narrow symmetry principles (Van Fraassen 1989, 2008; Okruhlik 2014).

Our preliminary stance leads to us include as narrow symmetries the least controversial example of transformations between DPMs that we might take to connect dynamically equivalent DPMs: spacetime transformations that do not change the action. By design this notion of narrow symmetry is too weak. It excludes not only the hidden symmetries mentioned above, which we might take as a positive feature but also the Galilean boosts of Newtonian Mechanics, which we might reasonably want to include since they are also 'spacetime symmetries'. However, what may seem like a fault in our analysis is in fact an important positive feature. In particular, we take the non-invariance of the action principle of Newtonian Mechanics under Galilean boosts to be indicative of the fact that inertial frames are *non-redundant structures* of the theory. On our view, if ones wishes to include Galilean boosts within the class of narrow symmetries and thus render inertial frames redundant structures, then the physically well-motivated approach is to modify the nomic structure such that they become narrow symmetries under the preliminary stance. We will return to this important example of our general approach to symmetry and structure in our discussion of Barbour–Bertotti theory in §8.5.

5.3.4 Regular and Irregular Nomic Structure

The final and most important point we will make in this context relates to the key formal feature of the nomic structure that is oddly not considered in Belot's rightly lauded analysis; that is, the issue of the well-posedness of the evolution equations. Nomic structure, as already noted, has two important functions: partitioning and projecting. From the perspective of the variational problem or canonical equations of motion, the most fundamental sub-division is then between those theories in which the nomic structure *enforces* a projection as a requirement for a well-posedness and those theories in which the nomic structure which allows for independent well-posed equations of motion for all models which are DPMs.

In the context of Lagrangian action principles this is a distinction which is (almost) entirely between what are usually called 'regular' Lagrangians and 'irregular' Lagrangians. We will conduct a full analysis of this distinction, including a highly important exception, in Chapters 7 and 8. For the time being, still operating on an abstract level, we can introduce the following distinction:

Definition 5.14 *Regular Nomic Structure is nomic structure that enforces a partition but not a projection.*

For a theory with regular nomic structure it is in principle possible to treat all DPMs as *prime facie* dynamically distinct. However, there are typically good physical reasons to apply a projection following the preliminary stance.

Definition 5.15 *Irregular Nomic Structure is nomic structure that enforces a partition and a projection.*

For a theory with irregular nomic structure it is mandatory to classify at least some DPMs as dynamically equivalent in order to construct a well-posed initial value problem. It is in principle then possible to treat all DPMs which are independently well-posed as dynamically distinct. However, again there are typically good physical reasons to apply a projection following the preliminary stance.

Significantly, irregular nomic structure carries with it a notion of dynamical equivalence such that all the narrow symmetries it implies will also be symmetries in the more traditional sense (i.e. variational, divergence, or Hamiltonian symmetries). The converse does not hold, however, since there are many straightforward examples of invariances of the variational problem which are associated with regular nomic structure. There is also an important connection between irregular nomic structure and Noether's second theorem that will be examined in detail later.

Finally, and crucially, in the canonical context, the existence of irregular nomic structure is precisely the motivation for introducing the *constrained* Hamiltonian formalism which is, of course, the arena of canonical quantum gravity and the problem of time. We will return to these connections in Chapter 7. Chapter 6 will focus exclusively on regular nomic structure and an exhaustive analysis of how the space of DPMs of a regular theory can be parameterized via *absolute charges*.

5.4 Chapter Summary

The structure of a theory is specified in terms of three types of structure: constitutive structure, nomic structure, and spatiotemporal structure. Constitutive structure is the most basic structure of a theory. This is the structure that one must assume in order to build the space of kinematically possible models. Nomic structure is the structure that represents the dynamical laws of a particular theory. Nomic structure serves to pick out a partition in the space of kinematically possible models and provides an equivalence relation between dynamically possible models. The latter role can be formalized in terms of a projection map that serves to 'project out' distinctions between dynamically possible models according to some dynamical equivalence principle. A broad symmetry is any map which transforms between dynamically possible models that are constitutively equivalent. A narrow symmetry is a subset of the broad symmetries which correspond to maps between dynamically possible models which are constitutively equivalent and dynamically equivalent. Under the preliminary stance on narrow symmetries, all broad symmetries which are variational symmetries of the full system under study are narrow symmetries. Regular nomic structure is nomic structure that enforces a partition but not a projection. For a theory with regular nomic structure it is in principle possible to treat all dynamically possible models as *prime facie* dynamically distinct. However, there are typically good physical reasons to apply a projection based at least on following the preliminary stance. Irregular nomic structure is nomic structure that enforces a partition and a projection. For a theory with irregular nomic structure it is mandatory to classify at least some dynamically possible models as dynamically equivalent in order to construct a well-posed initial value problem.

6

Conservation and Geometry

This chapter considers the geometric structure of theories with regular nomic structure within the framework of Hamiltonian mechanics. In this context, broad symmetries facilitate the identification of dynamically isolated substructures of a theory. In particular, the space of conserved charges and the Poisson sub-algebras of this space give representations of the interesting dynamically stable structures of a theory. However, there are good reasons to avoid a naïve reductionism in which the isolated structures of a theory are removed. Rather, once the broad symmetries are parameterized so that the isolated structures and conserved charges are identified, one can then proceed to identify the narrow symmetries in a second step based upon physical reasoning. The preliminary stance, that narrow symmetries are all broad symmetries which are variational symmetries of the full system under study are narrow symmetries, is the basic starting point in this approach.

6.1 Canonical Mechanics

The ideas articulated in the previous chapter were expressed such that they could be applied to virtually any formulation of a physical theory. In particular, the individual models that make up the space D might be expressed in terms of the solutions of a set of partial differential equations, whether or not those equations are derived from an action principle, or, moreover, have a corresponding Hamiltonian or canonical formulation. In this section we will specialize our analysis to the case of canonical systems in which a Hamiltonian analysis is well-defined.[1]

Consider the action functional:

$$S[\gamma] = \int_{t_0}^{t_1} L(q, \dot{q}, t) \mathrm{d}t, \tag{6.1}$$

where q represents the collection of generalized position coordinates, $\dot{q}$ the generalized velocities, t is a time parameterization, L is the Lagrangian which typically takes the form $L = T - V$ with T kinetic energy and V potential energy, and indices are suppressed. Hamilton's principle of least action states that the motion of a mechanical

---

[1] For the most part we will follow the classic textbook treatment of Arnold (2013), supplemented by the more detailed discussion of Belot (2007) and Arnold and Givental (2001). The problem of discerning whether or not a particular system of differential equations admits a variational formulation is called the 'inverse problem'. See Olver (1991, pp. 377–8) for the discussion of the history of the problem and an array of relevant references. Given a problem with a variational formulation, there are then further restrictions required to guarantee a well-defined canonical analysis, most importantly that the Legendre transform is well-defined. We will return to this point explicitly shortly.

system, γ_{cl}, correspond to the stationary points of the action functional, i.e. those such that $\delta S[\gamma]|_{\gamma_{\mathrm{cl}}} = 0$.

For regular nomic structure, which we have assumed, the space of DPMs, D, can then be specified directly in terms of the curves γ_{cl} or equivalently via the solutions to the Euler–Lagrange equations:

$$\frac{\mathrm{d}}{\mathrm{d}t}\left(\frac{\partial L}{\partial \dot{q}}\right) - \frac{\partial L}{\partial q} = 0 \,. \tag{6.2}$$

For our purposes it makes sense to further specialize to Lagrangian systems that can be made to be equivalent to a corresponding Hamiltonian system using a Legendre transform. This can be achieved first by restricting to Lagrangian systems that admit a Cauchy problem, and then by imposing an appropriate set of spatial boundary conditions on the variational problem. We will not need to specify the explicit form of such spatial boundary conditions for our analysis.

The key objects in the Hamiltonian formulation are the generalized momentum coordinates $p = \frac{\partial L}{\partial \dot{q}}$. For the case of an n-dimensional system a *phase space* can then be defined as a manifold $\Gamma = \mathbb{R}^{2N}$ with coordinates $p_1, \dots p_N; q_1 \dots q_N$. Lagrangian-Hamiltonian equivalence can be understood, in simple terms, via the equivalence between the Euler–Lagrange equations, which we can write as $\dot{p} = \frac{\partial L}{\partial q}$, and Hamilton's equations:

$$\dot{p} = -\frac{\partial H}{\partial q} \qquad\qquad \dot{q} = \frac{\partial H}{\partial p} \,, \tag{6.3}$$

where $H(p, q, t) = p \cdot \dot{q} - L(q, \dot{q}, t)$ is the Hamiltonian function.

Foreshadowing what follows, an alternative, less standard way of writing Hamiltonian mechanics for an n-dimensional system, in terms of an *extended phase space* $\Gamma' = \mathbb{R}^{2N+1}$ with coordinates $p_1, \dots p_N; q_1 \dots q_N; t$. We can then write Hamilton's equation (equation 6.3) as the 'vortex lines' of the one-form:

$$\theta' = p\mathrm{d}q - H\mathrm{d}t \,. \tag{6.4}$$

The one-form θ' is called the Poincaré-Cartan one-form, and will be the key object behind the theory of canonical transformations that we will consider shortly. For the proof of the equivalence with Hamilton's equations that proceeds via Stokes' lemma, see Arnold (2013, §44).

Returning to the more standard non-extended formalism, we can isolate the key geometric object as the symplectic potential given by $\theta = p\mathrm{d}q$. The exterior derivative of this structure gives the symplectic two-form:

$$\omega = \mathrm{d}\theta = \mathrm{d}p \wedge \mathrm{d}q \,. \tag{6.5}$$

The phase space Γ is then a symplectic manifold (Γ, ω). Symplectic manifolds are even-dimensional manifolds equipped with a closed, non-degenerate two-form, such as ω. They are generic to mechanical contexts since the cotangent bundle of a configuration space, $\mathcal{C}$, is typically given by a symplectic manifold, the phase space, $\Gamma = T^\star \mathcal{C}$.

Hamilton's equations can then be expressed in terms of the *Hamilton vector field* v_H defined implicitly via

$$\mathrm{d}H = \iota_{v_H}\omega\,,\tag{6.6}$$

where the *iota map*, ι, is the interior product on the exterior algebra of Γ. Hamilton's equations then state that the DPMs, γ_{cl}, are the integral curves of v_H:

$$\dot{\gamma}_{\mathrm{cl}} = v_H(\gamma_{\mathrm{cl}})\,.\tag{6.7}$$

By Frobenius' theorem, these integral curves define a regular foliation of any region in Γ where v_H is everywhere non-zero.

To find a way to characterize the space of DPMs, we must find a sub-manifold of Γ that is transverse to the integral curves γ_{cl}. A convenient way to do this is to find a function τ whose level surfaces are normal to v_H:

$$\mathcal{L}_{v_H}\tau = \{\tau, H\} = 1\,,\tag{6.8}$$

where the *Poisson bracket* is defined as $\{f, g\} = \iota_{v_f}\iota_{v_g}\omega$ and v_f and v_g are the Hamilton vector fields of f and g. Using a theorem due to Darboux theorem,[2] we know that, in a finite (topologically trivial) region around these level surfaces, a canonical set of coordinates $(\tau, Q^a; H, P_a)$ (for $a = 1, \ldots, N-1$) can be found whose Hamilton vector fields span the tangent distribution of Γ in that region and whose non-zero Poisson brackets are

$$\{\tau, H\} = 1 \qquad\qquad \{Q^a, P_b\} = \delta^a_b\,.\tag{6.9}$$

In this way, Darboux's theorem trivializes the geometry of Γ in a finite local patch and we can work as if (τ, H, Q^a, P_b) are independent coordinates on $\mathbb{R}^{2N}$. In particular, the tangent distribution of the level surfaces $D_t = \{p \in \Gamma | \tau(p) = t\}$ of τ is spanned by $(v_\tau, v_{Q^a}, v_{P^a})$, which is trivially transverse to v_H. Since all DPMs are tangent to v_H, the surfaces D_t thus intersect all DPMs in the local patch where Darboux's theorem applies.

Since this method for constructing the space of DPMs makes use of Darboux's theorem, it will generally not work globally in Γ. This can occur, for example, if the level surfaces of H change topology as in the closed $(E < 0)$ and open $(E > 0)$ solutions of the Kepler problem. However, by patching together finite local sets of Darboux coordinates that satisfy (6.9), one can always globally construct the space of DPMs. In practice, it may be difficult to find functions τ that satisfy the condition (6.8) even in a local finite region.

When such a function τ is found, it serves as a convenient time coordinate along the integral curves $\gamma_{cl}(t)$. We can thus think of the variable τ as a privileged clock variable. We will have much more to say about such clock variables later. See Fig. 6.1 for illustration.

The final key piece of machinery for studying canonical mechanical systems is the Hamilton–Jacobi equation. This equation follows fairly trivially from the formal

---

[2] For a proof of Darboux's theorem see §43.B of Arnold (2013). The proof presented there can straightforwardly be adapted to the construction of local coordinates that satisfy the conditions (6.9). As we note below, doing this in practice for concrete system can be very cumbersome.

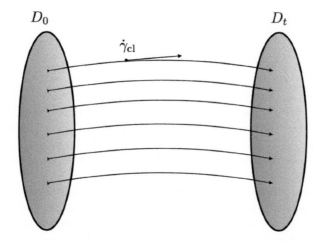

Fig. 6.1: The integral curves γ_{cl} of v_H and their intersection with the level surfaces of τ, which are transverse to this flow. The clock function τ can then be used as a time coordinate along γ_{cl}.

ingredients that we already have. In particular, we can identify the differential of the action (6.1) for a fixed initial point with the Poincaré–Cartan one-form (6.4):

$$\mathrm{d}S = p\mathrm{d}q - H\mathrm{d}t. \tag{6.10}$$

We then have from the definition of the total derivative that:

$$\frac{\partial S}{\partial t} = -H(p,q,t) \qquad\qquad p = \frac{\partial S}{\partial q}, \tag{6.11}$$

from which the Hamilton–Jacobi equation immediately follows:

$$\frac{\partial S}{\partial t} + H\left(\frac{\partial S}{\partial q}, q, t\right) = 0. \tag{6.12}$$

This is a first-order partial differential equation whose solution can be used to construct the space of DPMs when the system is well-posed. We will not perform a detailed analysis of this equation but study some simple cases where it can be partially integrated.

The standard ansatz for solving the Hamilton–Jacobi equation when the Hamiltonian has no explicit dependence on t is to require that the *principal function* S be expressed in terms of the time-independent *characteristic function* W as:

$$S = W - Et, \tag{6.13}$$

where the separation constant E exists for any time-independent system, and is interpreted as the total energy.

The ansatz (6.13) leads directly to the reduced Hamilton–Jacobi equation:

$$H\left(\frac{\partial W}{\partial q}, q, t\right) = E.$$ (6.14)

As will be familiar to many readers, the relationship between equations (6.12) and (6.14) is formally directly analogous to that between the time-dependent Schrödinger equation and the time-independent Schrödinger equation. This relationship, as well as the existence of the conserved quantity E, are absolutely fundamental to the analysis of the problem of time provided in the second volume. We will return to this connection in Chapter 14.

A particularly instructive way of thinking of the Hamilton–Jacobi approach to mechanics is in terms of canonical transformations. These correspond to any (smooth) transformation on phase space that preserves the symplectic two-form ω, and are also called *symplectomorphisms* (i.e. diffeomorphisms that preserve the symplectic structure).

The important point for our purposes is that one seeks to identify a canonical transformation with special phase space coordinates in which the dynamics are trivialized. In particular, we can look for local transformations $g : \mathbb{R}^{2N} \to \mathbb{R}^{2N}$ that transform the original phase space coordinates (p, q) to a new set of coordinates (P, Q) in which the Hamiltonian vanishes.

This can be achieved by defining a type-2 generating function $g(q, P, t)$ such that

$$d(S - g + PQ) = PdQ - Kdt$$ (6.15)

and then setting $K = 0$.[3]

Adding exact forms of this kind to S can always be done without changing the DPMs of a theory because such terms do nothing more but add a boundary term to the action, which leave the stationary points fixed. Expanding the differential of g and rearranging (6.15) leads to the following relations:

$$Q = \frac{\partial g}{\partial P} \qquad\qquad p = \frac{\partial g}{\partial q} \qquad\qquad H(q, p) + \frac{\partial g}{\partial t} = 0.$$ (6.16)

Because the transformed Hamiltonian K is set to vanish, Hamilton's equations say that (Q, P) are just initial data:

$$\dot{Q} = 0 \qquad\qquad\qquad \dot{P} = 0.$$ (6.17)

When the first two relations of (6.16) can be inverted for (q, p) in terms of the initial data (Q, P) and t, then the system (6.16) simply reproduces the main equations (6.11) of the Hamilton–Jacobi formalism with $g = S$. Thus, we can think of Hamilton's principal function S as a type-2 generating function of a canonical transformation that parameterizes the dynamics of the system in time t.

---

[3] Type-2 generating functionals by definition depend on the old coordinates and the new momenta. The extra exact term $d(PQ)$ in (6.15) is required for g to be a type-2 generating functional.

Using similar methods, it is possible to define a type-2 generating function for a canonical transformation that transforms the Hamiltonian to a function $K(\bar{P})$ only of the transformed momenta $\bar{P}$. Hamilton's equations then immediately imply that the $\bar{P}$'s are constants of motion. Such a transformation is always defined for finite dimensional time-independent systems when the constant energy surfaces $H = E$ are compact. These systems satisfy the conditions for the *Liouville–Arnold theorem*. In this case, the coordinates $\bar{Q}$ parametrise N-dimensional tori and are called *action angle variables*. The constants of motion $\bar{P}$ are then the maximum number of Poisson-commuting variables that can be found in a Hamiltonian system. These systems will provide a useful example to illustrate certain implications of our formalism below. For more details on how to construct such variables, see Chapter 10 of Arnold (2013).

6.2   Contact Mechanics

It will now prove highly insightful to introduce some less standard mathematical machinery, that is the theory of *contact geometry*.[4] As already mentioned above, symplectic structures are only well-defined on even dimensional manifolds. Since physical systems are typically constrained to constant energy sub-manifolds (hyperplanes) of an even dimensional phase space there is obvious physical salience to the problem of defining a suitable 'odd dimensional twin' to the symplectic geometry for manifolds of one dimension less than the even dimensional phase space. This odd dimensional twin is contact geometry, and the twin of the symplectic potential is the contact one-form.

The particular relevance of contact geometry to our analysis is as follows. The structure of symplectic geometry is well suited to the parameterization of broad symmetries which map between DPMs with the same value of energy. In contrast, contact geometry provides a framework for us to treat energy as a variable and thus for the parameterization of broad symmetries which transform the energy. There is, furthermore, a close connection between contact geometry and the representation of time in phase space. Thus by considering contact geometry we will have available not only a formal tool to provide a complete parameterization of the broad symmetries in the relevant mechanical contexts but also a tool to represent the structure of time with maximum precision.

To understand the difference between a symplectic manifold Γ and a contact manifold $\mathcal{A}$, it is instructive to compare the symplectic potential θ of Γ with the contact one-form α of $\mathcal{A}$. The main difference regards the non-degeneracy of the volume-forms one can define in the respective spaces. Given an even-dimension manifold Γ of dimension $2N$ the necessary and sufficient condition on θ for Γ to be a symplectic manifold is that the volume-form $(d\theta)^N$ (and therefore the symplectic two-form) be non-degenerate. Given an odd-dimension manifold $\mathcal{A}$ of dimension $2N + 1$ the analogous necessary and sufficient condition on α is that the volume-form $\alpha \wedge (d\alpha)^N$ be non-degenerate. A simple application of the Frobenius theorem can be used to show that this condition implies that α defines a foliation of $\mathcal{A}$ by codimension-1 sub-manifolds whose tangent distribution, $\{Y^i\}$, is not in the kernel of $d\alpha$. This defines a symplectic substructure

---

[4] For an introductory discussion, see Arnold and Givental (2001, §4). For modern introductions see Bravetti (2018) and de Leon and Lainz Valcázar (2019). For a discussion of the pivotal role of contact geometry in geometric optics see Arnold (1990).

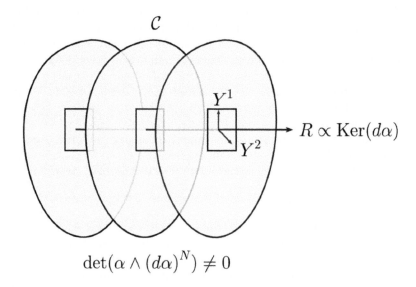

$$\det\!\left(\alpha \wedge (d\alpha)^{N}\right) \neq 0$$

Fig. 6.2: Sheaf of codimesnion-1 symplectic sub-manifolds and tangent distribution of contact manifolds.

of $\mathcal{A}$. The remaining non-symplectic direction called the *Reeb vector*, R, spans the kernel of $d\alpha$ and is normalized to satisfy $\iota_R \alpha = 1$, see Fig. 6.2. When $\alpha \wedge (d\alpha)^{N}$ is non-degenerate, it is straightforward to show that the Reeb vector, R, is unique.

To illustrate these structures more concretely, consider a set of local Darboux coordinates on $\mathcal{A}$ such that the $2N$ coordinates (q, p) parametrise the symplectic directions and the single coordinate z is chosen to parametrise the direction along the Reeb field. In these coordinates, the contact one-form can be written as:

$$\alpha = dz - p\,dq\,, \tag{6.18}$$

where the Reeb vector field takes the form $R = \frac{\partial}{\partial z}$. Just as a canonical transformation (or symplectomorphism) on a symplectic manifold Γ is a diffeomorphism that preserves the volume-form $(d\theta)^{N}$, a contact transformation (or *contacto*morphism) on a contact manifold $\mathcal{A}$ is a diffeomorphism that preserves the volume-form $\alpha \wedge (d\alpha)^{N}$.

Let us now study the geometric structure of D_t. This is the family of surfaces that parametrise the flow of v_H in Γ. We focus our attention on the tangent distribution of D_t. Because of how we have defined D_t using Darboux coordinates satisfying (6.9), we know that the basis vectors $\{v_{Q^a}, v_{P_a}, v_\tau\}$ are both tangent to D_t and invariant under the Lie flow of v_H. This basis is particularly convenient because it spans the tangent distribution of D_t and because it is in terms of a maximum set of (local) invariants of the dynamical flow. However, since D_t is an odd-dimensional space, it is a contact, rather than a symplectic, manifold. We can obtain a preferred contact one-form on D_t by taking the restriction of the symplectic one-form θ onto D_t. Up to an exact form, $\theta = \tau dH + PdQ$. Its restriction to D_t (i.e. the level surfaces of τ) is therefore $\alpha_t = \theta|_{\tau=t} = tdH + PdQ$. By inspecting (6.18), we see that, in these coordinates,

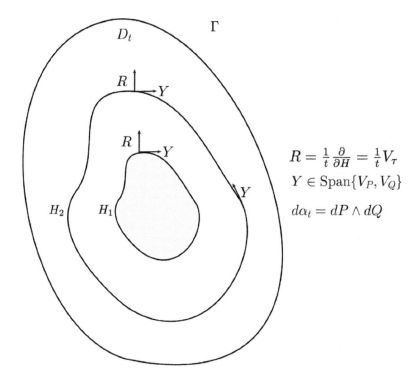

$$R = \frac{1}{t}\frac{\partial}{\partial H} = \frac{1}{t}V_\tau$$

$$Y \in \mathrm{Span}\{V_P, V_Q\}$$

$$d\alpha_t = dP \wedge dQ$$

Fig. 6.3: Contact geometry of the space of DPMs

the Reeb vector field is $\frac{1}{t}\frac{\partial}{\partial H}$ and the symplectic substructures have a symplectic form given by $d\alpha_t = dP \wedge dQ$.

We can now use this basis for the tangent distribution of D_t to understand the geometric structure of the space of DPMs. The Reeb vector $\frac{1}{t}\frac{\partial}{\partial H}$ generates flow between DPMs with different energies, where the value, E, of the energy is given by the coordinate H. These constant-H surfaces define symplectic sub-manifolds of D_t with the symplectic structure $dP \wedge dQ$ (see Fig. 6.3.).

These should be compared with the symplectic solution-space for Hamiltonian mechanics defined in standard texts such as Woodhouse (1997, §2.3). The difference between this and standard treatments is that the space D_t also includes all DPMs with different values of the conserved energy E. As already noted, including such DPMs will be crucial for our analysis later.

Note also the similarities and differences between α_t and the Poincaré–Cartan one-form (6.4) that we defined on the odd-dimensional extended phase space. This too is a contact one-form but on the extended phase space where t is now used to parametrise the Reeb direction $\frac{1}{H}\frac{\partial}{\partial t}$. While θ' is convenient for describing the flow of v_H parametrised by t, α_t is convenient for describing the space of DPMs with different energies E.

6.3   Conserved Charges and Symmetries

The Darboux coordinates (H, Q^a, P_a) on the space $\mathcal{D}_t$ of DPMs give us a lot of formal information about the dynamical properties of a particular theory. The space of contact transformations on $\mathcal{D}_t$ gives us a means to transform any (initial data set for a) DPM to any other simply by changing the value of H or (Q^a, P_a). Such transformations fit the notion of broad symmetry introduced in the last chapter. Despite its significance, it is important to emphasize the limitations of such a broad notion of symmetry. While it is always formally possible to find a set of Darboux coordinates satisfying the requirements of (H, Q^a, P_a) using, for example, the procedure outlined in §43.B of Arnold (2013), this procedure can be extremely cumbersome and explicit expressions are generally difficult, if not practically impossible, to obtain. Moreover, the group of broad symmetries of this kind is infinite, which raises the question of whether such a definition is too broad to be fruitful. Belot, for example, refers to such a broad definition as 'fruitless' and a 'recipe for disaster' (Belot, 2013, §3).

It is important to recognize that the fact that broad symmetries are so plentiful does not imply that they are uninformative. An embarrassment of riches still constitute riches! After all, broad symmetries encode non-trivial algebraic and dynamic structure. In this regard, one fruitful feature of broad symmetries is that they allow us to identify dynamically isolated substructures of a theory. Because the surfaces $\mathcal{D}_t$ are tangent to v_H, the functions (H, Q^a, P_a) parametrise the space of conserved charges and the Poisson sub-algebras of this space give representations of the interesting dynamically stable structures of a theory. Moreover, because the conserved charges label a theory's DPMs, there will be formal relationships between a theory's charges and its symmetries. It will prove useful to take a moment to discuss these relationships in a bit more detail to illustrate the utility of our framework and to highlight the connection with important cases appearing in the literature.

First, consider an infinitesimal diffeomorphism on Γ (not necessary a symplectomorphism) generated by a vector field $V \in T\Gamma$. The integral curves of v_H, and therefore the classical curves $\dot{\gamma} = v_H(\gamma)$, will be preserved if

$$\mathfrak{L}_V v_H(\mathrm{d}t) = 0. \tag{6.19}$$

A simple calculation shows that this condition is satisfied if

$$\mathfrak{L}_V \theta = c\theta + \mathrm{d}\phi \qquad\qquad \mathfrak{L}_V H \mathrm{d}t = cH\mathrm{d}t, \tag{6.20}$$

where c is a constant and ϕ is some arbitrary phase space function. This implies that the Cartan–Poincaré form θ' is rescaled, up to an exact form, by a constant e^c:

$$\theta' \to e^c \theta' + \mathrm{d}\phi. \tag{6.21}$$

Such transformations will transform the action $S[\gamma] = \int_\gamma \theta'$ so that $S \to e^c S + \phi|_{t_0}^{t_1}$ and the stationarity condition $\delta S[\gamma]|_{\gamma_{\mathrm{cl}}} = 0$ is preserved. Since this is the most general form of transformation that preserves the stationarity condition, the vector field V generates the most general infinitesimal diffeomorphism that will preserve the DPMs, γ_{cl}, of the theory. Note that for $c \neq 0$, it may be the case that $\mathfrak{L}_V \mathrm{d}t \neq 0$, and therefore that the time parameterization of γ gets rescaled by a constant. Note also that our construction is local on phase space. Global issues can be handled on a case-by-case basis but will generally not concern us in this book.

In the simplest case one can restrict to transformations such that $c = 0$ and $\mathfrak{L}_V H = 0$ so that t is invariant and θ can only be shifted by an exact form so that the symplectic two-form, ω, is preserved. We will call transformations that satisfy these conditions *Noether symmetries*. Because these transformations are canonical, they can be written in terms of some phase space function f such that $df = \iota_V d\omega$.[5] The second condition then immediately implies that $\{f, H\} = 0$, and therefore that f is a constant of motion. This means that every symmetry generated by the flow of some vector field V that preserves ω has an associated conserved charge f. This result has been called the *symplectic Noether theorem* (Souriau, 1974, p. 357) and we will use that name henceforth to differentiate it from the related but formally distinct first theorem due to Noether.[6] The charges f are then *Noether charges* of the theory.

For time-independent Hamiltonians, the Hamiltonian itself is a Noether charge with H playing simultaneously the role of the generator of time translations and of evolution in time. The interpretation of this dual role as both symmetry and evolution generator will be the subject of intense study throughout this book. More straightforward is the interpretation of the functions (Q^a, P_a) on $\mathcal{D}_t$, which are standard Noether charges. Together with the total energy, they saturate the number of independent Noether charges a theory can have.

The converse of the symplectic Noether theorem does not hold for all conserved charges. The constant function m is the *central charge* of the Poisson algebra on Γ since it Poisson commutes with all the other elements of the algebra. It is defined by $dm = 0$, and therefore generates no Hamilton vector field and can have no associated symmetry. This function nevertheless commutes (by definition) with H and is therefore conserved. Such functions will be important when studying phase space representations of the Galilean group below.

The function τ is not a Noether charge because $\mathfrak{L}_{v_\tau} H \neq 0$. For this reason, transformations that change constant energy surfaces, which are generated by v_τ, are generally not Noether symmetries. This doesn't mean, however, that such transformations are not symmetries: they clearly *are* broad symmetries in the sense that they transform DPMs to DPMs. They are simply not Noether symmetries because they are not generated by a conserved charge (although they are still canonical transformations on Γ).

To cover the remaining cases, we must consider $c \neq 0$. We will call symmetries of this kind *dynamical similarities* because they are general scaling symmetries that reduce to symmetries of the same name first studied in the Kepler problem. In that context, they have been used as an example of broad symmetries that relate *empirically inequivalent* DPMs (Belot, 2013; Wallace, 2022). However, in the context of homoge-

---

[5] This follows from the non-degeneracy of ω and Cartan's magic formula which has the general form: $\mathfrak{L}_X \pi = \iota_X d\pi + d(\iota_X \pi)$ for a n-form π.

[6] The symplectic Noether theorem was derived in the 1970s by Souriau (1997, p. 103) and Kostant (1970). For discussion in the context of Noether's work and further references see Kosmann-Schwarzbach (2010, p. 108).

neous cosmology dynamical similarities are best understood as relating empirically equivalent DPMs see (Gryb and Sloan, 2021).[7]

Dynamical similarities rescale the symplectic two-form and thus correspond to a rescaling of the unit of phase space volume (i.e. the unit of angular momentum). As noted above, for them to be symmetries they must also be accompanied by either one or both of a global rescaling of the time parameter and the total energy. While we include dynamical similarities for completeness, a full treatment is beyond the scope of this book. We note briefly that because of their non-symplectic character, they are closely connected to the theory of contact Hamiltonian flows.[8] As is proved in recent work (Bravetti and Garcia-Chung, 2021), such flows are associated with conserved charges, but these charges are *not* standard Noether charges and do not generate dynamically similarities through their Hamiltonian flows.

To summarize, it is clear that there are formal relationships between conserved charges and symmetries, but that such relationships can be complicated. In many cases, symmetries can be shown to be generated by conserved charges according to Noether's theorem. But broad symmetries can also be generated by non-conserved charges as the example of τ and v_τ shows. It's also possible for a conserved charge, namely the central charge of the Poisson algebra on Γ, to have no associated symmetry at all. In further cases there are symmetries, namely dynamical similarities, that have associated charges but are not generated by them. Finally, and importantly for the purposes of this book, there is a charge, namely the total energy, that can be both understood as generating a symmetry, namely time translations of the DPMs, and dynamical evolution. The complex web of entangled relationships between symmetries and conserved charges is a subtle affair. Nevertheless, the geometric picture we have presented lays bare the essential structures in a transparent and unifying way.

6.4   Isolated Substructures and Narrow Symmetries

We now consider in a bit more detail what can be learn from broad symmetries and how we may use this to inform a definition of narrow symmetry. We already saw in the previous section that many of these symmetries imply the existence of conserved charges because of the dynamical information built into their definition. The symplectic structure on Γ suggested further relationships between certain symmetries and conserved quantities through the symplectic Noether theorem. But broad symmetries encode even more non-trivial information due to the symplectic structure and, more generally, the Lie algebra of vector fields on Γ. In particular, we will now argue that broad symmetries can be used to identify the dynamically isolated substructures of a theory. These structures are powerful heuristic tools because they are stable structures that can, for example, be used as reference structures or even be completely removed from the theory without affecting the dynamics of the remaining structure within the theory.

---

[7] In this context, dynamical similarities are interesting for potentially providing a notion of classical singularity resolution (Koslowski, Mercati and Sloan, 2018; Sloan, 2019; Mercati, 2019) and an explanation for the arrow of time (Barbour, Koslowski and Mercati, 2014; Gryb, 2021).

[8] See for example discussions in Sloan 2018; Bravetti 2018; de Leon and Lainz Valcázar 2019.

To unpack these claims, consider a collection of vector fields $X = \{V^i\}$ that all satisfy the condition (6.19) and are therefore infinitesimal generators of broad symmetries. If the set X satisfies $[X, X] \subseteq X$ (i.e. the vector fields are involutive) then, by Frobenius' theorem, X arises as the tangent distribution of a regular foliation of Γ. The condition (6.19) further guarantees that all flows along the leaves of this foliation preserve the integral curves of v_H. It is then always possible to perform a quotient of Γ by the action of X using the leaves of the foliation to define equivalence classes on Γ. On the quotiented space, the preservation of the integral curves of v_H means that evolution of the remaining quantities is unaffected by the quotient (up to a choice of global time parameter). The symmetries generated by X have been *reduced* in the sense that the directions along X have been eliminated from the theory without affecting dynamics along the directions transverse to X.

This procedure is very general, although explicitly performing a quotient can be awkward in many cases. The procedure can even be applied in the case where the reduced space is not a symplectic manifold.[9] The end product is an integrable system that is effectively decoupled from the degrees of freedom along the X directions.

This reduction procedure becomes more transparent when we consider what it means for structures, such as phase space functions, that can be used to represent physical quantities. The relationship between the V^i and phase space quantities, again, passes through Frobenius' theorem. Since X arises as a regular foliation, we know there exists local coordinates v^i on this foliation such that $\iota_{V^i} \mathrm{d}v^j = \delta^{ij}$. Reducing the V^i directions is then the same as eliminating the quantities v^i from the theory. Care must be taken when applying this procedure globally, but this should always be possible provided the vector fields are globally involutive.

If the vector fields in X are the generators of Noether symmetries, the procedure can be extended. Each V^i field will have a corresponding conserved charge $P = \{p^i\}$ such that $\mathrm{d}p^i = \iota_{V^i}\omega$. The symplectic structure (and its inverse) can then be used to convert functions to vector fields (and vice versa). The condition $[X, X] \subseteq X$ immediately implies that the momenta p^i form a Poisson subalgebra $\{P, P\} \subseteq \{P, m\}$ on Γ, where m is the central charge of the Poisson algebra on Γ. Furthermore, the definition of v^i then implies: $\{v^i, p^j\} = \delta^{ij}$. Locally, one can then also construct the vector fields U^i that obey $\iota_{U^i} \mathrm{d}p^j = \delta^{ij}$ using $\mathrm{d}p^i = \iota_{U^i}\omega$. Since the momenta in P are conserved quantities by the symplectic Noether theorem, it may be useful to eliminate these from the theory and thus obtain a symplectic flow in the quotient space.[10] This is possible because the coordinates (v^i, p^i) label symplectic leaves of Γ. However, in general the fields U^i will not satisfy (6.19) because the v^is will be dynamical. This means that the dynamics on the larger quotient will usually depend explicitly on the conserved charges p^i.

Significantly, the primary motivation for the identification of conserved charges in mathematical and physical practice is *not* as a scaffold for a definition of narrow symmetry. Rather it is in the context of a broad notion of symmetry that charges and

---

[9] An example for how to do this in the case where X generates a dynamical similarity can be found in Sloan (2018) and (Gryb and Sloan, 2021).

[10] For more on symplectic reduction procedures including the crucial idea of momentum maps see Butterfield (2007).

their associated flows are identified as a means to *reduce the order* of the system of equations that need to be solved to explicitly construct the DPMs. By identifying the dynamically isolated substructures of a theory we can find the autonomous substructures whose evolution equations can be integrated without knowledge of the behaviour of the remaining substructures of the theory. This is the core pragmatic and interpretively neutral notion of symmetry which is relevant to mathematical and physical practice. The 'reduction' in question does not imply an identification of symmetry-related models as dynamically equivalent, rather it simply implies that the solutions in question do not, in general, require us to solve independent equations.

Where does this leave us with respect to the identification of narrow symmetries? Should we by default always remove isolated structures and thus define narrow symmetry transformations as any transformation between DPMs generated by involutive vector fields X satisfying (6.19)? One could then prescribe the dynamical identification of such DPMs and removal of the isolated structures along the lines of the procedure outlined above.

In our view there are two good reasons to avoid such a brand of naïve reductionism. First, and most obviously, as noted in our discussion of the arguments of Belot (2013), on such a view we will certainly end up identifying as dynamically equivalent DPMs that we have good reason to treat as distinct. In fact, the conserved quantities (E, Q^a, P_a) are all dynamically isolated in the relevant sense. But because they label all independent DPMs (up to a time parameterization), such a definition of narrow symmetry would lead to no distinct DPMs in any Lagrangian theory. Second, and relatedly, such a naïve reductive approach does not take into account relevant information we may have about the 'target system' which we intend to use our system of equations to model. In particular, when we know that the target system is a quasi-isolated subsystem of the universe, then there may be *direct empirical significance* to the isolated structure and thus good reason not to remove it.[11]

Our settled position, therefore, is that narrow symmetries should *not* simply be read off from the isolated structures of a theory. Making such an identification would mean that there would be no difference between a narrow and a broad symmetry, and this would have a trivialising effect on our conception of symmetry. On our view, the identification of isolated structures and conserved charges should, first and foremost, be thought of as a *heuristic tool* for identifying, for example, the 'reference structures' of a theory. Once the broad symmetries are parameterised so that the isolated structures and conserved charges are identified, one can then proceed to identify the narrow symmetries. The preliminary stance, that all the broad symmetries which are variational symmetries of the full system under study are narrow symmetries, is the basic starting point but is not intended to produce a definitive classification schema. In the following section we will illustrate our approach by studying the example of Newtonian Mechanics.

---

[11] For further discussion we again direct the reader to complementary discussion that can be found in Gryb and Sloan (2021). Further discussions are Kosso 2000; Brading and Brown 2004; Greaves and Wallace 2014; Friederich 2015; Gomes 2020.

6.5 Nomic Structure of Newtonian Mechanics

In this section we give a brief outline of N-particle Newtonian Mechanics (NM) in the presence of conservative forces. We then apply the framework introduced above to analyse the nomic structure and associated conserved charges. This simple example will serve to both illustrate the power of the formalism introduced so far, and also provide an intuitive example for the reader to keep in mind when we introduce more elaborate formalism in the next chapters.

6.5.1 Formulating Newtonian Mechanics

A Newtonian N-particle system describes a set of N idealized point particles propagating in time in a 3-dimensional Euclidean space. This motion is represented by the $3N$ particle position functions $q_I^i(t) : \mathbb{R} \to \mathbb{R}^{3N}$, where $i = (1, 2, 3)$ parametrises a spatial index and $I = (1, \ldots, N)$ parametrises the particles. These $3N$ Cartesian coordinates therefore carve out trajectories $\gamma(t)$ whose images are curves in a $3N$-dimensional configuration space $\mathcal{C} = \mathbb{R}^{3N}$, and where t takes values on the real line $\mathbb{R}$. The curves $\gamma(t)$ are therefore representations of entire histories of a system, and provide a complete description of the motion of that system. The tangents to γ at time t represent the instantaneous velocities $\dot{q}_I^i(t)$ of the Ith particle. These are functions from $\mathbb{R}$ to the tangent space $T\mathcal{C}$.

The configurations $q_I^i(t)$ and velocities $\dot{q}_I^i(t)$ at any time t are sufficient to determine the future evolution of the system provided the relevant constitutive and nomic structures are appropriately specified. For a conservative Newtonian N-particle system, this involves fixing the masses m_I of the Ith particle and the form of the potential energy function $V(q) : \mathcal{C} \to \mathbb{R}$.[12] In general, $V(q)$ will also depend on the value of certain coupling constants c_α such as Newton's constant G.

The dynamically possible histories of a system γ_{cl} are solutions to the second-order partial differential equation

$$ m_I \frac{\mathrm{d}^2}{\mathrm{d}t_N^2} q_I^i = \delta^{ij} \frac{\partial V}{\partial q_I^j} \,, \tag{6.22} $$

which is simply Newton's second law. The quantity δ^{ij} is the inverse of the flat Euclidean metric δ_{ij} in 3-dimensions. It is standard to represent this system using a privileged Cartesian coordinate system in which $\delta_{ij} = \text{diag}(1, 1, 1)$ and a *privileged temporal coordinate* t_N such that Newton's second law takes the precise form given by equation (6.22). We will return to the analysis and interpretation of the Newtonian time parameter in §10.2 and thus we will defer an extensive discussion of the structure of Newtonian time to the introduction of our new framework for the analysis of symmetry and structure. We ask the reader to bear with us up to that point since the formalism which we will introduce between now and then will prove highly instructive.

Formally, one can define a family of *inertial frames* by those sets of coordinate systems where Newton's second law can be written in this privileged form. These frames are therefore spanned by the transformations that leave the precise form of (6.22)

---

[12] We will ignore imposing specific continuity conditions on $V(q)$ and simply assume that it is smooth enough to integrate the equations of motion in the domain relevant to the particular application.

invariant. A subgroup of these can be found by considering time-independent transformations. In this case both the right-hand side and left-hand side transform as vectors under arbitrary coordinate transformations. The most general time-independent
Lie group to leaves δ^{ij} invariant is the Euclidean group, ISO(3) = SO(3) $\ltimes$ $\mathbb{R}^3$, of
translations and rotations, which has the fundamental representation:

$$q_I^i \rightarrow R^i{}_j(\theta_i)q_I^j + a^j \,, \tag{6.23}$$

where $R \in SO(3)$ while $a^i \in \mathbb{R}^3$ and θ^i are parameters of the translations and rotations
respectively. The equations of motion (6.22) are invariant under such transformations
provided the potential $V(q)$ is also invariant under Euclidean transformations. Given
we impose the requirement of the invariance of the potential, then the time derivatives
on the LHS of (6.22) are also invariant under the time-dependent Galilean boosts and
time translations:

$$q_I^i \rightarrow q_I^i + v^i t_N \qquad\qquad\qquad t_N \rightarrow t_N + a^0 \,, \tag{6.24}$$

where v^i are time-independent velocities and a^0 is a constant time offset.[13]   Using
the fundamental representation given by (6.23) and (6.24), it is relatively easy to
show that the (inhomogeneous) *Galilean group*, Gal(3),[14] parametrised by (v^i, a^μ, θ^i)
satisfies the semi-direct product structure Gal(3) = $\big(SO(3) \ltimes \mathbb{R}^3\big) \ltimes \mathbb{R}^{3,1}$. This can be
seen from its representation in terms of the boost, C^i, translation, $P^\mu = (E, P^i)$, and
rotation, L^i, and Hamiltonian, H, operators with non-zero commutators given by:

$$[L_i, P_j] = \epsilon_{ij}{}^k P_k \qquad\qquad [L_i, C_j] = \epsilon_{ij}{}^k C_k \tag{6.25}$$

$$[L_i, L_j] = \epsilon_{ij}{}^k L_k \qquad\qquad [C_i, H] = P_i \,. \tag{6.26}$$

All inertial frames are then related by elements of Gal(3) so defined. This, of course,
does not say anything about how such frames can be identified physically.

NM contains a privileged Cartesian coordinate chart and Newtonian time parameter. However, non-standard charts and non-inertial observers are not only admissible
in the theory but, in fact, even sometimes pragmatically useful for solving practical problems. In such charts, (6.22) picks up extra terms that can be pushed to the
right-hand side and interpreted as 'fictitious' forces.

The equations of motion (6.22) can alternatively be derived using Hamilton's Principle. Defining the kinetic energy:

$$T = \sum_I \frac{m_I}{2} \delta_{ij} \, \dot{q}_I^i \dot{q}_I^j \,, \tag{6.27}$$

the action for NM is

---

[13] Corollary 6 of Newton's Principia suggests that any extra terms added to the LHS of (6.22) for
arbitrary time-dependent boosts can be compensated by an observable total (linear) acceleration of
the universe. The correct broad symmetry group for NM should then plausibly include arbitrary $v^i(t)$
(Saunders, 2013). Though such transformations can be naturally accommodated within out analysis
framework, we will not do so here in order to keep the treatment as simple as possible.

[14] Note that some reference reserve Gal(3) for the *homogeneous* Galilean group of boosts and
rotations.

$$S_N[\gamma] = \int_\gamma \mathrm{d}t \left(T - V(q_I^i) \right) , \tag{6.28}$$

where the curves γ are bounded by the endpoint at (t_1, t_2). For variations where the configurations q_I^i are fixed on the endpoints, the stationary points of $S_N[\gamma]$ are the classical curves that satisfy (6.22).

The kinetic energy T can be rewritten using the identity $\mathrm{d}(q \cdot \dot{q}) = (q \cdot \ddot{q} + \dot{q} \cdot \dot{q}) \, \mathrm{d}t$ in terms of a total differential and a term that is invariant under the Galilean transformations (6.23) and (6.24). The action (6.28) is therefore invariant under the Galilean transformations up to a boundary term. Because this preserves the stationary points of S_N, we recover our previous result that the Galilean transformations are (broad) symmetries of NM.

The space of dynamically possible models (DPMs) of NM, $\mathcal{D}_N$, is then the collection of all γ satisfying (6.22). To parametrise, $\mathcal{D}_N$, and to understand the structure of the conserved quantities that generate the Galilean symmetries, it is convenient to construct a first-order initial value formulation of (6.22) using the Hamiltonian formalism. The Hamiltonian is defined via Legendre transform as

$$H = \sum_I p_i^I q_I^i - L = T + V , \tag{6.29}$$

where the kinetic energy

$$T = \sum_I \frac{\delta^{ij} p_i^I p_j^I}{2m_I} \tag{6.30}$$

is expressed in terms of the momenta

$$p_i^I = \frac{\partial L}{\partial \dot{q}_I^i} = m_I \delta_{ij} \dot{q}_I^j \tag{6.31}$$

and the Lagrangian is $L = T - V$.

Following the treatment of Hamiltonian mechanics via symplectic geometry introduced above, we can express Hamilton's equations in terms of the integral curves of the Hamilton vector field v_H defined via $\mathrm{d}H = \iota_{v_H} \omega$, where $\omega = \sum_I \mathrm{d}p_i^I \wedge \mathrm{d}q_I^i$ is the symplectic two-form on the phase space Γ coordinatised locally by (q_I^i, p_i^I). In these coordinates, we have

$$v_H = \sum_I \left[\frac{p_j^I \delta^{ij}}{m_I} \frac{\partial}{\partial q_I^i} - \delta_{ij} \frac{\partial V}{\partial q_I^j} \frac{\partial}{\partial p_i^I} \right] . \tag{6.32}$$

Newton's second law (6.22) is therefore equivalent to Hamilton's equations:

$$\dot{q}_I^i = v_H(q_I^i) = \delta^{ij} \frac{p_j^I}{m_I} \qquad \dot{p}_i^I = v_H(p_i^I) = -\delta_{ij} \frac{\partial V}{\partial q_I^j} . \tag{6.33}$$

We can identify the solution space of NM with the $2N-1$ dimensional level surfaces $\mathcal{D}_{\tau_E}$ of a *Newtonian clock observable* $\tau_E : \Gamma \to \mathbb{R}$ which is such that $\tau_E = t_N$ and

$$\{\tau_E, H\} = 1 \qquad\qquad \{\tau_E, Q\} = \{\tau_E, P\} = 0 \qquad (6.34)$$

where (Q, P) are constants of motion $\{Q, H\} = \{P, H\} = 0$ that locally parametrise the symplectic sub-manifold of $\mathcal{D}_{\tau_E}$. Again, we ask the reader interested in a more extensive discussion of the structure of time in NM to bear with us until §10.2 when we will return to the analysis of internal time parameters in NM in detail.

The Reeb vector for this system is easily seen to be expressed in these coordinates as

$$R = \frac{1}{t_N} \frac{\partial}{\partial H} \qquad (6.35)$$

so that the total energy $E = T + V$ is a conserved charge. The whole solution space is therefore parametrised by the $2N - 1$ constants of motion (Q, P, E). Changes in the Newtonian clock observable τ_E will read out changes in the privileged temporal coordinate t_N. Explicit expressions for τ_E as well as the constants of motion (Q, P) in terms of the original coordinates (q_I^i, p_i^I) involve solving partial differential equations similar to the Hamilton–Jacobi equation following the algorithm described in §43.B of Arnold (2013). In general, such explicit expressions are very difficult to obtain even for simple potentials $V(q_I^i)$. Their existence however is guaranteed, at least locally, by Darboux's theorem. Any explicit expressions will only be defined up to contact transformations on $\mathcal{D}_{\tau_E}$.

6.5.2   Identifying Conserved Charges

A tractable method to find explicit representations of a closed subset of the full set of (locally) conserved charges Q and P for arbitrary potentials is to consider the conserved charges that generate the space-time Galilean symmetries discussed above. Consider the following charges:

$$P_i = \sum_I p_i^I \qquad L_i = \sum_I \epsilon_{ij}{}^k q_I^j p_k^I \qquad C^i = \sum_I \left(m_I q^i - \tau_E \delta^{ij} p_j^I \right), \qquad (6.36)$$

It is straightforward to show that together with the Hamiltonian H, these charges generate the vector fields $(v_P^i, v_L^i, v_C^i, v_H)$ which form representations of the Galilean group:

$$[v_L^i, v_P^i] = \epsilon^{ij}{}_k v_P^k \qquad\qquad [v_L^i, v_C^j] = \epsilon^{ij}{}_k v_P^k \qquad (6.37)$$

$$[v_L^i, v_L^i] = \epsilon^{ij}{}_k v_L^k \qquad\qquad [v_C^i, H] = v_P^i, \qquad (6.38)$$

where $[,]$ is the Lie bracket for vector fields on Γ. It is also a simple exercise to show that the charges of (6.36) are preserved along the integral curves of v_H. Importantly, one needs to use the fact that the Newtonian clock τ_E must Poisson commute with all conserved charges of the theory except the Hamiltonian, to which it is canonically conjugate.

The charges (P_i, L_i, C_i, H), however, do *not* form infinitesimal representations of Gal(3). Instead, in addition to the Poisson brackets (6.25), we have the non-zero Poisson bracket

$$\{C_i, P_j\} = M\delta_{ij}, \tag{6.39}$$

where $M = \sum_I m_I$ is the total mass of the system. These are infinitesimal representations of the Bargmann group, Barg(3, 1) with semi-direct product structure $\text{Barg}(3,1) = (\text{SO}(3) \ltimes \mathbb{R}^3) \ltimes (\mathbb{R}^{3,1} \otimes \text{U}(1)_M)$. Here, M arises as a central charge transforming under U(1). Barg(3, 1) is the *central extension* of the universal cover of $Gal(3)$.[15] That the central extension appears in the Poisson algebra of the conserved charges is permissible because of the fact that the Poisson algebra of functions on Γ is itself a representation of the central extension of the group of Hamilton vector fields on Γ. Because the constant functions M are such that $\mathrm{d}M = 0$, they generate no Hamilton vector field, and therefore the charges (6.36) only generate vector fields that themselves are ordinary representations of Gal(3).

Significantly, the vector fields generating the Gal(3) symmetries are involutive, because they are representations of Gal(3), and obey the condition (6.19). In principle they could be eliminated by performing the general quotienting procedure introduced in the previous section. To get a symplectic Hamiltonian system one would need to remove the associated Galilean conserved charges as well. This is straightforward for the translations and boosts, but because the kinetic terms of NM depend on the angular velocities, the DPMs will depend on the conserved charges associated with angular momentum. This fact is often interpreted as suggesting that angular velocity is absolute in NM. However, the reduced dynamics can nevertheless be formulated by treating the components of the angular momentum in inertial frames as conserved quantities. For a discussion of the pros and cons of performing a reduction with respect to rotations, see Gomes and Gryb (2020) and Belot (1999) respectively.

This is where it is useful to appeal to our preliminary stance on narrow symmetries in order to differentiate between the status of different symmetries. Recall that following the preliminary stance, the narrow symmetries of a theory should be identified with broad symmetries which are variational symmetries of the full system under study. As is easily checked, the boost symmetries and constant rotations are *not* variational symmetries of NM, and therefore are *not* narrow symmetries by our definition. We therefore expect such symmetries to require special treatment. On our preliminary stance, DPMs related by boosts and constant rotations are taken to be dynamically distinct. Physically this is of course well justified by the privileged empirical role that boosts and rotations, as well as their conserved charges, play in the theory.

By contrast, consider the sub-group of transformations generated by the charges P_i and L_i corresponding to the time-independent Euclidean ISO(3) transformations of the spatial coordinate system,

$$x^i \rightarrow R^i{}_j(\theta)x^j + a^i, \tag{6.40}$$

which *are* variational symmetries. So too are time-independent shifts of the total linear momentum, P_i, of the system. These symmetries and the conserved charges P_i

---

[15] As a quick check, one can recover Gal(3) from Barg(3, 1) by setting the central charge $M = 0$.

correspondingly play no privileged empirical role and no dynamical information is lost in eliminating them form the theory.

The first stage of our analysis of NM is thus complete. We can define the projection map π_N through the equivalence relation relating the leaves of the foliation generated by the time-independent ISO(3) transformations and translations of the conserved charges P_i. Boosts and rotations will simply not be considered narrow symmetries in standard formulations of NM. We will, however, see that such symmetries can be *promoted* to narrow symmetries by a modification of the variational principle. Thus, the narrow symmetries of a theory can be adapted to the particular context in which the theory is being applied, but at the cost of explicitly changing the variational principle used to define the solutions of the theory. Such modifications will have concrete consequences that will be explored throughout the course of this book.

6.5.3 Structure Summary

The constitutive structures of NM are split into: matter, manifold and geometric constitutive structures. From the above we can identify these as:

Manifold: A spatial manifold $\mathbb{R}^3_x$ for representing spatial positions of particles in 3D. A temporal manifold $\mathbb{R}_t$ for representing instants. A spacetime manifold that is the product space of the temporal and spatial manifolds $\mathbb{R}_t \times \mathbb{R}^3_x$.

Matter: A configuration space, $\mathcal{C}$, of representations of N-particle positions in 3D: $\mathcal{C} = \mathbb{R}^{3N}$. A configuration velocity space, $T_q\mathcal{C}$, consisting of the tangent space to $\mathcal{C}$ at a point $q \in \mathcal{C}$. This represents the velocity of a particle at a particular position. The image of a curve $\gamma : \mathbb{R}_t \to T\mathcal{C}$ on the tangent bundle $T\mathcal{C}$ represented by the pair $\left(q^i_I(t), \dot{q}^i_I(t)\right) \forall t \in (t_1, t_2)$. The particle masses $m_I \in \mathbb{R}^+$ and other coupling constants c_α that enter the potential.

Geometric: There are two geometric structures needed: i) A temporal metric $g_t(t_1, t_2) = t_2 - t_1 \, \forall t_2 > t_1$ inherited from the natural metric on the temporal manifold $\mathbb{R}$. ii) A spatial metric $g_{ij} = \delta_{ij}$, where δ is the flat Euclidean metric on the spatial manifold $\mathbb{R}^3$.

Putting this all together we get the full specification of the space of kinematical structures:

$$\mathcal{K} = \left\{ \mathbb{R}_t \times \mathbb{R}^{3N}_x, g_t, \delta, \gamma : \mathbb{R}_t \to T\mathcal{C}, m_I, c_\alpha \right\} . \tag{6.41}$$

The kinematically possible models of NM are then given by the collection of all possible (images of) curves γ in $T\mathcal{C}$ conjoined with a specification of all the particle masses m_I and couplings c_α of the theory.

$$K = \{\gamma : \mathbb{R}_t \to T\mathcal{C}, m_I, c_\alpha\} \tag{6.42}$$

Partitioning map $n : K \to D$: We defined three equivalent procedures: i) Project that space of all images of γ (defined above) onto the space of γ_{cl}, represented by the pair $\left(q^i_I(t), \dot{q}^i_I(t)\right)$ that satisfy Newton's second law (10.6). ii) Project the space of images of γ onto the space γ_{cl} that are stationary points of the action functional. iii) Perform the Legendre transform (eq) of the curves γ and project these onto the space of curves satisfying Hamilton's equations (6.33). Undo the Legendre transform to get γ_{cl}. iv) Solve the Hamilton–Jacobi equation.

Projection Map π_N: Following the preliminary stance on narrow symmetries, the projection map of NM is specified by identifying all possible values of the charges P_i and the leaves of the foliation generated by the time-independent ISO(3) symmetries of the system as these are variational symmetries.

6.6 Chapter Summary

This chapter has considered the geometric structure of theories with regular nomic structure within the framework of Hamiltonian mechanics. In this context, broad symmetries allow us to identify dynamically isolated substructures of a theory. In particular, the space of conserved charges and the Poisson sub-algebras of this space give representations of the interesting dynamically stable structures of a theory. However, there are good reasons to avoid a naïve reductionism in which we remove the isolated structures of a theory. Rather, once the broad symmetries are parameterised so that the isolated structures and conserved charges are identified, one can then proceed to identify the narrow symmetries in a second step based upon physical reasoning. The preliminary stance, that narrow symmetries are all broad symmetries which are variational symmetries of the full system under study, is the basic starting point in this approach.

We have provided a concrete illustration of these idea by considering the nomic structure of Newtonian Mechanics. Here we find that boost symmetries and constant rotations are not variational symmetries of Newtonian Mechanics and therefore are not narrow symmetries under the preliminary stance. We therefore expect such symmetries to require special treatment. On our preliminary stance, dynamically possible models related by boosts and constant rotations are taken to be dynamically distinct. Physically this is of course well justified by the privileged empirical role that boosts and rotations, as well as their conserved charges, play in the theory. By contrast, the sub-group of transformations corresponding to the time-independent Euclidean (rigid) transformations of the spatial coordinate system are variational symmetries as are time-independent shifts of the total linear momentum of the system. These symmetries and the relevant conserved charges correspondingly play no privileged empirical role, and no dynamical information is lost in eliminating them form the theory through a projection procedure as described.

7
Irregular Nomic Structure

Irregular nomic structure is characterized by Lagrangians where the associated Hessian is not invertible. For irregular Lagrangians it will be true that at least some of the components of acceleration at a given time cannot be solved uniquely in terms of the configuration and velocity variables at that time. Irregularity of the Lagrangian cannot, however, be used in general as a sufficient condition for the diagnosis of degeneracy in the solutions to the relevant Euler–Lagrange equations. In the Hamiltonian formulation of theory with an irregular nomic structure, relations between the canonical variables are found generically: a constrained Hamiltonian formalism. The principal aim of this chapter is to show that the Dirac analysis of constrained Hamiltonian systems does not provide a fully satisfactory methodology to connect canonical first-class constraints, generalized Bianchi identities, and initial value constraints.

7.1 Irregular Lagrangians

Following Sudarshan and Mukunda (1974), Sundermeyer (1982), and Henneaux and Teitelboim (1992), the standard starting point for the constraint Hamiltonian analysis of mechanics is the fundamental formal distinction between *regular* and *singular* Lagrangians.[1] Consider a finite dimensional system with action:

$$I = \int_{t_1}^{t_2} dt\, L(q^i, \dot{q}^i, t) \tag{7.1}$$

where the index $i = 1, ..., N$ runs over the degrees of freedom of the system. It is instructive to write the Euler–Lagrange equations in the form:

$$\alpha_i(q, \dot{q}, t) \equiv \frac{\partial L}{\partial q^i} - \frac{\partial}{\partial t}\frac{\partial L}{\partial \dot{q}^i} = W_{ij}(q, \dot{q})\ddot{q}^j - K_i = 0 \tag{7.2}$$

where α_i is the Euler–Lagrange operator, W_{ij} is the Hessian matrix:

$$W_{ij} = \frac{\partial^2 L}{\partial \dot{q}^i \partial \dot{q}^j} \tag{7.3}$$

and

---

[1] For a modern treatment of singular Lagrangian systems focusing on Lagrangian constraints see (Díaz *et al.*, 2014; Diaz and Montesinos, 2018; Díez *et al.*, 2020). We will discuss our specific divergence from an important detail of that analysis in the next chapter.

$$K_i = \frac{\partial L}{\partial q^i} - \frac{\partial^2 L}{\partial \dot{q}^i \partial q^j} \dot{q}^j \, . \tag{7.4}$$

Writing the Euler–Lagrange equations in this form makes clear that the accelerations *at any particular time* are uniquely determined by the velocities *at that time* if and only if the Hessian can be inverted. This is, of course, equivalent to the condition that the determinate of the absolute value of the Hessian is non-zero.

We define a *regular* Lagrangian as a Lagrangian that leads to Euler–Lagrange equations that can be uniquely determined in terms of the initial configurations and velocities, and therefore satisfies $\det|W_{ij}| \neq 0$. This implies that the Euler–Lagrange equations are autonomous, integrable and of second-order. Furthermore, if the Lagrangian is regular then we are guaranteed by the inverse function theorem that there exists a one-to-one transformation between the configuration-velocity variables $(q, \dot{q}, t)$ and the configuration-momentum variables (q, p, t). This, in turn, means that the coordinates and momenta are independent variables (Sundermeyer, 1982, p. 11). Systems with regular Lagrangians have regular nomic structure in the sense that the equations of motion, both the second-order Euler–Lagrange equations and the first-order Hamiltonian equations, will be well-posed for all degrees of freedom.

By contrast, for an *irregular* (or *singular*) Lagrangian we have that $\det|W_{ij}| = 0$. This implies that at least some of the components, $\ddot{q}^i$, of acceleration at a given time cannot be solved uniquely in terms of the configuration and velocity variables at that time (Sundermeyer, 1982, p. 39).

We can explicitly represent the breakdown of integrability by focusing on the kernel of the Hessian. In particular, if the rank of the Hessian is $R < N$, then we can consider the $N - R$ null eigenvectors $\lambda^i_{(a)}(q, \dot{q})$:

$$\lambda^i_{(a)}(q, \dot{q}) W_{ij}(q, \dot{q}) = 0 \tag{7.5}$$

for $i, j = 1, ..., N$ and $a = 1, ..., N - R$. This should be compared to the case of a regular Lagrangian system where the non-null directions are used to solve the Euler–Lagrange equations by integrating (7.2).

Our discussion here concerns regular and irregular *Lagrangians*, not regular and irregular *physical systems*. The same physical system may be described by a regular or irregular Lagrangian depending upon the approach taken to the representation of the system's degrees of freedom. It is true, however, that the regularity and irregularity, as indicated by the condition on the Hessian, is a *coordinate invariant* statement (Sundermeyer, 1982, pp. 38–9) and is thus not a mere representational artefact which can be transformed away by a more perspicuous choice of configuration variables. Moreover, the irregularity of the Lagrangian, and the consequent existence of the null vectors of the Hessian, cannot be used as a sufficient condition for the diagnosis of degeneracy in the solutions to the relevant Euler–Lagrange equations in the simple sense that the irregularity of the Lagrangian may merely indicate that there is no stationary point of the action. We will return to the analysis of irregular Lagrangians in the context of the geometric analysis provided in the following chapter. Before then, it will be instructive to consider the Hamiltonian dynamics of a systems with an irregular Lagrangian description.

7.2   Constrained Hamiltonian Mechanics

As was noted above, regular systems are precisely those in which the transformation to the phase space variables is guaranteed to be unique and the coordinates and momenta will then be independent variables. It is thus not a surprise that when dealing with the Hamiltonian formulation of theory with an irregular nomic structure we generically find relations between the canonical variables: a *constrained* Hamiltonian formalism.

We know that the number of independent relations between canonical variables should be given by the dimension of the null space of the Hessian which above we took this to be $N - R$. We can introduce a set of explicit *canonical constraints* which describe the relations between positions and momenta via a relation of the form:

$$C_s = p_s - g_s(q, p_\alpha) \tag{7.6}$$

where the index $s = 1, ..., N - R$ runs over the canonical variables which are not independent, $\alpha = 1, ..., R$ runs over the remaining independent variables, and the functions g_s encode the relevant interdependencies. These relations are called *primary constraints* following the coinage of Anderson and Bergmann (1951). From these relations we can, at least locally, derive an explicit form of the dynamics of a constrained Hamiltonian by substituting (7.6) to eliminate the $N - R$ dependent momenta, deriving the analogue of Hamilton's equations, and then applying consistency conditions. In the interests of space we will neglect the detailed technical steps that lie behind this explicit approach; see Sundermeyer (1982, pp. 45–50) and Gitman and Tyutin (1990, pp, 13–16) for details.

Decompositions of the kind used in the explicit approach break down globally and for this reason it is standard to define the primary constraints *implicitly*. Following Henneaux and Teitelboim (1992) and Sundermeyer (1982), we enforce relations the form:

$$\phi_r(q, p) \approx 0 \tag{7.7}$$

for $r = 1, ..., R'$.

Two points regarding this implicit approach to Hamilton constraints are of immediate significance. First, we do not require in this implicit representation that the primary constraint equations (7.7) are independent. Rather, the number of independent constraint equations is given by the dimension of the null space of the Hessian which we took to be R above. Thus we will expect in general to find that $R < R'$ and that $R' - R$ of the constraints are not independent relations.

Second, in this equation we have introduced Dirac's notion of a 'weak equality' via the sign $\approx$. From an algebraic perspective, and falling back on the explicit constraint formalism again, the meaning of the weak equality symbol can be rigorously defined via the condition that two phase space functions, $F(q_i, p_i)$ and $G(q_i, p_i)$ for $i = 1, ..., N$, are weakly equivalent if and only if they are equivalent *after* applying the substitution (7.6); thus we have that

$$F(q_i, p_i) \approx G(q_i, p_i) \to F(q_i, p_\alpha, g_s(q, p_\alpha)) = G(q_i, p_\alpha, g_s(q, p_\alpha)) \tag{7.8}$$

where again $s = 1, ..., N - R$ and $\alpha = 1, ..., R$, and the expression is only guaranteed to be well-defined locally on phase space.

From the implicit perspective that we are following, the weak equality between two functions can be understood in terms of the functions being equal on a subspace within phase space defined by the constraints themselves—the primary constraint surface $\Sigma \subset \Gamma$. Thus, we can write:

$$F(q,p) \approx G(q,p) \rightarrow F(q,p)|_\Sigma = G(q,p)|_\Sigma \tag{7.9}$$

The significance of weak equality in the context of the constraints themselves is that whereas $\phi(q,p)_r = 0$ would indicate that the constraints are equal to zero throughout phase space and would thus have zero Poisson brackets with the canonical variables, $\phi_r(q,p) \approx 0$ indicates that the constraints are zero on the constraint surface and can have non-zero Poisson brackets with the canonical variables. The obvious implication for the explicit formalism is that we can write (7.6) as $C_s \approx 0$.[2]

So far our analysis has focused solely on the basic kinematical structure of a constrained Hamiltonian system. Let us now proceed to the construction of the Hamiltonian, the implications of dynamical consistency of the constrained system, and the all important distinction between first-class and second-class constraints.

We assume that the rank of the Hessian is constant throughout $(q, \dot{q})$ and that the implicit primary constraint conditions (7.7) define a smooth sub-manifold embedded in phase space, the primary constraint surface $\Sigma \subset \Gamma$. As per above, we assumed that there are R independent relations among the R' constraints. This means that the primary constraint surface will be of dimension $2N - R$.

The important next step is to impose regularity conditions on the R' (implicitly defined) constraint functions given by (7.7). The key requirement is that the $2N - R$ dimensional constraint surface is coverable by open regions, within each of which the constraint functions $\phi_{r'}$ $r' = 1, ..., R'$ can be split into independent constraints ϕ_r $r = 1, ..., R$ such that the following Jacobian matrix has finite matrix elements and is of rank $N - R$:

$$\frac{\partial \phi_r}{\partial(q^i, p_i)} \tag{7.10}$$

where $r = 1, ...R$. We then require that the remaining dependent constraints $\phi_{\bar{r}}, \bar{r} = R + 1, ..., R'$ are strongly zero as a consequence of the others.

The regularity conditions imply the following two important relations. First, for any smooth function which is weakly vanishing $G \approx 0$ there is some function $g^{r'}$ such that $G = g^{r'}\phi_{r'}$. Second, if $\lambda_n \delta q^n + \mu^n \delta p_n = 0$ for arbitrary variations δq^n and δq^n tangent to the constraint surface then, given some arbitrary function $u^{r'}$, we have that:

$$\lambda_n = u^{r'} \frac{\partial \phi_{r'}}{\partial q^n}\bigg|_\Sigma \tag{7.11}$$

$$\mu_n = u^{r'} \frac{\partial \phi_{r'}}{\partial p_n}\bigg|_\Sigma \tag{7.12}$$

For more on the regularity conditions including proofs of the above see Henneaux and Teitelboim 1992, pp. 7–8, cf. Sundermeyer 1982, p. 51.

---

[2] More geometrically, we use weak equalities when considering vector/tensor fields (and their flows) on Γ to indicate the *restriction* of these vector fields to the regular sub-manifolds Σ of Γ defined by the canonical constraints since these flows are not defined *intrinsically* as Hamiltonian flows on Σ.

The signifiant point is then that, following Henneaux and Teitelboim (1992, pp. 8–11), assuming regularity and considering the variation of the Hamiltonian function δH, leads to conditions on the Hamiltonian, the canonical variables, the primary constraints, and the arbitrary functions $u^{r'}$, that allow us to write Euler–Lagrange equations as generalized Hamiltons equations of the form:

$$\dot{q}^n = \frac{\partial H}{\partial p_n} + u^{r'} \frac{\partial \phi^{r'}}{\partial p_n} \tag{7.13}$$

$$\dot{p}_n = -\frac{\partial H}{\partial q^n} + u^{r'} \frac{\partial \phi^{r'}}{\partial q^n} \tag{7.14}$$

$$\phi_{r'}(q, p) = 0 \tag{7.15}$$

These equations can be shown to result from the variational principle:

$$\delta \int_{t_1}^{t_2} \dot{q}^n p_n - H - u^{r'} \phi_{r'} = 0 \tag{7.16}$$

with $\delta q^n(t_1) = \delta q^n(t_2)$ and the functions $u^{r'}$ now identified as Lagrange multipliers. It is important to not that this is *not* the original variational principle, rather it is a new regular variational principle which is defined on an extended space that contains the Lagrange multiplier functions.

The equations of motion implied by this variational principle can be rewritten as:

$$\dot{F} = \{F, H\} + u^{r'}\{F, \phi_{r'}\} \tag{7.17}$$

where F is an arbitrary phase space function and we have used the standard Poisson bracket.

The equation (7.17) provides a perspicuous means by which to formulate dynamical consistency conditions on the primary constraints. In particular, we can write the requirement that the constraints be persevered in time as:

$$\dot{\phi}_{r'} = \{\phi_{r'}, H\} + u^{r''}\{\phi_{r'}, \phi_{r''}\} = 0 \tag{7.18}$$

where r' and r'' index the primary constraints.

Secondary constraints then arise when satisfaction of this equation leads to a relation between the canonical variables independent of the multipliers $u^{r''}$. When they exist, dynamical consistency conditions must also be applied to the secondary constraints. These in turn may lead to tertiary constraints, and it is possible that the algorithm may in fact fail to terminate. We will comment further on the relation between secondary canonical constraints and Lagrangian constraints in the next chapter.

Assuming the Hamiltonian constraint algorithm does terminate, we end up with a complete set of constraints (primary, secondary, tertiary...) that we can write as ϕ_j with $j = 1, ... R' + K = J$, where R' is the number of (possibly dependent) primary constraints and K is the number of further constraints generated by the dynamical consistency algorithm. Eventually, this will lead to further modifications of the variational principle in the form of an extended Hamiltonian.

The next step is to consider the restrictions that the dynamical consistency conditions on the complete set of constraints place on the Lagrangian multipliers. In particular, we can formulate a condition on the Lagrange multipliers based upon the preservation of the full set of constraints in time. Following our expression (7.18) this takes the form:

$$\dot{\phi}_j = \{\phi_j, H\} + u^m\{\phi_j, \phi_m\} = 0 \qquad (7.19)$$

where $m = 1, ...R'$ and $j = 1, ...R' + K = J$. The general solution to these equations can be expressed in terms of a decomposition with the Lagrange multipliers written in terms of components that depend up the dynamical consistency conditions and components that are totally arbitrary (Henneaux and Teitelboim, 1992, pp. 13–15). Explicitly, this solution take the form:

$$u^m \approx U^m + v^a V_a^m \qquad (7.20)$$

where U^m and V_a^m are solutions to the inhomogeneous and homogenous forms of (7.19) respectively, and the multipliers v^a are *totally arbitrary*.

The index $a = 1, ...A$ is summed over the number of independent solutions of the homogenous equation $V_a^m\{\phi_j, \phi_m\} \approx 0$ and guaranteed to be constant on the constraint surface, given we assume the matrix $\{\phi_j, \phi_m\}$ to be of constant rank on that surface. The A arbitrary multipliers v^a, which are associated with the solutions to the homogenous equations, will prove of interpretational significance in the context of the degree of freedom counting.

With the general form of u^m now in hand we can return to the equations of motion (7.17) which we can now write as:

$$\dot{F} \approx \{F, H_T\} \qquad (7.21)$$

with the total Hamiltonian defined to be:

$$H_T = H + U^m\phi_m + v^a V_a^m \phi_m \qquad (7.22)$$

The final piece of machinery that we need to implement the constrained Hamiltonian formalism is the crucial distinction between first-class and second-class constraints. This classification is distinct from that between primary and secondary constraints and far more significant for the diagnosis of dynamical redundancy in systems with canonical constraints. The classification takes the following simple form:

First-Class Constraints. A constraint is first-class if and only if it has a Poisson bracket with every other constraints that is weakly vanishing; that is, a constraint ϕ_1 is first-class if and only if:

$$\{\phi_1, \phi_i\} \approx 0 \qquad (7.23)$$

for all $i = 2, ...J$

Second-Class Constraints. A constraint is second-class if and only if it has a Poisson bracket with at least one other constraint is not weakly vanishing; that is, a constraint ϕ_1 is second-class if and only if:

$$\{\phi_1, \phi_2\} \not\approx 0 \qquad (7.24)$$

For at least one $\phi_2 \in \{\phi_i\}, i = 2, ...J$. All constraints that fail to be first-class are necessarily second-class.

These definitions can be applied equally to any function on phase space. Thus we can talk about first- and second-class functions in general, whether or not they are constraints.

The first significant application of the first-class concept relates to the total Hamiltonian. If we write the total Hamiltonian (7.22) as a linear sum of two pieces, $H_T = H' + v^a \phi_q$, where $H' = H + U^m \phi_m$ and $\phi_a = V_a^m \phi_m$, then we can show that H_T and ϕ_a are individually first-class. The total Hamiltonian is thus the sum of a first-class function H' together with the first-class primary constraints ϕ_a multiplied by arbitrary constants v^a.[3]

In the next section, we will consider the interpretation of primary first-class constraints as generating functions of infinitesimal contact transformations which lead to changes in the canonical variables that do not affect the physical state. This interpretation of first-class constraints as indicative of dynamical redundancy, and therefore 'gauge generating', follows from a theorem, for primary constraints, and a conjecture, for secondary constraints, both due to Dirac (1964). It is within these fairly straightforward classical results that seeds of the problem of time is sown and they will thus be worthwhile considering in some depth.

7.3   Dirac's Theorem

What is often called 'Dirac's theorem' refers to an argument (not in theorem form) that occurs in the first lecture within his 1964 *Lectures of Quantum Mechanics* which gives a typically terse 'textbook' presentation of the theory of constrained Hamiltonian dynamics, excluding general relativity, based upon lectures he delivered at Yeshiva University, New York, in 1964. Of particular relevance is the argument leading to the statement:

We come to the conclusion that the ϕ_a's which appeared in theory in the first place as the primary first-class constraints, have this meaning: *as generating functions of infinitesimal contract transformations that lead to changes in the q's and p's that do not affect the physical state.*[4]

On the next page, after some analysis of the status of first-class secondary constraints he concludes:

The final result is that those transformations of the dynamical variables which do not change physical states are infinitesimal contact transformations in which the generating function is a primary first-class constraint or possibly a secondary first-class constraint ... I think it might be that all the first-class secondary constraints should be included among the transformations which do not change the physical state, but I have not been able to prove it. Also, I have not found any example for which there exist first-class secondary constraints which do generate a change in the physical state.[5]

Let us consider the relevant formal arguments in a little more detail, following the influential discussion of Henneaux and Teitelboim (1992, pp. 19–20) which is based on (Dirac, 1964, p. 20–1). The Dirac argument can be formulated explicitly as a theorem as follows.

---

[3] As noted by Henneaux and Teitelboim (1992, pp. 16) the splitting of the total Hamiltonian is not unique since renaming of the arbitrary functions allows for different decompositions between the two pieces.

[4] Dirac (1964, pp. 21), italics in original.

[5] Dirac (1964, pp. 22–4).

Theorem 7.1 *Let us assume that:*

   i. *We have a constrained Hamiltonian theory with a dynamically closed and irreducible set of first-class primary constraints ϕ_a, with associated totally arbitrary multipliers v^a, a time dependent phase space function $F(t) : (q,p)_t \to \mathbb{R}$, and external time parameter, t.*

  ii. *The physical state of the system at an initial time S_t can be specified by a full set of canonical variables $(q,p)_{t_1}$; that is, although we must allow for physical states to not be uniquely determined by specified of canonical variables, we can assume that canonical variables uniquely determine physical states. S_t thus supervenes on $(q,p)_t$.*

 iii. *Whenever we make a specification of the physical state at an initial time, the equations of motion (7.21) then fully determine the physical state at other times.*

  iv. *Define any transformation of the canonical variables that does not change the physical state as a gauge transformation.*

Then,

   D. *The first-class primary constraints, ϕ_a, generate gauge transformations.*

The proof follows from a straightforward calculation based upon two applications (7.21) with different specifications of the arbitrary multipliers.

Proof Specify F at some initial time t_1 as F_1. Specify the change in the function to some infinitesimally later time $t_2 = t_1 + \delta t$ as $\delta F = F_2 - F_1$. Calculate δF via (7.21). Since the multipliers v^a are totally arbitrary we can repeat that calculation with a different multipliers $\bar{v}^a$ leading to $\delta \bar{F}$ again via (7.21). Define the difference between these two as $\triangle F = \delta \bar{F} - \delta F$.

   By i to iv $\triangle F$ must be a gauge transformation. This is because the physical state of the system was specified at t_1 and will thus by iii be uniquely determined at t_2 independently of the choice of the multipliers. This means that any difference between $\delta \bar{F}$ and δF must therefore correspond to a transformation of the canonical variables that does not change the physical state, which is a gauge transformation by iv.

   The explicit form of $\triangle F$ is given by:

$$\triangle F = (v_a - \bar{v}_a)\delta t \{F, \phi_a\} \tag{7.25}$$

By assumption we have that $(v_a - \bar{v}_a)\delta t$ is a small and arbitrary multiplier and we can make the identification $\epsilon^a = (v_a - \bar{v}_a)\delta t$. We then have that:

$$\triangle F = \epsilon^a \{F, \phi_a\} \tag{7.26}$$

which implies that ϕ_a is a generating function of a gauge transformation. $\square$

   A simple corollary then follows concerning the definition of a gauge-invariant observables or Dirac observables:

Corollary 7.1 *Consider a theory with primary constraints only. Define a phase space function which is invariant under gauge transformations as a Dirac observable. Then, the Dirac observables are given by the functions which have a weakly vanishing Poisson bracket with the full set of primary first-class constraints.*

The corollary follows directly from Dirac's theorem combined with the geometry of the constraint surface since a weakly vanishing Poisson bracket implies constancy along the orbits of the vector field generated by the constraints and thus gauge invariance.

An important terminological observation regarding 'Dirac's theorem' is that in Dirac's original formulation the word 'gauge' does not feature. Rather, he phrases his conclusion in the following terms: 'the primary first-class constraints [are] generating functions of infinitesimal contact transformations, they lead to changes in the q's and p's that do not change the physical state' (Dirac, 1964, p. 21). The use of the term 'contact' transformation may be misleading. Plausibly, Dirac is using the term interchangeably with canonical transformation, as was then standard in the physics literature, rather than specifically pointing to the fact the relation with contact manifolds and the odd dimensional structure of the energy surface (Goldstein, 1980, p. 383). We can, therefore, take his framing of the implications of the theorem to be faithfully preserved within the modern rendering of Henneaux and Teitelboim (1992) where the words 'gauge transformation' do occur.

More significantly, introducing the notion of gauge transformation in these terms is arguably both too general and too specific. It is too general since it opens up a vagueness as to what we mean by 'does not change the physical state'. In particular, such a notion of gauge would, under some interpretations of the physical state, include rigid global transformations, like uniform spatial translations, which have nothing to do with first-class constraints nor ill-posed initial value problems. As was discussed Chapter 6, from a formal perspective, it is perfectly legitimate to interpret these transformations as changing the physical state—and in some physical applications such an interpretation may be perspicuous. However, in other contexts, in particular when considering transformations of the universe as a whole, there are strong physical reason for taking these transformations to not change the physical state. This is in contrast to 'gauge transformations' whose formal treatment is necessarily different, and whose interpretation is not tied to the physical context.

Isolating a more solid basis for distinguishing 'gauge transformations' than simply 'does not change the physical state' will be the project of the next chapter. To foreshadow our approach, it is worth quoting Sundermeyer's ever-perceptive comment on the definition of gauge transformation:

[O]ne should consider two histories which evolve from a given initial state under the influence of arbitrary functions as physically equivalent. Let me define a transformation mediating between such histories as a 'gauge transformation'. Observe that although the notion of gauge transformation here is motivated by the form of the equation [(7.21)] they were present on the Lagrangian level too, where certain accelerations were found to depend on arbitrary functions and their derivatives.[6]

The principal virtue of setting things up in terms of histories, rather than focusing on transformations that do not change the physical state, is that such a perspective simultaneously makes clear both the difference between gauge and rigid global transformation and the connection to ill-defined initial value problems becomes clear.[7] The

---

[6] Sundermeyer (1982, p. 89).

[7] Interestingly, as pointed out by Pons (2005), a definition of gauge transformation in terms of histories has a strong historical precedent going back all the way to the work of Anderson and Bergmann (1951). See also Gitman and Tyutin (1990) and treatments based on the gauge generator

problem, however, is that one would like to be able to move back and forth between the symmetries defined at the level or histories and the *dynamical redundancy* which exists at the level of the initial value problem.

This leads us to a key problem for the Dirac's theorem itself. As noted by Barbour and Foster (2008), and as made explicit in our rendition of the theorem, the conclusion of the argument presented by Dirac, and repeated by Henneaux and Teitelboim, is evidently limited in scope by the assumption that the constrained Hamiltonian theory in question contains an external time variable. There is a degree of ambiguity with regard to whether or not Dirac intended to explicitly acknowledge this limitation.[8] Henneaux and Teitelboim, however, do make this restriction clear, even though, as we shall see, they believe that the conclusion that first-class primary constraints generate gauge transformations *does* apply to the case of first-class primary constraints in theories without an external time parameter in any case. We will return to these issues also in detail in the next chapter.

Finally, it is worth us now commenting briefly upon the complex and subtle issue of the interpretation of first-class *secondary* constraints. Following on from his discussion of the 'theorem', Dirac then makes the 'conjecture' that secondary first-class constraints might also generate transformations of the physical variables that do not change physical states. Pons (2005), perceptively, rephrases this as 'all first-class constraints generate gauge transformations at a given time' (p. 511). There is then a large literature discussing the extent to which the conjecture can in fact be proved, and the relevance of 'pathological' counter-examples.[9] We will return to the relevance of secondary constraints in the following chapter.

7.4 Noether's Second Theorem

The second of Emmy Noether's two landmark theorems has garnered less attention within foundational discussions of symmetries. However, from a foundational perspective, it is arguably the more important of the two. The second theorem is considerably more complex to both state and interpret than the first, and thus warrants a more extended treatment. Here we will consider the standard analysis, in the following chapter we will return to the theorem and its implications in the context of our own novel approach.

We saw in the last chapter that in the context of canonical mechanics the implications of Noether's first theorem can be understood in terms of the 'symplectic Noether theorem' that for a dynamical system with Hamiltonian H, for every one-parameter group of transformations which leaves v_H invariant there is an associated conserved quantity K that generates the one-parameter group.

The second theorem, by contrast, defies such a simple canonical rendering. Let us start by framing Noether's second theorem in words. Following Sundermeyer (1982, pp. 25–6), the theorem can be stated concisely as follows:

(Castellani, 1982; Mukunda, 1980; Pons, Salisbury and Shepley, 1997; Pitts, 2013; Pooley and Wallace, 2022).

[8] We will discuss this point at length in our historical analysis of the problem of time to be provided in Volume II of this work.

[9] See Lusanna 1990; Lusanna 1991; Pons 2005 and the more complete lists of references therein.

Noether's Second Theorem. If an action is invariant under infinitesimal transformations of an infinite continuous group parameterized by r arbitrary functions then there exists r independent algebraic or differential *generalized Bianchi identities* which hold irrespective of the satisfaction of the Euler–Lagrange equations. The converse relation also holds.

This follows Noether's own statement of the theorem fairly closely (Kosmann-Schwarzbach, 2010, p. 6). Later in the book we will consider the two important illustrations that Noether provides in her original treatment: the Bianchi identities of general relativity and the Weierstrass condition in reparameterization invariant mechanics.

A little more formally, the theorem can be stated as follows.[10] First, define a set of transformations of the variables via:

$$\bar{t} = t + U(f)$$
$$\bar{q}^i = q^i + T^i(f) \tag{7.27}$$

where the transformations act on (t, q^i)-space and depend upon an arbitrary continuously differentiable function $f \in C^{n+2}(\Omega)$ of order $n+2$ on the space-time region Ω in which the variation is being performed. We have also introduced the linear differential operators $T^i, i = 1, \ldots, r$ and U defined up to order n.

Second, write the invariance of the action under the transformation 7.27 as the condition:

$$\int_a^b \left[L\left(t, \bar{q}(\bar{t}), \frac{d\bar{q}(\bar{t})}{d\bar{t}}\right) \frac{d\bar{t}}{dt} - L\left(t, q(t), \frac{dq(t)}{dt}\right) \right] dt = 0 . \tag{7.28}$$

Note that for fields over space-time, the Euler–Lagrange operator α_i defined earlier generalises to:

$$\alpha_i \equiv \frac{\partial L}{\partial q^i} - \partial_\mu \frac{\partial L}{\partial(\partial_\mu q^i)} , \tag{7.29}$$

where μ is a space-time index.

Let us also define the *adjoint* $\tilde{O}$ of a differential operator O. To do this we require an integration measure to define square integrable functions. The natural choice is the volume form vol_g of the space-time metric g_{ab}. In terms of this measure, the adjoint is then defined as

$$\int_\Omega hT(f) \, \mathrm{vol}_g = \int_\Omega f\tilde{T}(h) \, \mathrm{vol}_g , \tag{7.30}$$

where $h \in C^n(\Omega)$ and both $(f, h) \in \mathbb{L}^2(\Omega, \mathrm{vol}_g)$. This definition will put an additional restriction on the domain of $\tilde{O}$, as we will see below.

---

[10] Here we provide the more straightforward formulation of the theorem since the details which our relevant for our analysis are already contained in this formulation. Our treatment is partially based on that of Logan (1977) and Neuenschwander (2017). For detailed discussion of Noether's second theorem in a historical context see Kosmann-Schwarzbach (2010). For a rigorous formal overview see Olver (1991). For a philosophical discussion see Brading and Brown (2003). We will return to important limitations in the analysis of Brading and Brown in the next chapter.

Following (Logan, 1977) the generalized Bianchi identities, which are implied by the invariance (7.28) under Noether's second theorem, can then be proven to take the simple form:

$$\tilde{T}^i(\alpha_i) - \tilde{U}(\dot{q}^i \alpha_i) = 0. \tag{7.31}$$

These are off-shell 'gauge identities' which hold whether or not the equations of motion are satisfied.

At this point one would like to be able to tell a story about the formal and physical interrelation between Noether's second theorem and the associated generalized Bianchi identities, on the one hand, and the irregularity of the Lagrangian, the existence first-class constraints, and the well-posedness of the initial value problem, on the other. However, the Dirac analysis does not provide a direct methodology to connect the first-class constraints, generalized Bianchi identities, and initial value constraints. In the following chapter we will provide a geometric formalism that allows us to establish such connections, and in doing so provide a diagnostic framework for the *dynamical redundancy* associated with initial value constraints.

7.5 Chapter Summary

Irregular nomic structure is characterized by Lagrangians where the associated Hessian is not invertible. For irregular Lagrangians it will be true that at least some of the components of acceleration at a given time cannot be solved uniquely in terms of the configuration and velocity variables at that time. Irregularity of the Lagrangian cannot however be used in general as a sufficient condition for the diagnosis of degeneracy in the solutions to the relevant Euler–Lagrange equations. In the Hamiltonian formulation of a theory with an irregular nomic structure, we generically find relations between the canonical variables: a constrained Hamiltonian formalism. We have reconstructed an explicit form of Dirac's theorem that associates (first-class) Hamiltonian constraints with gauge transformations together with a corollary which establishes a suitable connection with observables and gauge invariance. We have then offered sustained critical commentary on the scope and interpretation of the theory with particular reference to the limitation that the theory explicitly assumes an external time variable and thus does not apply to time reparameterization invariant theories. Finally, we have discussed Noether's second theorem that relates a class of infinitesimal symmetries to the existence of generalized Bianchi identities or gauge identities. Our principal conclusion is that the Dirac analysis of constrained Hamiltonian systems *does not* provide direct methodology to connect the canonical first-class constraints, generalized Bianchi identities, and *initial value constraints*. We are looking for a general framework for identifying dynamical redundancy associated with initial value constraints and this requires us to go beyond the standard analysis in terms of irregular Lagrangians, Dirac's constraint algorithm, and Noether's second theorem.

8
Diagnosing Dynamical Redundancy

This chapter will provide a formal framework to resolve the key interpretative issue encountered in the last chapter for the case of theories with fixed-time parameterization; that is, on the assumption of a fixed-time parameterization, we will demonstrate that it is the existence of initial value constraints that is indicative of the specific species of underdetermination problem associated with dynamical redundancy within the analysis of theories with irregular nomic structure. As such, it is the identification of initial value constraints that is the fundamental basis for the isolation of gauge degrees of freedom. This criterion thus supersedes both Noether's second theorem and the Dirac criterion (based on the existence of first-class canonical constraints) as the fundamental diagnostic tool for interpreting a gauge theory.

8.1 Variational Symmetries of Histories

The first step in our analysis is to return to the formulation of the variational principle paying particular attention to the boundary terms. First consider the theory defined by the action functional

$$S[q^i(t)] = \int_{t_1}^{t_2} L\left(q^i, \dot{q}^i, \dots, \tfrac{\mathrm{d}^k}{\mathrm{d}t^k} q^i\right) \mathrm{d}t, \tag{8.1}$$

depending on the configuration variables $q^i(x)$, where the Lagrangian L depends on k derivatives of q^i. In general, we allow for the possibility for i to range over a continuous spatial index x of n-dimensions. In that case, the Lagrangian is itself a functional over spatial configurations. For most applications, we will be considering spacetimes of the form $(\mathcal{M}, g)$ and it will be most convenient to use the natural volume form $\mathrm{vol}_g = \sqrt{-|g|}\mathrm{d}^{k+1}x$ induced by g as the integration measure for S, where x is some set of spatiotemporal coordinates. Integration by parts can then easily be performed using the metric compatible covariant derivative ∇_g. In that case, we have

$$S[q^i(x)] = \int_\Omega \mathrm{vol}_g\, \mathcal{L}(q^i, \nabla_\mu q^i, \dots, \nabla^{(k)}_{\mu\cdots\nu} q^i), \tag{8.2}$$

where Ω is some connected spacetime region and $\mathcal{L}$ is the *Lagrangian density* and the q^i can also include the spacetime metric g.

We will be interested in formulating the equations of motion in regions where $\Omega \subseteq \mathcal{M}$ is globally hyperbolic so that we can understand the conditions under which they can be expressed as a well-posed Cauchy problem. If we consider a foliation of

Ω into spacelike (Cauchy) hyper-surfaces Σ_t of constant time t, then (8.2) reduces to (8.1) when

$$L(t) = \int_{\Sigma_t} \mathrm{vol}_{\bar{g}} N \mathcal{L} \,, \qquad (8.3)$$

where $\bar{g}$ is the restriction of g to Σ_t and N is the *lapse* function such that $\sqrt{-|g|} = N\sqrt{|\bar{g}|}$ for t coordinates adapted to Σ_t. The discussion below therefore applies generally to the field theory case with variational derivatives adapted accordingly and with boundary terms supplemented by the appropriate contribution from the time-like or null boundary of Ω.

Classical solutions are then solutions to the variational equation

$$
\delta S[q^i(t); \delta q^i(t)]\Big|_{q^i=q^i_{\mathrm{cl}}} = \int_{t_1}^{t_2} \left[\frac{\partial L}{\partial q^i} \delta q^i + \frac{\partial L}{\partial \dot{q}^i} \frac{\mathrm{d}}{\mathrm{d}t}(\delta q^i) + \ldots + \frac{\partial L}{\partial\left(\frac{\mathrm{d}^k}{\mathrm{d}t^k} q^i\right)} \frac{\mathrm{d}^k \delta q^i}{\mathrm{d}t^k} \right] \mathrm{d}t
$$

$$
= \int_{t_1}^{t_2} \delta q^i(t)\, \mathrm{d}t \left[\frac{\partial L}{\partial q^i} - \frac{\mathrm{d}}{\mathrm{d}t}\left(\frac{\partial L}{\partial \dot{q}^i}\right) + \ldots + (-1)^k \frac{\mathrm{d}^k}{\mathrm{d}t^k}\left(\frac{\partial L}{\partial\left(\frac{\mathrm{d}^k}{\mathrm{d}t^k} q^i\right)}\right) \right]_{q^i=q^i_{\mathrm{cl}}}
$$

$$
+ \left[\frac{\partial L}{\partial \dot{q}^i} \delta q^i + \frac{\partial L}{\partial \ddot{q}^i} \frac{\mathrm{d}}{\mathrm{d}t}(\delta q^i) - \frac{\mathrm{d}}{\mathrm{d}t}\left(\frac{\partial L}{\partial \ddot{q}^i}\right)\delta q^i + \ldots \right]_{t_1}^{t^2} = 0, \quad (8.4)
$$

which must be satisfied for smooth variations $\delta q(t) \in C^k[t_1, t_2]$ that are otherwise arbitrary except (possibly) on the boundary. In our notation, the first argument of the variation δS indicates the functional dependence of S and the second argument the smearing functions used to perform the variation.[1] The equations (8.4) are then the Euler–Lagrange (EL) equations. The set of dots in the boundary term indicates an increasingly complicated series of boundary contributions that can be straightforwardly obtained by integrating the first line by parts.

The EL equations can be usefully split into a local piece and a boundary piece by defining the EL 'vectors'

$$
\alpha_i(t) \equiv \frac{\delta S}{\delta q^i(t)} = \frac{\partial L}{\partial q^i} - \frac{\mathrm{d}}{\mathrm{d}t}\left(\frac{\partial L}{\partial \dot{q}^i}\right) + \ldots + (-1)^k \frac{\mathrm{d}^k}{\mathrm{d}t^k}\left(\frac{\partial L}{\partial\left(\frac{\mathrm{d}^k}{\mathrm{d}t^k} q^i\right)}\right), \qquad (8.5)
$$

which is essentially the variation δS smeared with a *Dirac* delta function. Strictly speaking, this definition holds only for t inside the region (t_1, t_2) of variation since the δ-function is not well-defined on the boundary. The local equations

$$\alpha_i(t) = 0 \qquad (8.6)$$

must then be supplemented with a boundary condition of the form

$$
\left[\frac{\partial L}{\partial \dot{q}^i} \delta q^i + \frac{\partial L}{\partial \ddot{q}^i} \frac{\mathrm{d}}{\mathrm{d}t}(\delta q^i) - \frac{\mathrm{d}}{\mathrm{d}t}\left(\frac{\partial L}{\partial \ddot{q}^i}\right)\delta q^i + \ldots \right]_{t_1}^{t^2} = 0 \,, \qquad (8.7)
$$

---

[1] The index structure of the arguments is mostly for illustrative purposes.

which can be seen as a restriction either on the boundary variation or on the quantities $\frac{\partial L}{\partial \left(\frac{d^k}{dt^k} q^i \right)}$ (i.e. on the domain of the functional S).

The boundary conditions should rightfully be thought of as specifying the independent initial (and boundary) data that must be used for solving the $(k+1)^{\text{th}}$ order differential equations (8.6). Only initial data satisfying the boundary equation (8.7) can be well-posed even if solutions exist. Note that, on the boundary or on a slice Σ_t, all functions and their derivatives should be considered *independent*. To do a proper characterization of the initial data, it is therefore helpful to reformulate the system as a first-order system, whose integrability can be easily assessed. We will return to this point shortly when considering a degree of freedom count for irregular Lagrangian systems.

For variations of the form

$$\delta q^i = M^{ij} \alpha_j \,, \tag{8.8}$$

where $M^{ij} = -M^{ji}$ are anti-symmetric functions of the q^i, the local term of the EL equation is automatically satisfied:

$$\delta S_{\text{loc}} = \int_{t_1}^{t_2} dt \, \alpha_i M^{ij} \alpha_j = 0 \,. \tag{8.9}$$

The vanishing of the boundary term requires $M_{ij} = 0$ on the boundary for unrestricted Lagrangians and when $\alpha_i \neq 0$.[2]

Let us then define the full set of (broad) *symmetries-over-histories* as any transformation of the form:

$$q^i(t) \rightarrow q^i(t) + \delta_\epsilon q^i(t), \tag{8.10}$$

where ϵ stands for an arbitrary parameter $\epsilon_\alpha(t)$ defined over an entire history, such that if q^i is a solution, then $q^i + \delta_\epsilon q^i$ is also a solution.[3] The broad symmetries therefore induce arbitrary diffeormorphims on the space of DPMs. We will give a local characterization of this space later using the Hamiltonian formalism.

The symmetries-over-histories under the preliminary stance will then be the subset of the broad symmetries-over-histories that exactly preserve the action. Significantly such symmetries leave the action invariant—even *off-shell*. Formally, we can write this as

$$\delta S[q^i, \delta_\epsilon q^i] = 0 \,, \tag{8.11}$$

with $\delta_\epsilon q^i$ unconstrained on the boundary and q^i not necessarily satisfying $\alpha_i = 0$.[4]

---

[2] Note that, while these transformations are formally symmetries of the action off-shell, they are *trivially* zero on-shell (i.e. when $\alpha_i = 0$). They are thus trivial symmetries that exactly vanish when the classical equations of motion are satisfied. We include them here for completeness, and because they can be used, in general, to relate different representations of non-trivial narrow symmetries—see (Henneaux and Teitelboim, 1992, §3.1.5) for a more complete discussion. To find the interesting narrow symmetries of a theory, we must thus quotient the full group of gauge symmetries of S by these trivial ones.

[3] Requiring the preservation of the solution space automatically excludes the trivial transformations discussed above, but below we will be considering identities that happen to hold off-shell, so it will be valuable to include the trivial transformations in our analysis.

[4] We are assuming here that the symmetries satisfying (10.6) are known in advance. When the narrow symmetries are not fully known in advance, it is possible to formulate (10.6) in terms of a formal condition on $\delta_\epsilon q^i$.

Taking the functional derivative of (8.11) and using the EL equations (8.4) to eliminate the term depending on the functional derivative of $\delta_\epsilon q^i$, we find the on-shell relation

$$\delta^2 S[q_{\text{cl}}^i; \delta q^i, \delta_\epsilon q^j] = 0\,. \tag{8.12}$$

We have used a notation where the expression is to be evaluated at the value of the first argument; that is, $F[q_0; \ldots] = F[q; \ldots]\big|_{q=q_0}$ for some functional F of q. The vanishing of the second variation above is equivalent to the local expression

$$\int_{t_1}^{t_2} \mathrm{d}t'\, \frac{\delta^2 S}{\delta q^i(t)\delta q^j(t')}\bigg|_{q^i=q_{\text{cl}}^i}\, \delta q^i(t)\delta_\epsilon q^j(t') = 0\,, \tag{8.13}$$

where the variations $\delta q^i(t)$ are arbitrary except at the boundary, where they vanish.

Given this arbitrariness, the narrow symmetries are found to be the null eigenvectors, in the functional sense, of the second functional derivative of S when the EL equations are satisfied. A generating set for these symmetries can then be found by finding a basis for the on-shell kernel of $\delta^2 S$.

Note that each element of a generating set of the narrow symmetries is a functional degree of freedom, spanned by the independent gauge parameters ϵ (which we will define more explicitly below), defined over a complete history. The definition (8.13), though mathematically elegant, is usually not of much practical use however since it is a functional eigenvalue equation. A more useful definition will be given later in terms of the degeneracy of EL equations themselves.

8.2   Noether's Second Theorem Revised

The formalism we have introduced allows for a perspicuous reevaluation of the implications of Noether's second theorem. First, let us bring (8.11) into a more recognizable form. Let us write the variations $\delta_\epsilon q^i$ in terms of an nth order differential operator $T_\alpha^i(\epsilon^\alpha)$ of the form

$$\delta_\epsilon q^i = T_\alpha^i(\epsilon^\alpha) \equiv \sum_{a=0}^{n} T_{(a)\alpha}^i \frac{\mathrm{d}^a}{\mathrm{d}t^a}\epsilon^\alpha\,. \tag{8.14}$$

We can then define the adjoint $\tilde{T}_\alpha^i(f_i)$ such that

$$\int_{t_1}^{t_2} \mathrm{d}t\, f_i T_\alpha^i(\epsilon^\alpha) = \int_{t_1}^{t_2} \mathrm{d}t\, \epsilon^\alpha \tilde{T}_\alpha^i(f_i)\,, \tag{8.15}$$

for all continuously differentiable functions $f_i \in C^n(t_1, t_2)$ and $\epsilon^\alpha \in C^{n+k}(t_1, t_2)$. We will see below that the boundary conditions used to define the adjoint will place further restrictions on the functions f_i (i.e. the domain of $\tilde{T}_\alpha^i$). Noether's second theorem (8.11) then reads

$$\delta S[q^i; \delta_\epsilon q^i] = \int_{t_1}^{t_2} \mathrm{d}t\, \alpha_i T_\alpha^i(\epsilon^\alpha) = \int_{t_1}^{t_2} \mathrm{d}t\, \epsilon^\alpha \tilde{T}_\alpha^i(\alpha_i) = 0\,. \tag{8.16}$$

For arbitrary gauge parameters ϵ^α this reduces to the local equation

$$\tilde{T}_\alpha^i(\alpha_i) = 0\,. \tag{8.17}$$

Which is just the generalized Bianchi identities (7.31) for the case in which $\bar{q}^i(t) = q^i(t) + \delta_\epsilon q^i(t)$ and $\bar{t} = t$.

This local equation, while necessary, is not sufficient for the vanishing of (8.11). We must supplement the differential equations (8.17) with the appropriate boundary conditions, which must vanish for arbitrary ϵ_α. These boundary conditions must be satisfied on *all* final boundaries Σ_{t_2}. This leads to local (in time) constraints between the components of T^i and α_i—one for each derivative of ϵ_α appearing in T^i. These constraints enforce that each of the time derivatives of the gauge parameters ϵ_α can be independently specified at any time, and are precisely the Lagrangian constraints of the theory. Mathematically they specify the domain of the adjoint operator $\tilde{T}_\alpha$ as indicated above.

Let us now consider the interpretation of Noether's second theorem. We will focus on the explicit expressions for the simplest (but important) non-trivial case where the symmetries are first-order differential operators in t. In this case, $n = 1$ and integration by parts of the left-hand side of (8.15) gives

$$\tilde{T}_\alpha^i = \left(T_{(0)\alpha}^i - \dot{T}_{(1)\alpha}^i \right) - T_{(1)\alpha}^i \frac{\mathrm{d}}{\mathrm{d}t} \tag{8.18}$$

and the boundary term

$$\left[\left(T_{(0)\alpha}^i \frac{\partial L}{\partial \dot{q}^i} + T_{(1)\alpha}^i \alpha_i \right) \epsilon^\alpha + T_{(1)\alpha}^i \frac{\partial L}{\partial \dot{q}^i} \dot{\epsilon}^\alpha \right]_{t_1}^{t_2} = 0 \,. \tag{8.19}$$

The adjoint defined in (8.18) leads to the explicit form of Noether's second theorem, $\tilde{T}_\alpha^i \alpha_i = 0$ when $k = 1$. Because the boundary terms must hold at-an-instant, the functions ϵ^α and their time derivatives $\dot{\epsilon}^\alpha$ should be treated as arbitrary independent functions. We thus get two constraints. The first:

$$T_{(1)\alpha}^i \frac{\partial L}{\partial \dot{q}^i} = 0 \,, \tag{8.20}$$

arises from the vanishing of the $\dot{\epsilon}^\alpha$ term and is called the *primary constraint* of the theory. It holds off-shell. A second constraint, which holds only on-shell (i.e. when $\alpha_i = 0$), results from the ϵ^α term and is given by

$$T_{(0)\alpha}^i \frac{\partial L}{\partial \dot{q}^i} \approx 0 \,, \tag{8.21}$$

where we have used the '$\approx$' sign to indicate an equation that is only required to hold on-shell. The condition above doesn't historically have a specific name since versions of it were originally derived using the Dirac methodology where it can play different roles. Note that in deriving (8.21), we have assumed that ϵ_α has non-trivial time dependence — an assumption we will revisit below when deriving Noether's first theorem.

Let us then investigate the consequences of the degeneracies of the variational principle that result from Noether-2 symmetries for the solvability of the equations of motion. To do this, note that $\frac{d}{dt} = \dot{q}^i \frac{\partial}{\partial q^i} + \ddot{q}^i \frac{\partial}{\partial \dot{q}^i}$ when $k = 1$. Thus,

$$\alpha_i = \frac{\partial L}{\partial q^i} - \frac{d}{dt}\left(\frac{\partial L}{\partial \dot{q}^i}\right) = \frac{\partial L}{\partial q^i} - \dot{q}^j \frac{\partial^2 L}{\partial q^j \partial \dot{q}^i} - \ddot{q}^j \frac{\partial^2 L}{\partial \dot{q}^i \partial \dot{q}^j}. \tag{8.22}$$

The EL equations $\alpha_i = 0$ can then be seen as differential equations relating the accelerations $\ddot{q}^i$ to the velocities $\dot{q}^i$ and configurations q^i at any given time t. To solve them, the Hessian of the Legendre transform:

$$W_{ij} = \frac{\partial^2 L}{\partial \dot{q}^i \partial \dot{q}^j} \tag{8.23}$$

must be invertible as a matrix. The primary constraints (8.20), however, immediately imply that W_{ij} is *not* invertible because they imply that the $T_{(1)}{}^i{}_\alpha$ lie in the kernel of W_{ij}:

$$T_{(1)}{}^i{}_\alpha W_{ij} = 0. \tag{8.24}$$

This means that the primary constraints prevent the variational principle from generating well-posed equations of motion in the precise sense that the accelerations cannot be solved uniquely in terms of the configurations and their velocities at a given time. It is this form of underdetermination that, we will argue, should be part of a good definition of gauge symmetry — although we will also see that this can arise in a more diverse set of circumstances than we have considered here.

The condition (8.24) can be used in combination with the identity $T_{(1)}{}^i{}_\alpha \alpha_i = 0$, which is trivially satisfied on-shell, to produce the on-shell constraint

$$T_{(1)}{}^i{}_\alpha K_i \approx 0, \tag{8.25}$$

where we have defined

$$K_i = \frac{\partial L}{\partial q^i} - \dot{q}^j \frac{\partial^2 L}{\partial q^j \partial \dot{q}^i}. \tag{8.26}$$

The constraints (8.25) are usually called the *Lagrangian constraints* of the theory, which follow from the primary canonical constraints and the equations of motion.

Normally the Lagrangian constraints are derived directly from the Lagrangian using Dirac's methodology. In the Dirac approach, one starts by reading off the primary canonical constraints directly from the form of the Lagrangian and then derives additional constraints by consistency with the equations of motion. If one were to do that here, one would want to impose that the time derivative of (8.25) vanish when $\alpha_i = 0$. Using the fact that the $T_{(1)}{}^i{}_\alpha$ are in the kernel of W_{ij} off-shell, Noether's second theorem, which is also valid off-shell, then tells us that

$$\frac{d}{dt}\left(T_{(1)}{}^i{}_\alpha K_i\right) = \frac{d}{dt}\left(T_{(1)}{}^i{}_\alpha \alpha_i\right) = T_{(0)}{}^i{}_\alpha \alpha_i. \tag{8.27}$$

The condition $T_{(0)}{}^i{}_\alpha \alpha_i \approx 0$ then arises as a closure condition for the Dirac algorithm.

From the point of view taken here, the Noether methodology assumes the existence of a Noether-2 symmetry. In this case, the on-shell condition (8.21) can be used, in combination with $\alpha_i = 0$ to derive the two on-shell conditions

$$T_{(0)}{}^i{}_\alpha W_{ij} \approx 0 \qquad\qquad T_{(0)}{}^i{}_\alpha K_i \approx 0, \tag{8.28}$$

in the same way that the analogous conditions were derived for $T_{(1)}{}^{i}{}_{\alpha}$.[5] Together, these conditions imply $T_{(0)}{}^{i}{}_{\alpha}\,\alpha_{i} \approx 0$, which closes the Dirac algorithm. We thus see that the assumptions of the Noether methodology are mutually consistent: when a Noether-2 symmetry exists the Dirac algorithm is guaranteed to close. We also see that this closure condition can be expressed as resulting from an off-shell identity; namely Noether's second theorem. The various distinctions between 'primary' and 'Lagrangian' constraints arise simply from the formal distinctions between $T_{(0)}{}^{i}{}_{\alpha}$ and $T_{(1)}{}^{i}{}_{\alpha}$. In the Noether methodology, these arise as one tight, self-consistent package.

Let us note here that the amount of underdetermination in the equations of motion depends explicitly on the form of the infinitesimal generator of the Noether-2 symmetry in question. This can be read-off directly from the dimension of the kernel of W_{ij}. When $T_{(1)}{}^{i}{}_{\alpha}$ is non-zero there are off-shell null vectors of W_{ij} for each value of α.

The same is *nearly* true when $T_{(0)}{}^{i}{}_{\alpha}$ is non-zero, although only on-shell. The arguments leading to (8.21) assume that the gauge parameter ϵ_{α} is a non-trivial function of t. If this is *not* the case (i.e. if the gauge parameters are constant functions of the independent variables), then $T_{(1)}{}^{i}{}_{\alpha} = 0$ (because $\dot{\epsilon} = 0$ definition) and (8.19) has the solution

$$\frac{d}{dt}\left(T_{(0)}{}^{i}{}_{\alpha}\frac{\partial L}{\partial \dot{q}^{i}}\right) = 0. \tag{8.29}$$

This implies, therefore, that $T_{(0)}{}^{i}{}_{\alpha}\frac{\partial L}{\partial \dot{q}^{i}}$ need only be a constant of motion not necessarily equal to zero. This is Noether's first theorem restricted to the case where the action is not allowed to have explicit dependence on the independent variables. We see now why the presence of Noether-1 symmetries, whose infinitesimal generators are constant functions of the independent variables, leads to constants of motion. Importantly, the Noether-1 symmetries are still required to keep the action invariant but *do not* lead to underdetermination in the equations of motion. They thus provide an important example of why invariance of the action is, in general, not a reliable way to implement a notion of gauge symmetry based on matching underdetermination in the equations of motion with underdetermination of representations by phenomena.

8.2.1   Counting Degrees of Freedom

We will argue below that for first-class constrained systems it can be shown that the number of primary canonical constraints will match the number of independent generalized Bianchi identities. Secondary (and higher-order) canonical constraints will then correspond to the Lagrangian constraints, which are the time derivatives of the higher-order constraints. From this perspective, as already noted, the generalized Bianchi identities simply express the closure of the Dirac algorithm, which must occur because we have assumed the theory generated by S has the symmetries of the form (8.14). Further complications in the relevant relations are introduced by considering systems with second-class canonical constraints. Analysis of the relations between generalized

---

[5] For simplicity, we have assumed here that $\frac{\partial T_{(0)}{}^{i}{}_{\alpha}}{\partial \dot{q}^{i}} = 0$ so that these expressions are valid only for classical symmetries that are not generalized symmetries. The form of the primary canonical and Lagrangian constraints, however, does *not* require such an assumption.

Bianchi identities, Lagrangian constraints, and Hamiltonian constraints can be found in Sundermeyer 1982; Lusanna 1990; Lusanna 1991; Chaichian and Martinez 1994; Díaz, Higuita and Montesinos 2014.

Let us now explicitly derive the degree of freedom count as implied by the different relations we have derived. The local expression for the generalized Bianchi identities (8.17) results from the existence of narrow symmetries-over-histories. There is, therefore, one identity for every function $\epsilon^\alpha(t)$ defined over a complete history. On the other hand, the Lagrangian constraints (8.20) result from compatibility of the narrow symmetries with the variational principle of the theory *at a time*. Each constraint reflects the freedom to vary a derivative of ϵ^α independently of the others. There is thus one Lagrangian constraint for each non-zero value of $T^i_{(a)\alpha}$ for $a \geq 1$. Of course, there is no *a priori* reason why these degree-of-freedom counts should match. In general they will not: one is related to the functional degrees of freedom over a history that relate equivalent DPMs, while the other reflects the number of free functions that can be specified at a given time to relate the independently specifiable initial data for the evolution equations.

A simple example of where this degree of freedom count does not match occurs when only $T^i_{(0)\alpha}$ is non-zero but is constant in time. In this case, the generalized Bianchi identity reduces to Noether's first theorem and there is no Lagrangian constraint at all. And because ϵ^α is not independently specifiable at any given time, the constraint (8.21) does not impose a reduction in the freely specifiable initial data of the system. Thus, the generalized Bianchi identities enforce a global restriction at the level of a history, but not at any given time.

Given these observations, we can provide a more nuanced analysis of the relationship between the existence of generalized Bianchi identities and 'underdetermination' within a system of equations. In particular, we can diagnose an ambiguity that has been introduced into the philosophical literature by Brading and Brown (2003), owing to a failure to clearly differentiate between the inequivalent conditions for underdetermination at the level of histories and underdetermination in the initial value problem. Within this influential analysis we find the following statement:

Noether's second theorem tells us that in any theory with a local Noether symmetry there is always a *prima facie* case of underdetermination: more unknowns than there are independent equations of motion ... This underdetermination means that there are, in general, as many identities involving the field as there are arbitrary functions involved in defining the local symmetry transformations.[6]

Now, if interpreted as a statement with regard to underdetermination at the level of histories, then Brading and Brown are correct; that is, Noether's second theorem indeed implies identities involving the fields *at the level of histories* and thus underdetermination *at the level of histories*. However, typically, in the context of gauge theories we are interested in identifying underdetermination *at the level of instantaneous states* and thus formulating gauge identifies *at the level of instantaneous states*. For such purposes satisfaction of Noether's second theorem is necessary but not sufficient for the identification of underdetermination. What we require is a precise prescription for the diagnosis of *underdetermination-at-a-time* from *symmetries-over-histories*. For

---

[6] Brading and Brown (2003, p. 104).

this purpose neither Dirac's Theorem nor Noether's Second Theorem are fit for purpose. The novel analysis of the following section will provide formal tools for resolving this important formal and interpretative issue.

8.2.2   Example: Electromagnetism

The electromagnetic field tensor can be written as $F_{\mu\nu} = \partial_\mu A_\nu - \partial_\nu A_\mu$. The action for matter-free electromagnetism is the functional

$$S[A^\mu] = \int_\Omega \mathrm{d}^4 x \, F^{\mu\nu} F_{\mu\nu} \qquad (8.30)$$

of the vector potential $A^\mu(x)$, which depends on the spacetime point x, with μ and ν spacetime indices. For simplicity, we have used a flat Minkowski metric in Cartesian coordinates so that the volume form is trivial and indices can be lowered and raised straightforwardly. The theory is manifestly invariant under the Noether-2 symmetries

$$A^\mu \to A^\mu + \partial^\mu \epsilon = \begin{pmatrix} A^0 - \dot\epsilon \\ A^a + \partial^a \epsilon \end{pmatrix}, \qquad (8.31)$$

where a is a spatial index.

In what follows we use a notation whereby the index i ranges over all the fields and therefore becomes a spacetime index μ and a continuous spatial coordinate x. Moreover, we assume there is only one gauge parameter indexed over space. We then have $T_\alpha^i \to T^\mu(x)$. The non-zero components of T_α^i can be read off from (8.31), and in this notation are

$$T_{(1)}^0 = -1 \qquad\qquad T_{(0)}^a = \partial^a . \qquad (8.32)$$

The adjoint can then be calculated using (8.18) and takes the form

$$\bar{T}^\mu = \begin{pmatrix} \frac{\mathrm{d}}{\mathrm{d}t} \\ \partial^a \end{pmatrix} . \qquad (8.33)$$

The EL vectors for electromagnetism are

$$\alpha_\mu(x) = \partial^\nu F_{\mu\nu} = \begin{pmatrix} \partial^a E_a \\ -\dot{E}_a + \epsilon_{abc}\partial^b B^c \end{pmatrix} . \qquad (8.34)$$

We thus find that the generalized Bianchi identities are

$$\tilde{T}^\mu \alpha_\mu = \frac{\mathrm{d}}{\mathrm{d}t}\left(\partial^a E_a\right) - \partial^a \dot{E} + \epsilon_{abc}\partial^a \partial^b B^c = 0 , \qquad (8.35)$$

which is trivially satisfied and equivalent to the covariant equation $\mathrm{d}F = 0$.

The non-zero expansion coefficients of T_α^i in (8.32) lead to primary constraints

$$T_{(1)}{}^\mu{}_\alpha \frac{\partial L}{\partial \dot{A}^\nu} = -\frac{\partial L}{\partial \dot{A}^0} = 0 , \qquad (8.36)$$

which indicate that A^0 is a Lagrange multiplier in the theory, and the Lagrangian constraint (after using the primary constraint)

$$T_{(1)}^\mu K_\mu = \alpha_0 = \partial^a E_a \approx 0 \,, \tag{8.37}$$

which is the so-called *Gauss constraint* of the matter-free theory. From (8.21), we obtain the Gauss constraint more directly because

$$T_{(0)}{}^\mu{}_\alpha \frac{\partial L}{\partial \dot{A}^\mu} = \partial^a E_a \approx 0 \,. \tag{8.38}$$

The second condition of (8.28), $T_{(0)}{}^\mu{}_\alpha K_\mu = \partial^a K_a \approx 0$, then says, after a straightforward calculation, that the Gauss constraint is propagated in time $\partial^a \dot{E}^a = 0$ guaranteeing the closure of the Dirac algorithm.

8.3 Initial Value Constraints

While a full degree of freedom count can, in principle, be derived from the Lagrangian formalism through a careful treatment of the boundary conditions of the variational principle, a richer picture is gained from explicitly studying the integrability properties of the equations of motion directly. To do this, it is most convenient to move to a first-order formalism, where the integrability and well-posedness of the equations of motion can be most easily assessed. Moreover, in first-order form it is possible to express the variational principle and resulting equations of motion directly in terms of geometric quantities, with the subtleties of the boundary terms contained in constraints between the configuration variables and their velocities.[7] We will show here a procedure for converting a second-order system to a first-order system, but this procedure can easily be generalized to higher-order systems.[8] Thus, our analysis applies in general to systems of arbitrary order in derivatives of t.

The key idea is to double the variables of the system by introducing the velocity variables v^i, and impose additional constraints on these variables so that the resulting first-order system is equivalent to the old system when $v^i = \dot{q}^i$. An elegant way of achieving this is to use the slightly modified action

$$S_1 = \int_{t_1}^{t_2} dt \left(\frac{\partial L(q^i, v^i)}{\partial v^i} (\dot{q}^i - v^i) + L(q^i, v^i) \right) , \tag{8.39}$$

where we replace $\dot{q}^i \to v^i$ in the Lagrangian L. When the quantities $\frac{\partial L}{\partial v^i}$ are linearly independent (i.e. they don't satisfy constraints), then the extra term will enforce $\dot{q}^i = v^i$ on-shell. As we will see, this requirement on $\frac{\partial L}{\partial v^i}$ will be precisely the requirement that the resulting EL equations for the system are well-posed in the variables q^i and v^i.

To unpack these claims more carefully, we can take advantage of the first-order form of the equations of motion of the system to express the formalism in terms of

---

[7] An enlightening analysis of the geometry of the Euler-Lagrange equation which complements our own approach in a number of respects can be found in (Curiel, 2010).

[8] See, Woodhouse (1997, § 2.3, p. 24) for details.

more geometric quantities. First, we define the velocity phase space $\Gamma = T\mathcal{C}$ equipped with local coordinates (q^i, v^i), and define the one-form

$$\theta_L = \frac{\partial L}{\partial v^i} dq^i \tag{8.40}$$

on Γ and the Hamiltonian function

$$H = v^i \frac{\partial L}{\partial q^i} - L . \tag{8.41}$$

We can then consider a curve induced by the map $\gamma : \mathbb{R} \to T\Gamma$ on Γ and the tangent vector, X, which in coordinates reads

$$X = \dot{q}^i \frac{\partial}{\partial q^i} + \dot{v}^i \frac{\partial}{\partial v^i} , \tag{8.42}$$

where the derivatives are with respect to the time parameter $t \in \mathbb{R}$ along γ. Using these geometric quantities, the action $S_1[\gamma]$ then reads

$$S_1[\gamma] = \int_{t_1}^{t_2} dt \, (\iota_X \theta_L - H) = \int_\gamma (\theta_L - H dt) . \tag{8.43}$$

In a slight abuse of notation, we define $dt = \gamma^* dt$ as the differential along the image of γ induced by the pullback of the differential dt on its domain. We thus have $\iota_X dt = 1$ and $\iota_{Y_i} dt = 0$ for all Y_i such that $[Y_i, X] = 0$, which can be taken as a more geometric definition of the tangent vector $X = \frac{d}{dt}$ to the curve.

Let us restrict to the case where the action has no explicit dependence on the independent variables. Because of this, any variation due to a change in the independent variables can be expressed as a variation of the dependent variables by pulling these back by γ. Without loss of generality, we can thus express the variational derivative of the action $S[\gamma]$ in terms of infinitesimal variations generated by a vector field $u \in T\Gamma$. Using our previous notation, we can express such variations as $\delta q^i \to \mathfrak{L}_u q^i$. For our analysis later, it will be convenient to split these variations into those that lie tangent to γ, and therefore could correspond to variations of t and are parallel to X, and those that are transverse to γ, and therefore satisfy $\mathfrak{L}_u t = 0$ and are spanned by the Y_i vectors above. But for a general u, the variation of S_1 can be written in terms of the Lie drag of S by the vector field u:

$$\delta S_1[\gamma; u] \equiv \mathfrak{L}_u S[\gamma] = \int_\gamma \left(\iota_u d\theta_L - (\mathfrak{L}_u H) dt - H \mathfrak{L}_u dt + d \left(\iota_u \theta_L \right) \right)$$

$$= -\int_{t_1}^{t_2} \left(\iota_u \left(\iota_X \omega_L + dH \right) dt + H \mathfrak{L}_u dt \right) + \iota_u \theta_L \Big|_{t_1}^{t_2} , \tag{8.44}$$

where we have defined the exact two-form $\omega_L = d\theta_L$. Note that while ω_L is closed, we will now see that it will be degenerate when $S[\gamma]$ has narrow symmetries-over-histories so that it is a *pre*-symplectic rather than symplectic two-form in this case.

Each term in (8.44) is significant. First let us consider (8.44) as an equation for fixing the classical solutions X for arbitrary u. For arbitrary variations u transverse to X, $\mathfrak{L}_u dt = 0$ and the integrand is zero only if Hamilton's equations

$$\iota_X \omega_L + dH = 0 \tag{8.45}$$

are satisfied. When ω_L is non-degenerate, there is a unique vector field X over Γ that solves this equation. Using

$$\omega_L = d\theta_L = d\left(\frac{\partial L}{\partial v^i} dq^i\right) = \frac{\partial^2 L}{\partial q^i \partial v^j} dq^i \wedge dq^j + \frac{\partial^2 L}{\partial v^i \partial v^j} dv^i \wedge dq^j \tag{8.46}$$

and

$$dH = d\left(v^i \frac{\partial L}{\partial v^i} - L\right) = \left(v^j \frac{\partial^2 L}{\partial v^j \partial q^i} - \frac{\partial L}{\partial q^i}\right) dq^i + \frac{\partial^2 L}{\partial v^j \partial v^i} v^j dv^i \tag{8.47}$$

as well as (8.42), Hamilton's equations (8.45) become

$$\iota_X \omega_L + dH = \left[(v^j - \dot{q}^j)\frac{\partial^2 L}{\partial v^i \partial q^j} + \frac{d}{dt}\left(\frac{\partial L}{\partial v^i}\right) - \frac{\partial L}{\partial q^i}\right] dq^i + (v^j - \dot{q}^j)\frac{\partial^2 L}{\partial v^i \partial v^j} dv^i . \tag{8.48}$$

The dv^i term enforces $v^i = \dot{q}^i$ (for non-degenerate $\frac{\partial^2 L}{\partial v^i \partial v^j}$) and the dq^i term enforces the EL equations, as expected, when this condition is satisfied. The variational principle (8.44) is therefore equivalent to Hamilton's principle—at least when solutions exist and are well-posed.

In general, θ_L will have no kernel so that the first-order variables can be varied freely in which case the vanishing of the boundary terms requires $u(t_1) = u(t_2) = 0$. Significantly, variations tangent to X are forbidden in the variation if the parameterization on γ is fixed since these are not valid variations of the classical curves (these variations change the parameterization). The important case where arbitrary parameterizations of γ are allowed will be treated separately in the context of our analysis of reparameterization invariant mechanics and the problem of time in Chapter 11.

With the variations tangent to X ruled out, all remaining variations will be transverse to X. The geometry of the situation then provides us with a precise means to connect the transverse variations to initial value constraints. In particular, we can use (8.44) to deduce the formal consequences of S having narrow symmetries-over-histories in the case of transverse variations.

Let us treat (8.44) as an equation for the tangent vector fields u_i for an arbitrary X (not necessarily satisfying (8.45)). As above, we first consider the case where the u_i are transverse to X and form a closed Lie algebra $[u_i, u_j] = f_{ij}{}^k u_k$. In this case we again have $\mathfrak{L}_{u_i} dt = 0$, and for arbitrary X we must have separately that

$$\mathfrak{L}_u \theta = 0 \qquad\qquad \mathfrak{L}_{u_i} H = 0 . \tag{8.49}$$

The first of these equations takes the form

$$\iota_u \omega_L = -d\left(\iota_u \theta_L\right) , \tag{8.50}$$

which means that, in genera, u is a canonical transformation generated by the phase space function $\iota_u \theta_L = 0$. However, on-shell we can use (8.45) to deduce that

$$\iota_u \iota_X \omega_L = -\iota_u dH = 0, \tag{8.51}$$

because of the second independent equation of (8.49). Thus, on-shell, we must have separately that

$$\iota_u \omega_L = 0 \qquad\qquad \mathcal{L}_u H = 0 \tag{8.52}$$

and

$$\iota_u \theta_L = 0. \tag{8.53}$$

Thus, the u_i generate narrow symmetries-over-histories of S if they are in the kernel of ω_L and are invariances of the Hamiltonian.

We will refer to $\iota_u \theta_L = 0$ as *Initial Value Constraints* (IVCs) because they are constraints on the boundary data. Note that we obtained them from $d(\iota_u \theta_L) = 0$ since, in general, u will depend on a time-dependent arbitrary parameter. However, if u is time-independent (i.e. if the symmetry is 'global') then no IVC is generated. We will study this case explicitly in the context of transverse and tangential variations in Chapter 11.

Crucially, it is the existence of initial value constraints that is indicative of the specific *underdetermination-at-a-time* problem associated with dynamical redundancy within the formalism. As such, it is the identification of IVCs that is the fundamental basis for the isolation of gauge degrees of freedom. This criterion thus supersedes both Noether's second theorem and the Dirac criterion (based on the existence of first-class canonical constraints) as the fundamental diagnostic tool for interpreting a gauge theory.

Returning to our more formal analysis it is worth noting that $\iota_X \iota_u \omega_L = -\iota_X d(\iota_u \theta_L) = -\frac{d}{dt}(\iota_u \theta_L)$. We thus have, on-shell, that

$$\mathcal{L}_u H = -\frac{d}{dt}(\iota_u \theta_L). \tag{8.54}$$

The condition $\mathcal{L}_u H = 0$ can then be seen as a condition for the propagation in time of the IVCs. In many cases, the conditions $\iota_u \theta_L = 0$ will be automatically satisfied by the form of the Lagrangian. In this case, these constraints are what were called the primary constraints within the Dirac language and the conditions $\mathcal{L}_u H = 0$ can be used to generate the secondary constraints which are usually required only to vanish on-sell. In §8.4 we will give explicit representations for the symmetries u and see which parts of u are responsible for the primary and higher-order constraints. We will also see that the generalized Bianchi identities can be used to show that the $(n+1)^{\text{th}}$ order constraints are given by the time derivative of the nth order constraints, which completes the connection.

We see from the derivation above that while the symmetries generated by primary and secondary (or higher-order) constraints are all off-shell symmetries, only the primary constraints generate IVCs and degeneracies of ω_L off-shell. Off-shell, the symmetries generated by secondary constraints are canonical transformations of the action where the boundary variation exactly cancels the boundary term generated by the canonical transformation of the local action. But only on-shell are the secondary (and higher-order) constraints required to vanish.

The primary constraint surface is pre-symplectic owing to the fact ω_L is degenerate there. This defines the image of the *Legendre transform* $\mathbb{L} : TC \to T^*C$ which, in local Darboux coordinates (q^i, p_i), takes the form

$$p_i \equiv \frac{\partial L}{\partial v^i} \, . \tag{8.55}$$

This transformation is bijective when the Hessian, W_{ij}, of $\mathbb{L}$

$$W_{ij} = \frac{\partial^2 L}{\partial v^i \partial v^j} \tag{8.56}$$

has no kernel. Since we are assuming no degeneracies among the configurations variables q^i (i.e. no second-class constraints), the kernel of ω_L, which takes the form

$$\omega_L = \mathrm{d}\theta_L = \mathrm{d}\left(p_i \mathrm{d}q^i\right) = \mathrm{d}p_i \wedge \mathrm{d}q^i \, , \tag{8.57}$$

maps bijectively to the kernel of W_{ij}. We can thus understand the degeneracies of ω_L as the failure of the invertibility of $\mathbb{L}$ that occurs because of the restriction of $\mathbb{L}$ to the primary constraint surface.

We can extend $\omega_L \to \omega_e$ and $\theta_L \to \theta_e$ canonically off the primary constraint surface using a canonical symplectification procedure. Using (8.50), we see that ω_e can be used to define the flow of u everywhere on the extended phase space in terms of a canonical transformation generated by the IVCs. Restricting this flow to the constraint surface, then gives precisely the symmetry generators u for the theory.

Since we are assuming there are no independently specified constraints among the configuration variables q^i (i.e. we are assuming there are no independent second-class constraints), the kernel of W_{ij} can be used to construct the kernel of ω_L, and indicates $\mathrm{Dim}(\ker(W))$ constraint relations among the p_i.

The link between the null directions of ω_L and a set of constraints among the p_i can be established as follows. Consider the *extended* symplectic two-form ω_e on Γ that can be expressed in local coordinates as

$$\omega_e = \mathrm{d}p_i^e \wedge \mathrm{d}q^i \, , \tag{8.58}$$

where p_i^e match $\frac{\partial L}{\partial v^i}$ when these quantities exist and are otherwise canonical Darboux coordinates corresponding to $\mathrm{d}q^i$. Thus, ω_e is equal to ω_L along all non-degenerate directions but extends ω_L to a canonical symplectic two-form when ω_L is degenerate. Note that Γ may have a non-trivial topology inherited from the global structure of ω_L that ω_e will respect. But because ω_e is non-degenerate, the u_i define the functions h_i, up to a constant, according to

$$\iota_{u_i}\omega_e = \mathrm{d}h_i \, . \tag{8.59}$$

We can find solutions to (8.45) when ω_L is degenerate by restricting the flow X_e defined by

$$\iota_{X_e}\omega_e = \mathrm{d}H \tag{8.60}$$

to the surface $C = \{(q_i, p^i) | h_i = 0\}$. On this surface, $\mathrm{d}h_i = 0$, and the u_i become degenerate directions of the restriction $\omega_e|_C = \omega_L$ as required. The restricted flow

$X_e|_C = X$, however, is well-defined. It is rather curious that, while ω_e is perhaps *the* central structure used in canonical analysis to find solutions to (8.45) in the degenerate case (the Poisson bracket on Γ, for example, is not defined without it), most standard textbooks do not define it explicitly or comment on how it is related to the two-form ω_L obtained from the variational principle.

To fix the constant offset of the h_i, one needs to know more than simply the null directions u_i of ω_L. When u is transverse to X, this information is provided by the $\text{Dim}(\ker(W))$ independent relations (8.53). We then have

$$h_i = \lambda_i^j \iota_{u_j} \theta_L \,, \tag{8.61}$$

for some invertible matrix λ_j^i. It can then be shown from (8.50) that $\lambda_j^i = \delta_j^i$.

We note here that our point of departure was to *assume* that one has constructed an action S that is known to be invariant under some set of gauge symmetries, as this is consistent with physical practice. Under this assumption, there must exist an invertible matrix λ_i^j that satisfies the above requirements—otherwise the action S would not have the available symmetries. This is a slightly different attitude than in the usual Dirac algorithm where one does not assume that the full set of Lagrangian symmetries are yet known or are even consistent (and therefore require extension or modification).

The procedure above for finding solutions X from X_e is not unique because any transformation along the flow of the u_i also satisfies (8.45). In other words, for any solution X_1 of (8.45) there is another solution X_2 related by

$$X_2 = X_1 + \xi^i \iota_{\omega_c^{-1}} \mathrm{d} h_i \,, \tag{8.62}$$

where the ξ^i are *completely free* functions. Since the functions ξ^i are complete free, the theory defined by S can only be predictive for phase space functions that do not depend on the ξ^i. This can be achieved by restricting to phase space functions invariant under the flow of h_i; that is, the functions O_D satisfying

$$\iota_{\omega_e^{-1}} \mathrm{d} O_D \wedge \mathrm{d} h_j \Big|_C = \{O_D, h_i\} \Big|_C = 0 \,, \tag{8.63}$$

where we have defined the Poisson bracket $\{f, g\} = \iota_{\omega_e^{-1}} \mathrm{d} f \wedge \mathrm{d} g$.

It is worth noting here that the set of functions O_D with this property would be equivalent to the Dirac observables if one includes in u_i tangent vectors parallel to X. We will show in Chapter 11 that the argument just presented fails for that case.

8.4 Dynamical Redundancy

We finally have all the ingredients in place to complete our analysis and provide a fullly rigorous prescription for the diagnosis of dynamical redundancy within systems with singular Lagrangians and fixed-time parameterization. Our aim is to rewrite our symmetry conditions using the explicit expressions (8.14) for the narrow symmetries of S. This will allow us to connect our formalism to more familiar expressions from the Lagrangian and canonical constraint analysis. We begin by writing u in terms of the quantities of (8.14). Since the u generate infinitesimal transformations on velocity phase space, we have:

$$u = \delta_\epsilon q^i \frac{\partial}{\partial q^i} + \delta_\epsilon \dot{q}^i \frac{\partial}{\partial v^i} \,. \tag{8.64}$$

Since we have assumed that we have been able to bring our theory in first-order form, the terms of (8.14) for $a > 1$ can be rewritten using field redefinitions as terms of order $a \leq 1$ and are therefore unnecessary (including such terms explicitly is not problematic but only complicates the explicit form of the equations).

It is important to note that u is the pullback of U by γ and, as such, gives an independent vector along each point t along X. At any given time t, the functions ϵ^α are therefore independent of their derivatives. We will see, however, that requiring consistency between ϵ^α and its derivatives will imply one further important (and well-known) relationship between the initial value constraints and the symmetry conditions on H. For the moment it is useful to separate u into pieces that depend on the independent time derivatives of ϵ^α. Using (8.14) in (8.64) we find that $u = \sum_a u_\alpha^a \frac{d^a}{dt^a} \epsilon^\alpha$ where

$$u_\alpha^0 = T_{(0)}{}^i{}_\alpha \frac{\partial}{\partial q^i} + \dot{T}_{(0)}{}^i{}_\alpha \frac{\partial}{\partial v^i} \tag{8.65}$$

$$u_\alpha^1 = T_{(1)}{}^i{}_\alpha \frac{\partial}{\partial q^i} + \left(T_{(0)}{}^i{}_\alpha + \dot{T}_{(1)}{}^i{}_\alpha \right) \frac{\partial}{\partial v^i} \tag{8.66}$$

$$u_\alpha^2 = T_{(1)}{}^i{}_\alpha \frac{\partial}{\partial v^i} \,. \tag{8.67}$$

We can now write the initial value constraints $\iota_u \theta_L$ in a more explicit and familiar form. Since each vector u_α^a implies an independent equation that must be satisfied on phase space, we find, using $\theta_L = \frac{\partial L}{\partial v^i} dq^i = p_i dq^i$,

$$\pi_\alpha^2 \equiv T_{(0)}{}^i{}_\alpha p_i = 0 \qquad\qquad \pi_\alpha^1 \equiv T_{(1)}{}^i{}_\alpha p_i = 0 \,. \tag{8.68}$$

The vector u^3 contains no $\frac{\partial}{\partial q^i}$ component, and therefore contributes no independent constraint. Above, we have defined the *primary constraints* π^1 and *secondary constraints* π^2 for reasons we will explain below.[9]

We can write explicit expressions for the kernel of ω_L and the symmetries u such that $\mathfrak{L}_u H = 0$ of the Hamiltonian. For this we must use the explicit expressions for (8.46) and (8.47). The simplest expressions to calculate are those for u^3 since these only involve contractions with dv^i. Both $\iota_{u^2}\omega_L = 0$ and $\mathfrak{L}_{u^2} H = 0$ imply

$$T_{(1)}{}^i{}_\alpha \frac{\partial L}{\partial v^i \partial v^j} = T_{(1)}{}^i{}_\alpha W_{ij} = 0 \,, \tag{8.69}$$

where we recall that W_{ij} is the Hessian of the Legendre transform. We therefore find that the $T_{(1)}{}^i{}_\alpha$ are null vectors of the Hessian W_{ij}.

---

[9] Note that the index of π increases as the ϵ-derivative index of T decreases. The reason for this will become clear in a moment.

The next easiest terms to calculate are those due to $\iota_{u^0}\omega_L = 0$ and $\mathcal{L}_{u^0}H = 0$. From $\iota_{u^0}\omega_L = 0$, one obtains a one-form equation with independent dq^i and dv^i components. The dv^i component is easily seen to imply

$$T_{(0)}{}^i{}_\alpha W_{ij} = 0, \tag{8.70}$$

which implies that $T_{(0)}{}^i{}_\alpha$ are further independent null vectors of W_{ij}. The dq^i term gives

$$\dot{T}_{(0)}{}^i{}_\alpha W_{ij} = T_{(0)}{}^i{}_\alpha F_{[ij]}. \tag{8.71}$$

The contraction of this equation with $T_{(0)}{}^j{}_\alpha$ is zero. Thus, inverting this equation for $\dot{T}_{(0)}{}^i{}_\alpha$ can only give the component of $\dot{T}_{(0)}{}^i{}_\alpha$ that is independent of the kernel of W_{ij}. This is usually sufficient since, as we have seen above and will see further below, the kernel of W_{ij} can be known through the other means. When combined with $\mathcal{L}_{u^0}H = 0$, (8.71) can be used to eliminate the terms depending on $\dot{T}_{(0)}{}^i{}_\alpha$ to obtain

$$T_{(0)}{}^i{}_\alpha K_i = 0, \tag{8.72}$$

where we have defined

$$K_i = \frac{\partial L}{\partial q^i} - v^j \frac{\partial^2 L}{\partial v^i \partial q^j}. \tag{8.73}$$

which is equivalent to our earlier definition (7.4) for $v^i = \dot{q}^i$.

The equation (8.72) is the standard definition of the Lagrangian constraint associated with the null eigenvector $T_{(0)}{}^i{}_\alpha$ of W_{ij} (Sudarshan and Mukunda, 1974; Sundermeyer, 1982). However, for $T_{(0)}{}^i{}_\alpha$ the corresponding Lagrangian constraint is often dynamically uninformative because the transformation only affects the potential term of the Lagrangian. Historically, this Lagrangian identity, which is equal to (8.21) and can be derived from the generalized Bianchi identity and the other Lagrangian constraints, has not been called a Lagrangian constraint.

It is then easy enough to guess that one will obtain the more recognizable Lagrangian constraint for $T_{(1)}{}^i{}_\alpha$ by combining $\iota_{u^1}\omega_L = 0$ and $\mathcal{L}_{u^1}H = 0$. Indeed this is precisely what one finds after following the procedure outlined above and making use of the fact that $T_{(0)}{}^i{}_\alpha W_{ij} = T_{(1)}{}^i{}_\alpha W_{ij} = 0$. Explicitly,

$$T_{(1)}{}^i{}_\alpha K_i = 0, \tag{8.74}$$

which is just equivalent to (8.20) because $T_{(1)}{}^i{}_\alpha$ is a null vector of W_{ij}. We have thus shown that the coordinate-free expressions $\iota_u\omega_L$ and $\mathcal{L}_uH = 0$ are equivalent to the existence, in particular coordinates, of Lagrangian constraints due to non-invertibility of the Legendre transform arising from narrow symmetries of the action.

The local form of the generalized Bianchi identities (8.17) is satisfied automatically given these constraints since $\alpha_i = \ddot{q}^j W_{ij} - K_i$ so that

$$T_{(0)}{}^i{}_\alpha \alpha_i - \frac{d}{dt}\left(T_{(1)}{}^i{}_\alpha \alpha_i\right) = T_{(0)}{}^i{}_\alpha K_i - \frac{d}{dt}\left(T_{(1)}{}^i{}_\alpha K_i\right) = 0. \tag{8.75}$$

Alternatively, one could view the generalized Bianchi identities as a statement of the propagation in time of the Lagrangian constraint associated with $T_{(1)}{}^i{}_\alpha$.

Note that we can derive the generalized Bianchi identity directly from our variational principle (8.44) by using $u = \sum_a u_\alpha^a \frac{\mathrm{d}^a}{\mathrm{d}t^a} \epsilon^\alpha$ and performing integration by parts on the integrand. The resulting local identity is

$$\sum_a (-1)^a \frac{\mathrm{d}^a}{\mathrm{d}t^a} \left(\iota_{u_\alpha^a} \alpha \right) = 0 \,, \tag{8.76}$$

where α is the one-form $\alpha = \iota_X \omega_L + \mathrm{d}H$. It is straightforward to show that this reproduces the explicit form (8.75) of Noether's second theorem. Noether's theorem is useful in this context, even though it is weaker than the full set of constraints derived here, because it provides an off-shell relationship between the derivatives of the $T_{(a)}{}^i{}_\alpha$. This can provide a useful link between the standard Dirac algorithm and the picture presented here.

To establish this link more fully, we must first derive a relationship between the Lagrangian constraints $T_{(a)}{}^i{}_\alpha K_i = 0$ and the initial value constraints $\pi_\alpha^{(2-a)} = T_{(a)}{}^i{}_\alpha p_i = 0$ for $a = \{0,1\}$. To do this, we will adapt the analysis of Sundermeyer (1982, p. 55) to our purposes. First, we require the vanishing of the time derivative of $\pi_\alpha^{(2-a)}$:

$$\dot{\pi}_\alpha^{(2-a)} = \dot{T}_{(a)}{}^i{}_\alpha p_i + T_{(a)}{}^i{}_\alpha \frac{\partial p_i}{\partial q^j} \dot{q}^j = 0 \,, \tag{8.77}$$

which holds because $T_{(a)}{}^i{}_\alpha W_{ij} = 0$. We then use Hamilton's equations (8.45) to rewrite the Lagrangian constraint as

$$T_{(a)}{}^i{}_\alpha K_i = T_{(a)}{}^i{}_\alpha \left(\dot{q}^j \frac{\partial p_i}{\partial q_j} - \dot{p}_i \right) . \tag{8.78}$$

Inserting (8.77) into this expression we find

$$T_{(a)}{}^i{}_\alpha K_i = - \left(\dot{T}_{(a)}{}^i{}_\alpha p_i + T_{(a)}{}^i{}_\alpha \dot{p}_i \right) = -\dot{\pi}_\alpha^{(2-a)} \,. \tag{8.79}$$

This relation only holds on-shell. Thus, from this perspective, the Lagrangian constraints are required for the consistent propagation of the initial value constraints. Using these expressions, the generalized Bianchi identity in the form (8.75) then implies that

$$\dot{\pi}_\alpha^1 = \pi_\alpha^2 \,, \tag{8.80}$$

where we have set an integration constant to zero by requiring the consistent propagation of π_α^1. This means that the Lagrangian constraint for $a = 1$ is simply the secondary constraint π_α^2, and Noether's second theorem states that this secondary constraint is propagated in time. This occurs because we have assumed, from the beginning, that S has a consistent set of symmetries, and that we are simply exploring the consequences of these.

Let us briefly consider the contrast between the perspective developed here and the Dirac approach. In the Dirac algorithm, one starts with a variational principle that *need not be assumed* to obey all the consistency requirements $\mathcal{L}_u H = 0$ and initial value constraints $\iota_u \theta_L = 0$ for a particular symmetry group. Rather, one introduces gauge fields to implement some symmetry transformations in the action, and these imply the primary constraints $\pi^1_\alpha = 0$. The local symmetries of the action do not specify, on their own, how the variation should consistently be applied on the boundary. Thus, one checks the consistency of the naive boundary variations (imposing only primary constraints) by computing its time evolution using the equations of motion for X and $\mathfrak{L}_X \pi^1_\alpha$, and requiring $\dot{\pi}^1_\alpha$ be equal to zero. If this is not zero, then one 'discovers' that one must impose the secondary constraint to ensure the consistency of the algorithm. Further propagation of the secondary constraints could imply further constraints. However, in the case we are considering (i.e. no second-class constraints and $a \leq 1$), Noether's second theorem guarantees the closure of the Dirac algorithm at the level of the secondary constraints.

In general, the Dirac algorithm will close whenever there is a (satisfiable) generalized Bianchi identity corresponding to the original symmetry of the theory provided one uses a consistent variation on the boundary. In the context of the Dirac algorithm, this consistency is achieved by adding the secondary constraints to the naive variational principle. In practice, this is done by defining the so-called *extended Hamiltonian*. In our formalism, this arises when computing the explicit solutions of (8.45) using ω_e and then restricting to the full constraint surface C (which includes both primary and secondary constraints if these are present).

The value of our perspective is that it makes explicit that the initial value constraints are relations that must hold in order for the variational principle to have the required symmetries. Moreover, our perspective does not rely upon Dirac's theorem in order to *indirectly* connect the irregularity of the Lagrangian to redundancy in the phase space formalism via the canonical constraints. Rather, initial value constraints are explicitly constructed such that the independently specifiable degrees of freedom at a time are identified without scope for formal or interpretative ambiguity. Once this is achieved there is no further dynamical redundancy.[10]

8.5   Nomic Structure of Barbour–Bertotti Theory

Barbour–Bertotti theory (BB theory) is a simple gauge-theoretic approach to classical mechanics where time-dependent shifts are implemented as gauge symmetries using 'best matching' (Barbour and Bertotti, 1982; Barbour, 2009b). It provides a simple yet powerful illustration of the ideas just outlined. Here we will consider the nomic structure of a version of the theory that only implements time-dependent *translational shifts* as gauge symmetries and ignores rotational shifts and time reparameterizations.[11]

---

[10] The perspective developed here can be connected naturally to the idea of a 'manifest' symmetry and 'free' variation developed in detail in Gryb and Thébault (2014, 2016b). A full formal analysis connecting the two modes of analysis will not be presented in the interest of space. The interested reader will be able to draw the relevant connections straightforwardly, however, based upon the comparison between the construction of the Barbour–Bertotti theory in Gryb and Thébault (2016b, §3.2), and that given in the next section.

[11] Aficionados of the various models for 'neo-Newtonian' spacetimes will recognize this modified version of NM as the variational analogue of a Galilean spacetime. The rotations are the most controversial and complicated aspect of Barbour–Bertotti theory and we neglect their treatment here to avoid what would be a rather lengthly, and largely tangential, analysis of the relevant complications. For a recent comprehensive analysis—including a powerful generalization of the original Barbour–Bertotti

8.5.1 Best Matching Translational Shifts

The first step in the construction of our simplified Barbour–Bertotti model is to consider the form of the Lagrangian for NM. In particular, we can note that by assumption the potential $V(q)$ can be taken to be translation-invariant. This then means that it is only the kinetic term

$$T_{\mathrm{N}} = \sum_I \frac{m_I}{2} \delta_{ij} \dot{q}_I^i \dot{q}_I^j \,, \tag{8.81}$$

which is not invariant under the time-dependent translational shifts:

$$q_I^i \to q_I^i + a(t) \,. \tag{8.82}$$

Significantly the Newtonian Lagrangian $T_{\mathrm{N}} - V(q)$ does have the transformation (8.82) as a divergence symmetry. We have considered this feature already when we discussed Galilean boosts. These are simply the subset of the time-dependent translational shifts for which $a(t) = \epsilon t \hat{a}$. where ϵ is an infinitesimal parameter and $\hat{a} \in \mathbb{R}^3$. Recall that Galilean boosts are the primary example of a broad symmetry transformation which are not variational symmetries but which are divergence symmetries.

Following our preliminary stance on narrow symmetries we would thus *not* classify the transformations of as narrow symmetries of NM; that is, we would not take models related by Galilean boosts to be dynamically identical. Thus we would apply the same general categorization to time-dependent translational shifts. One might plausibly wish, however, to make include time-dependent translational shifts as narrow symmetries since, in particular, one has sound physical reasons to take the total linear momentum to be non-physical within the class of systems one wishes to describe. As per our earlier discussions, what we take to be the most salient approach in such circumstances is to proceed down the path of voluptuary redundancy and introduce the time-dependent translational shifts as explicit variational symmetries via the introduction of auxiliary variables.

To form an invariant action, we introduce the *velocity shift* fields $v(t)$ that transform as

$$v \to v + \dot{a} \tag{8.83}$$

under shifts. This field is of course precisely the auxiliary field artificially parameterizing the symmetry as discussed in general terms in our discussion above.

Let us then define a shift-invariant time derivative

$$D_t q_I^i = \dot{q}_I^i - v^i \tag{8.84}$$

which in turn can be used to define the shift-invariant Barbour–Bertotti (BB) Lagrangian

$$L_{\mathrm{BB}} = \sum_I \frac{m_I}{2} \delta_{ij} D_t q_I^i D_t q_I^j - V(q_I^i) \,. \tag{8.85}$$

---

treatment—see Gomes and Gryb (2020). The time reparameterization, by contrast, are central to the topic of this book. We do not include them here so that we do not forestall the relevant detailed discussions of the classical problem of time, for which interpretation of time reparameterization is key.

Hamilton's principle applied to this Lagrangian mandates that the variation of the BB action:

$$S_{\mathrm{BB}}[\gamma] = \int_{\gamma} L_{\mathrm{BB}}\, \mathrm{d}t \tag{8.86}$$

with respect to the shift fields v^i, will vanish.

Significantly, the velocity shift fields v^i can be interpreted as actively shifting the velocity frame of the system. Hamilton's principle therefore selects a dynamically privileged frame of reference that minimizes the kinetic energy. When this happens, the configurations in the resulting frame are said to be *best matched*.

Solving this variational problem explicitly allows us to understand the physical interpretation of the best matched frame. In particular, since S_{BB} is independent of $\dot{v}^i$ the vanishing of the v^i variation leads straightforwardly to the condition:

$$\frac{\partial L_{\mathrm{BB}}}{\partial v^i} = \sum_{I} m_I \delta_{ij} \left(\dot{q}_I^j - v^j \right) = 0 \,. \tag{8.87}$$

We note that the quantity

$$p_i^I = \frac{\partial L_{\mathrm{BB}}}{\partial \dot{q}_I^i} = m_I \delta_{ij} \left(\dot{q}_I^j - v^j \right) \tag{8.88}$$

is the definition of the momentum of the Ith particle in the frame defined by v^i. The best matching condition (8.87) therefore specifies that the dynamically privileged frame be that frame where the total linear momentum of the system is zero.

One can find the velocity-shift field $v^i(t)$ that solves the condition (8.87) for any trajectory $q_I^i(t)$ in configuration space:

$$v^i(t) = \frac{1}{M_{\mathrm{tot}}} \sum_{I} m_I \dot{q}_I^i \equiv \dot{q}_{\mathrm{cm}}^i \,, \tag{8.89}$$

where $M_{\mathrm{tot}} = \sum_{I} m_I$ is the total mass of the system of particles. Because $v^i(t)$ is just the centre-of-mass velocity of the system $\dot{q}_{\mathrm{cm}}^i$, we see that best matching defines a dynamically privileged centre-of-mass frame where the total linear momentum of the system is zero. The remaining Euler–Lagrange equations for q_I^i give:

$$m_I \frac{\mathrm{d}}{\mathrm{d}t} \left(\dot{q}_I^i - v^i \right) = -\delta^{ij} \frac{\partial V}{\partial q_I^j} \,, \tag{8.90}$$

which reproduces Newton's second law, but in a centre-of-mass frame.

We can analyse the role of the velocity shift fields v^i within BB theory by studying the integrability of the Euler–Lagrange equations. For this we can use the tools for studying irregular Lagrangian systems developed in §7.1, in particular the null eigenvectors of the Hessian.

The Hessian of BB theory is a slightly awkward mathematical object that is a function of both the generalized coordinates x and velocities $\dot{x}$ whose tensor components are spanned only by the $\mathrm{d}x$ directions. Let us therefore use a notation where

$x = (q_I^i, v^i)$ so that $W = W_{\alpha\beta}\, \mathrm{d}x^\alpha \otimes \mathrm{d}x^\beta$, where α and β range over all the components of x. Given this, and using a tensor products of the basis elements $\mathrm{d}q_i^I$ and $\mathrm{d}v^i$ for the tensorial components of the Hessian W, we find

$$W = \sum_{IJ} m_I \delta_{ij}^{IJ}\, \mathrm{d}q_I^i \otimes \mathrm{d}q_J^j. \tag{8.91}$$

This tensor is clearly degenerate in all $\mathrm{d}v^i$ components because the momentum $\pi_i = \frac{\partial L_{\mathrm{BB}}}{\partial v^i} = 0$. This reflects the fact that the shift fields v^i behave as Lagrange multipliers for this system.

A simple basis for the kernel of W is

$$\lambda_i = \frac{\partial}{\partial v^i}, \tag{8.92}$$

where we understand the index i to label independent null vectors. The existence of null directions in W of course indicates that the BB theory has *irregular nomic structure*.

We can find Lagrangian constraints for the system by computing the quantity K defined in (7.4). A short calculation using the previous basis elements $\mathrm{d}q^i$ and $\mathrm{d}v^i$ for the vectorial components of K leads to

$$K = \sum_I \left(\frac{\partial L_{\mathrm{BB}}}{\partial q_I^i} - \frac{\partial^2 L_{\mathrm{BB}}}{\partial \dot{q}_I^i \partial v^j} \dot{v}^j \right) \mathrm{d}q_I^i + \frac{\partial L_{\mathrm{BB}}}{\partial v^i}\, \mathrm{d}v^i \tag{8.93}$$

$$= \sum_I \left(-\frac{\partial V}{\partial q_I^i} + m_I \delta_{ij} \dot{v}^j \right) \mathrm{d}q_I^i - \sum_I m_I \delta_{ij} \left(\dot{q}_I^j - v^j \right) \mathrm{d}v^i, \tag{8.94}$$

where we have used the fact that $\frac{\partial L_{\mathrm{BB}}}{\partial v^i} = \frac{\partial^2 L_{\mathrm{BB}}}{\partial \dot{q}_I^i \partial q_J^j} = 0$. The Lagrangian constraints

$$\lambda_i(K) = -\sum_I m_I \delta_{ij} \left(\dot{q}_I^j - v^j \right) = 0 \tag{8.95}$$

therefore straightforwardly reproduce the BB constraint (8.87).

The fact that the v^i directions are null indicates that the Euler–Lagrange equations for the system do not fix the value of the variables v^i in terms of their initial values: they are arbitrary Lagrange multipliers. The Lagrangian constraints (8.95) then require that the centre-of-mass velocity of the system be equal to arbitrary value of these Lagrange multipliers. This implies that one vectorial degree of freedom—the centre-of-mass velocity—of the original Newtonian configurations q_I^i that would otherwise have to be fixed by initial data is now arbitrary. Thus, the nomic structure of the BB theory does not determine the dynamical behaviour of the centre-of-mass velocity of a Newtonian system. The nomic structure is, however, determinate with respect to the dynamical evolution of the remaining degrees of system. The BB theory is therefore most appropriate for describing systems where the value of the centre-of-mass is not interpreted as a physical variable—typically on the basis of not being observable. We will see in §10.3 that our full 'Nomic-AIR analysis' of the system will reproduce these intuitions.

In this context it is of course key to our classification schema to consider the variation of the action $S_{\mathrm{BB}}[\gamma]$ with respect to the shift fields v^i. The Euler–Lagrange equations for the system do not fix the value of the variables v^i *because the variation of the action is free with respect to these variables.* As per our definition of a gauge symmetry, we thus take v^i to be a product of dynamical redundancy in the formalism and, furthermore, we *must* categorize the gauge transformations (8.82) as narrow symmetries.

8.5.2 Unreduced Constrained Hamiltonian Formulation

The constrained Hamiltonian formulation of Barbour–Bertotti theory offers a number of formal and conceptual advantages. First, it allows us to demonstrate straightforwardly that the BB system does indeed pose well-defined evolution equations for the non-redundant degrees of freedom given suitably defined potentials $V(q)$. This is, of course, due to the fact that as a first-order formulation the integrability of the equations of motion can be more readily assessed. Second, and even more importantly for our purposes, the particular advantage of Hamiltonian dynamics, constrained or otherwise, is that it allows us to more easily characterize the space of dynamically possible models, the structure of which is key to our new framework for the analysis of spatiotemporal structure. This will be articulated in the following chapters drawing on the details of Barbour–Bertotti theory specified here.

The first step, as ever, toward the Hamiltonian formalism is to take the Legendre transform. In particular, to perform the Legendre transform of the BB theory defined by the Lagrangian (8.85), we compute the canonical momenta

$$p_i^I \equiv \frac{\partial L_{\mathrm{BB}}}{\partial \dot{q}_I^i} = m_I \delta_{ij} \left(\dot{q}_I^j - v^j \right) \qquad\qquad \pi_i \equiv \frac{\partial L_{\mathrm{BB}}}{\partial v^i} = 0 \,. \qquad (8.96)$$

The vanishing of π_i reflects the failure to define a bijective Legendre transform as a result of the null directions of W. These primary constraints prevent us from being able to express the generalized 'velocities' $\dot{v}$ (they are actually *accelerations* in space-time) in terms of the canonical momenta π_i.

To produce an integrable first-order system, we therefore must replace $\dot{v}^i$ with arbitrary functions we will call N^i. Using these functions and inverting $\dot{q}_i^I$ for p_i^I and v^i, we obtain

$$H_{\mathrm{BB}} = \dot{v}^i \pi_i + \sum_I \dot{q}_I^i p_i^I - L_{\mathrm{BB}} \qquad (8.97)$$

$$= \sum_I \frac{\delta^{ij} p_i^I p_j^I}{2 m_I} + V(q_I^i) + N^i \pi_i + v^i \sum_I p_I^i \qquad (8.98)$$

The Hamiltonian equations of motion

$$\dot{q}_I^i = \sum_I \frac{\delta^{ij}}{2 m_I} p_i^I + v^i \qquad\qquad \dot{p}_i^I = -\frac{\partial V}{\partial q_I^i} \qquad (8.99)$$

$$\dot{v}^i = N^i \qquad\qquad \dot{\pi}_i = -\sum_I p_i^I \qquad (8.100)$$

are consistent with the primary constraints $\pi_i = 0$ only when the secondary constraint

$$\sum_I p_i^I = 0 \tag{8.101}$$

is satisfied.

The equations of motion imply that $\sum_I p_i^I$ is a constant of motion. This is of course implementing the result of the Noether's first theorem of the original Newtonian system, only at the level of the BB equations of motion. This then means that if we choose initial data such that the secondary constraint (8.101) is satisfied then the secondary constraint will be satisfied at all times. We can implement such a requirement explicitly by, for example, choosing suitable values for the arbitrary (but bounded and continuous) functions N^i given any initial conditions for $\dot{q}_I^i$).

We can then show that in regions where $V(q_I^i)$ is bounded and continuous, the Hamiltonian system (8.99) is locally integrable by Picard–Lindelöf theorem (Arnold, 1992). Moreover, this system is equivalent to the original Lagrangian system under Legendre transform because the secondary constraint (8.101) is equivalent to the Lagrangian constraint (8.95) when the Hamiltonian equations of motion are satisfied.[12]

Significantly, since the functions N^i are freely specifiable, the system is invariant under the transformations

$$q_I^i \to q_I^i + a^i \qquad\qquad p_i^I \to p_i^I \tag{8.102}$$

$$v^i \to v^i + \dot{a}^i \qquad\qquad \pi_i \to \pi_i\,, \tag{8.103}$$

which corresponds to a redefinition of N^i of the form $N^i \to N^i + \ddot{a}^i$. This is just the invariance of the original BB action. It is important to emphasize, however, that there is no real sense in which this transformation is generated by any phase space function. One could write a *gauge generator* (Pons, Salisbury and Shepley, 1997; Pons, Salisbury and Sundermeyer, 2010), $\mathcal{G}_i$ of the form

$$\mathcal{G}_i = \sum_I p_i^I - \dot{\pi}_i\,, \tag{8.104}$$

which can be used to produce the appropriate gauge transformation up to a boundary term. However, because of the $\dot{\pi}_i$ term, which is needed to ensure that the shift of v^i is the time derivative of the shift of q_I^i, the 'gauge generator' is not a proper phase space function and, therefore, there is no Hamilton vector field associated with $\mathcal{G}_i$. This formulation of BB theory is therefore not well adapted to a rigorous phase space analysis.

8.5.3 Partially Reduced Constrained Hamiltonian Formalism

The issue isolated at the end of the last sub-section motivates us to move to a partially reduced constrained Hamiltonian formulation of the Barbour–Bertotti theory.

---

[12] It is worth noting that in our treatment here we have retained both the primary constraints and secondary constraints which are viewed as initial value constraints and thus we are not required to extend the Hamiltonian along the lines of the Dirac–Henneaux–Teitelboim schema described above.

By partially reducing the Hamiltonian formulation of the BB theory presented above we gain access to a formalism that is amenable to a straightforward canonical analysis of the symmetries via Dirac's theorem.

Starting with the BB Hamiltonian (8.97), we can add a gauge-fixing condition $F^i = v^i - \lambda^i \approx 0$ for the primary constraint $\pi_i \approx 0$. Since the gauge-fixing condition is canonically conjugate to the gauge-fixed constraint (i.e. $\{F^i, \pi_j\} = \delta^i_j$), the projection of the symplectic structure onto the gauge-fixed surface is constant along the gauge-fixed directions. Under such conditions, the Dirac bracket reduces to the usual Poisson bracket on the gauge-fixed surface according to a general argument first given by Dirac (1959). It is then possible to apply the gauge-fixing conditions as strong equations using the canonical Poisson bracket on the reduced phase space. The resulting Hamiltonian is

$$H^{\text{red}}_{\text{BB}} = \sum_I \frac{\delta^{ij} p^I_i p^I_j}{2m_I} + V(q^i_I) + \lambda^i \sum_I p^I_i \, . \tag{8.105}$$

We could also call this partially reduced formalism the 'shift-eliminated' Barbour–Bertotti Theory because the shift fields v^i and momenta π_i have been explicitly eliminated.

To study the effects of the partial reduction, let us consider the Lagrangian theory that corresponds to this reduced Hamiltonian theory. The Hamiltonian equations of motion resulting from (8.105) are:

$$\dot{q}^i_I = \frac{\delta^{ij}}{m_I} p^I_j + \lambda^i \qquad\qquad \dot{p}^I_i = -\frac{\partial V}{\partial q^i_I} \, . \tag{8.106}$$

However, since the λ^i's are no longer canonical variables, to undo the Legendre transform we must solve the remaining constraint $\sum_I p^I_i \approx 0$ in terms of the functions λ^i. This is straightforward to do in this version of BB theory because the required solution involves setting λ^i to the centre-of-mass velocity of the system. The resulting Lagrangian is:

$$L^{\text{red}}_{\text{BB}} = \sum_I \frac{m_I}{2} \delta_{ij} \left(\dot{q}^i_I - \dot{q}^i_{\text{cm}} \right) \left(\dot{q}^j_I - \dot{q}^j_{\text{cm}} \right) - V(q^i_I) \, , \tag{8.107}$$

where $q^i_{\text{cm}} = \frac{1}{M_{\text{tot}}} \sum_I m_I q^i_I$. One can easily verify that the Hamiltonian $H^{\text{red}}_{\text{BB}}$ results from a Legendre transform of this Lagrangian, where $\sum_I p^I_i$ arises as a primary constraint due to the null vectors of the Hessian along the q^i_{cm} directions.

This particular formulation of the theory is considerably more convenient in the Hamiltonian formalism because the constraint $\sum_I p^I_i \approx 0$ need only be solved as an initial value constraint in order to integrate the Hamiltonian system. This convenience becomes crucial in more complicated theories where explicit solution of the constraints is not possible.

In additional to this formal advantage, the partially reduced formalism provides a platform for a clear conceptual analysis of the gauge transformations and resulting dynamical redundancy of the Barbour–Bertotti theory. In particular, the first-class constraint $P^{\text{tot}}_i \equiv \sum_I p^i_I \approx 0$ can be shown to generate a Hamilton vector field

$$v_{P_{\mathrm{tot}}} = \sum_I \frac{\partial}{\partial q_I^i} \qquad (8.108)$$

whose integral curves correspond to the orbits parameterized by the gauge transformations

$$q_I^i \to q_I^i + a^i(t) \qquad\qquad p_i^I \to p_i^I . \qquad (8.109)$$

It is easy to verify that these transformations leave the Lagrangian $L_{\mathrm{BB}}^{\mathrm{red}}$ invariant. In this context there is clear and rigorous sense in which the first-class constraints of the shift-eliminated Hamiltonian theory are the generators of the gauge transformations of $L_{\mathrm{BB}}^{\mathrm{red}}$ in precisely the sense of Dirac's theorem.

It is then straightforward to construct a maximal independent set of Dirac observables for this theory. Consider the centre-of-mass coordinates:

$$q_{I,\mathrm{cm}}^i = q_I^i - q_{\mathrm{cm}}^i , \qquad (8.110)$$

These commute with the constraints $P_i^{\mathrm{tot}} \approx 0$ as do their conjugate momenta

$$p_i^{I,\mathrm{cm}} = p_i^I - P_i^{\mathrm{tot}} . \qquad (8.111)$$

Thus, these coordinates provide us with a set of functions invariant under the gauge symmetries generated by $P_i^{\mathrm{tot}} \approx 0$. The partially reduced Barbour–Bertotti theory provides us with a straightforward theory in which the logic behind Dirac's theorem is both formally and conceptually faultless. The transformations generated by the first-class constraint to not change the physical state and the physical observables of the theory are precisely those that commute with the constraint.

Significantly, however, the transformations in question are *not* the same as the gauge transformations of the original BB theory, which has the additional degrees of freedom in v^i. This evidences the powerful lesson that the true dynamical redundancy of a physical theory is not always presented to us in explicit form within either the Lagrangian or Hamiltonian formalisms. We must apply physical insight to interpret the degrees of freedom and variation in question and it is typically sensible to partially reduce our formalism such that the interpretation of the phase space action of constraints can be unambiguously established.

8.5.4 Narrow Symmetries and Dynamical Redundancy

The narrow symmetries for BB theory are

$$v^i \to v^i + \dot{a}^i \qquad (8.112)$$

$$q_I^i \to q_I^i + a^i . \qquad (8.113)$$

The α index of T_α^i then splits into a piece for the v^i variables, which we will call U_j^i, and one for the q_I^i variables, which we will call R_{Ij}^i. Using this notation, the non-zero components of the operators U and R can be read off from (8.112) giving

$$U^i_{(1)j} = \delta^i_j \qquad\qquad R^i_{(0)Ij} = \delta^i_j \,. \qquad (8.114)$$

The adjoint operators $\tilde{U}^i_j$ and $\tilde{R}^i_{Ij}$ can then be computed from this using (8.18). The result is

$$\tilde{U}^i_j = -\delta^i_j \frac{\mathrm{d}}{\mathrm{d}t} \qquad\qquad \tilde{R}^i_{Ij} = \delta^i_j \,. \qquad (8.115)$$

Calling $p^I_i = \frac{\partial L_{\mathrm{BB}}}{\partial \dot{q}^i_I}$, the v^i components of α are

$$\alpha^{(v)}_i = -\sum_I p^I_i \qquad (8.116)$$

while the q^i_I components are

$$\alpha^{(q)I}_i = -\frac{\partial V}{\partial q^i_I} - \dot{p}^I_i \,. \qquad (8.117)$$

The generalized Bianchi identities (8.17) are then

$$-\sum_I \frac{\partial V}{\partial q^i_I} - \frac{\mathrm{d}}{\mathrm{d}t}\sum_I p^I_i + \frac{\mathrm{d}}{\mathrm{d}t}\sum_I p^I_i = 0 \,, \qquad (8.118)$$

which simply expresses the fact that the potential V is invariant under translations of q^i_I. The Lagrangian constraint (8.20) only receives a contribution from $\tilde{U}^i$. It expresses the vanishing of the total momentum

$$\sum_I p^I_i = 0 \,, \qquad (8.119)$$

which is the usual BB constraint. Note that the generalized Bianchi identity expresses the propagation of this constraint, and therefore the closure of the Dirac algorithm.

This analysis can be easily repeated for the shift-eliminated BB theory. The only difference is that the v^i variables are not present so all contributions due to its transformation properties vanish. Using the invariance of V under translations of q^i_i, the generalized Bianchi identities read

$$\frac{\mathrm{d}}{\mathrm{d}t}\sum_I p^I_i = 0 \qquad (8.120)$$

and there is no Lagrangian constraint. In this theory, the usual BB constraint arises as a primary constraint (see below) and thus, the generalized Bianchi identity simply expresses that this constraint is propagated by the dynamics (and therefore that the Dirac algorithm closes).

8.5.5   Structure Summary

Let us conclude by setting out how the key formal structures of the partially reduced Barbour–Bertotti formalism can be specified in geometric terms. In particular, we can noted that the space of D of dynamically possible models (DPMs) of the partially reduced Barbour–Bertotti theory is the space of integral curves of $v_{H_{\text{BB}}^{\text{red}}}$ that lie on the constraint surface $P_i^{\text{tot}} = 0$. Significantly the restriction of the symplectic two-form $\omega = \mathrm{d}p_I^i \wedge \mathrm{d}q_I^i$ is degenerate on this surface. This means that one needs to add addition gauge-fixing conditions G^i such that $\det\left\{G^i, P_i^{\text{tot}}\right\} \neq 0$. Under this condition, the restriction of the symplectic two-form ω onto this gauge-fixed surface is indeed non-degenerate and the integral curves of $v_{H_{\text{BB}}^{\text{red}}}$ on this gauge-fixed surface are equivalent to the Hamiltonian equations (8.106). One can see this by choosing the gauge-fixing conditions $G^i = q_{\text{cm}}^i - a^i$ which satisfy $\left\{G^i, P_i^{\text{tot}}\right\} = 1$ and using $\lambda^i = \dot{a}^i$. The space of DPMs is therefore the space of all integral curves of $v_{H_{\text{BB}}^{\text{red}}}$ that lie on *any* of the gauge-fixed surfaces defined by the functions a^i. These geometric structures provide us with a rigorous means to define the *partitioning map* $n : K \to D$ which projects from the space of KPMs (roughly phase space curves) to the space of dynamical curves.

The second nomic structure is the *projection map* π_N which provides the definition of dynamical distinctness and dynamical identity between dynamically possible models. It thus projects to a space of distinct dynamically possible models (DDPMs) and thus has the action $\pi_N : D \to \tilde{D}$ from the space of DPMs to the space of DDPMs. For the partially reduced Barbour–Bertotti the space of DDPMs, $\tilde{D}$ can be defined as the space of *families* of integral curves related by different value of a^i. This space can be explicitly constructed (whenever $V(q)$ and a^i are bounded and continuous) in terms of the Dirac observables (8.110) and (8.111), which commute with P_i^{tot}. The space of $\tilde{D}$ is then isomorphic to the space of the projection of the integral curves of $v_{H_{\text{BB}}^{\text{red}}}$ onto the reduced phase space of Dirac observables.

Manifold: A spatial manifold $\mathbb{R}_x^3$ for representing spatial positions of particles in 3D. A temporal manifold $\mathbb{R}_t$ for representing instants. A spacetime manifold that is the product space of the temporal and spatial manifolds $\mathbb{R}_t \times \mathbb{R}_x^3$.

Matter: A configuration space, $\mathcal{C}$, of representations of N-particle positions in 3D: $\mathcal{C} = \mathbb{R}^{3N}$. A configuration velocity space, $T_q\mathcal{C}$, consisting of the tangent space to $\mathcal{C}$ at a point $q \in \mathcal{C}$. This represents the velocity of a particle at a particular position. The image of a curve $\gamma : \mathbb{R}_t \to T\mathcal{C}$ on the tangent bundle $T\mathcal{C}$ represented by the pair $\left(q_I^i(t), \dot{q}_I^i(t)\right) \forall t \in (t_1, t_2)$. The particle masses $m_I \in \mathbb{R}^+$ and other coupling constants c_α that enter the potential.

Geometric: There are two geometric structures needed. i) A temporal metric $g_t(t_1, t_2) = t_2 - t_1 \,\forall t_2 > t_1$ inherited from the natural metric on the temporal manifold $\mathbb{R}$. ii) A spatial metric $g_{ij} = \delta_{ij}$, where δ is the flat Euclidean metric on the spatial manifold $\mathbb{R}^3$.

Putting this all together we get the full specification of the space of kinematical structures which is identical to that we defined for NM:

$$\mathcal{K} = \left\{\mathbb{R}_t \times \mathbb{R}_x^{3N}, g_t, \delta, \gamma : \mathbb{R}_t \to T\mathcal{C}, m_I, c_\alpha\right\}. \tag{8.121}$$

The kinematically possible models of (partially reduced) Barbour–Bertotti theory are then given by the collection of all possible (images of) curves γ in $T\mathcal{C}$ conjoined with a specification of all the particle masses m_I and couplings c_α of the theory. The nomic structure is then given by:

Partitioning map $n : K \to D$: the nomic partition can be defined restricting to integral curves of the vector field $v_{H_{\mathrm{BB}}^{\mathrm{red}}}$ that lie on *any* of the gauge-fixed surfaces defined by the functions a^i.

Projection Map $\pi_N : D \to \tilde{D}$ from the space of DPMs to the space of DDPMs is then a projection onto the space given (equivalently) by: i) considering *families* of integral curves related by different value of a^i; or ii) considering the projection of the integral curves of $v_{H_{\mathrm{BB}}^{\mathrm{red}}}$ onto the reduced phase space of Dirac observables

8.6   Chapter Summary

One of the central innovations of this book is the construction of a new geometric framework for the diagnosis of dynamical redundancy in theories with irregular nomic structure. In this present chapter, our focus has been on theories with fixed time parameterization which in our framework corresponds to variations which are restricted to be transverse rather than tangential. The tangential variations that relate to time reparameterizations require a separate formal treatment which will be provided at length in our later discussions.

According to our approach, it is the existence of initial value constraints that is indicative of the specific species of underdetermination problem associated with dynamical redundancy within the analysis of theories with irregular nomic structure. In particular, it is the existence of initial value constraints that is indicative of the specific underdetermination-at-a-time problem associated with *prima facie* problematic dynamical redundancy within the formalism. Significantly, by refocusing the formal analysis on initial value constraints, rather than generalized Bianchi identities or first-class constraints, our framework supersedes approaches to 'gauge' symmetry based upon Noether's second theorem or the Dirac criterion. These criteria are then recoverable in the relevant contexts from our initial value constraints approach, which should form the primary analysis tool for the analysis of gauge theories.

The formal ideas presented in this chapter were then illustrated via a context case study of Barbour–Bertotti theory. This is a simple gauge-theoretic approach to classical mechanics where time-dependent shifts are implemented as gauge symmetries using best matching. We consider the special case of the theory in which we implement time-dependent translational shifts as gauge symmetries and ignore rotational shifts and time reparameterizations. Application of our framework to this theory leads to recovery of the relevant results including the relevant initial value constraints and associated definition of narrow symmetries and dynamical distinctness.

9

The New Framework

The Nomic-AIR framework for the analysis of symmetry and structure in physical theory is a powerful and general new formal framework that allows for the exhaustive analysis and classification of the transformation properties of nearly any type of structure in nearly any type of theory. The framework is also flexible enough to be compatible with different approaches to the definition of the narrow symmetries of a theory. The core idea of the framework is to track the behaviour of a given structure induced by transformations on the space of dynamically possible models in comparison to the induced transformations on the relevant nomic structure. This chapter will introduce the framework, present some basic theorems, and consider the idea of local and global structures.

9.1 The AIR Classification

Consider any non-constitutive (i.e. nomic or spatiotemporal) structure $\mathcal{S}$ in $\mathcal{T}$. Assume the non-constitutive structures to be represented by a map with the domain a space, U, and the codomain a space, V, so that we have $\mathcal{S} : U \to V$ or equivalently $\mathcal{S}(u) = v$ for $v \in V$ and $u \in U$.

A simple example of such a structure is the nomic structure given by the partitioning map $n : K \to D$. Here the domain is the space of kinematically possible models and the codomain the space of dynamically possible models. A more complex example, this time of a temporal structure, would be the proper time interval where the domain is paths in the events space $\gamma \in \mathcal{M}$ together with a metric tensor constructed via the relevant geometric objects $g_{\mu\nu} \in G_i$, and the codomain is the real line, $\mathbb{R}$. In general, we can take the domain and codomain of a structure to definable using the constitutive structures. We leave the detailed explanation how to do this to concrete applications provided in later chapters.

Let us then consider endomorphisms on the space of DPMs that is transformations $\phi : D \to D$ or equivalently $d' = \phi(d)$ for $d, d' \in D$. We assume a structure $\mathcal{S}$ has been defined via maps with the domain and codomain defined in terms of the constitutive structures, together with the basic mathematical objects such as the real numbers. It then makes sense to consider the *pullback of a structure by an endomorphism*, that is we can define the precomposition map as follows:

Definition 9.1 *A **Precomposition Map**, $\phi_d^\star \mathcal{S}$. is given by transferring the action of a particular endomorphisms ϕ_i to a particular structure $\mathcal{S}_a$ by pre-applying it to the latter's domain:*

$$\phi_d^\star \mathcal{S} = \mathcal{S} \circ \phi_d = \mathcal{S}(\phi(d)) \tag{9.1}$$

In intuitive terms, the precomposition corresponds to considering the implications of a particular transformation on a particular nomic or temporal structure by pre-applying the transformation onto the 'input' of the structure map. It is simply a more general formal way of representing the old Leibnizian idea of considering how our representation of laws or time change under a transformation of the contents of the universe. We will make this connection explicit in the context the representation of spatiotemporal structure in NM and Barbour–Bertotti theory in the next chapter.

It is worth commenting briefly on the conditions for Equation 9.1 to be well-defined. First, and most significantly, as shall be seen shortly, it will prove significant for our analysis to consider structures whose pullbacks will not necessarily be defined for all ϕ; that is, it will be an important part of our analysis to consider cases in which Equation 9.1 fails to give a well-formed expression. We should note also that we are using the word 'pullback' in a more general context than its more conventional meaning in terms of a linear map on the space of one-forms. For us, the notion of 'pullback' will apply to a variety of different structures. These will include tensor fields of arbitrary rank as well as other more general mathematical structures, such as ordering relations, provided these can be represented as maps with domain D. In these cases, it will serve merely as a shorthand for precomposition. We retain the term 'pullback' however due to the close relationship between our general notion and the more conventional notion of 'pullback' used extensively in the literature in reference to covariant tensor fields and diffeomorphisms.[1]

Consider then a theory, $\mathcal{T}$, equipped with manifold, geometric, and matter structures such that the space of kinematical models can be defined as $K = \{\mathcal{M}, G_i, T_\alpha\} \forall i, \alpha$. Consider first the space of transformations, Ψ, given by the endomorphisms ψ_i of K:

$$\Psi = \{\psi_i, \forall i \mid \psi_i : K \to K\}. \tag{9.2}$$

This theory is further equipped with nomic structure N playing the role of a partitioning function $n : K \to D$. The space of transformations, Φ, is the space of all endomorphisms ϕ_i of D:

$$\Phi = \{\phi_i, \forall i \mid \phi_i : D \to D\}. \tag{9.3}$$

For our purposes, K and D will mostly be a smooth differentiable manifold and we will restrict to ψ_i and ϕ_i that are diffeomorphisms of K and D, that is, $\psi_i \in \mathrm{Diff}(K)$ and $\phi_i \in \mathrm{Diff}(D)$. Clearly, we will have that $\Phi \subset \Psi$.

It is worth noting our definitions allow for a subset of diffeomorphisms of K which are *not* diffeomorphisms of D, that is, $\bar{\psi} \notin \Phi$, and can be such that they do not preserve the nomic partition, that is, $\bar{\psi}^\star n \neq n$. In what follows, our analysis will primarily focus on the pullback of structures under ϕ_i, however it will be necessary to discuss the special role of transformations $\bar{\psi}$ in some contexts.

The space Φ is partitioned into three non-overlapping subspaces, $\mathrm{Abs}(\mathcal{S})$, $\mathrm{Rel}(\mathcal{S})$, and $\mathrm{Inc}(\mathcal{S})$, which we define as follows:

---

[1] For more details of the standard differential geometry notion of pullback see Malament (2012, §1.5).

Definition 9.2 *Absolute Subspace*: $Abs(S)$: *the subspace of all $\phi_i \in \Phi$ such that the pullback of the structure under ϕ_i is trivial for all tokens of the structure:* $\phi_i^\star S_a = S_a, \forall a$. *Such structure S transforms* invariantly, *and is in this sense is* absolute *under ϕ_i.*

Definition 9.3 *Relative Subspace*: $Rel(S)$: *the subspace of $\phi_i \in \Phi$ such that the pullback of the structure under ϕ_i is well-defined and non-trivial:* $\phi_i^\star S_a = S_b$, *for some $a \neq b$. Such structure S transforms* covariantly, *and in this sense is* relative *under ϕ_i.*

Definition 9.4 *Incomplete Subspace*: $Inc(S)$: *the subspace of $\phi_i \in \Phi$ such that the pullback of the structure under ϕ_i does not have a closed action on the tokens of the structure: there exists at least one S_a s.t. $\phi_i^\star S_a \neq S_b, \forall b$. Since it is not closed under the action of ϕ_i, such structure S does not transform in an appropriately well-behaved way, and in this sense is* incomplete *under ϕ_i.*

As defined, the AIR regions are non-overlapping (i.e. mutually exclusive). Furthermore, the union of the three AIR regions must be Φ itself, and therefore that they exhaust the states in consideration. Finally, since the identity $\mathbb{1}$ is obviously also an element of Φ, $Abs(S)$ will always contain at least this element, and is therefore non-empty.

We collect these definitions below using elementary set theory

Definition 9.5 *Totality*. *The space of transformations Φ is the universe of all elements under study:*
$$\Phi = U = \neg\emptyset.$$

Definition 9.6 *No Overlap*. *The AIR regions are mutually exclusive:*
$$Abs(S) \cap Rel(S) = Abs(S) \cap Inc(S) = Inc(S) \cap Rel(S) = \emptyset.$$

Definition 9.7 *No Gaps*. *The AIR regions exhaust Φ:*
$$\Phi = Abs(S) \cup Rel(S) \cup Inc(S).$$

Definition 9.8 *Identity*. *There is at least one map under which all structures are invariant:*
$$\mathbb{1} \in Abs(S) \Rightarrow Abs(S) \neq \emptyset.$$

Definitions 9.5 and 9.6 immediately imply the following (rather trivial) theorem:

Theorem 9.1 *Each AIR region can be defined via the complement of the other two:*

i. $Abs(\mathcal{S}) = \neg\,(Rel(\mathcal{S}) \cup Inc(\mathcal{S}))$
ii. $Rel(\mathcal{S}) = \neg\,(Abs(\mathcal{S}) \cup Inc(\mathcal{S}))$
iii. $Inc(\mathcal{S}) = \neg\,(Abs(\mathcal{S}) \cup Rel(\mathcal{S}))$

Proof Consider i. By Definition 9.5 together with the complement laws we have that $\Phi = \mathrm{Abs}(\mathcal{S}) \cup \neg\mathrm{Abs}(\mathcal{S})$ which implies that $\mathrm{Abs}(\mathcal{S}) \cup \neg\mathrm{Abs}(\mathcal{S}) = \mathrm{Abs}(\mathcal{S}) \cup \mathrm{Rel}(\mathcal{S}) \cup \mathrm{Inc}(\mathcal{S})$ by Definition 9.7. From this it is straightforward to show that $\neg\mathrm{Abs}(\mathcal{S}) = \mathrm{Rel}(\mathcal{S}) \cup \mathrm{Inc}(\mathcal{S})$. Then, by involution we have that $\mathrm{Abs}(\mathcal{S}) = \neg\,(\mathrm{Rel}(\mathcal{S}) \cup \mathrm{Inc}(\mathcal{S}))$. The same follows *mutatis mutandis* for ii and iii. □

A further definition that is implicit in the way we described the nomic structure is that since the projection map is surjective, it must always transform in either an absolute or relative manner. Incompleteness of this structure would correspond to a failure of surjectivity. This leads us to define:

Definition 9.9 *Nomic Completeness*. *The projection map part of the nomic structure always transforms in a well-defined way:*

$$Inc(\pi_N) = \emptyset.$$

As a corollary to this definition we can recall that our formalism does allow for a representation of transformations that *do not* preserve the nomic structure but do preserve the kinematic structure. These would be a subset of the transformations $\bar{\psi}$ mentioned above. We will pick up their relevance again in §10.1.5.

It will also prove worthwhile at this stage to define two important special types of structure of a given theory $\mathcal{T}$ that can distinguished explicitly based upon the AIR regions. The first special type of structure is the *fixed structure* of $\mathcal{T}$. This is structure which has no well-defined covariance properties. Fixed structure always transforms invariantly, whenever it transforms in a well-defined way.

Definition 9.10 *Fixed Structure*. *The collection of all structures for some theory theory $\mathcal{T}$, for which the relative subspace of Φ is empty:*

$$Rel(\mathcal{S}) = \emptyset.$$

Our notion of fixed structure is similar in some respects to the that defined in the 'standard framework' for symmetries (discussed by, for example, Pooley 2017). In particular, fixed structure in that context is structure that is identical in all KPMs. This structure will necessarily be fixed in our sense also. However, the converse clearly does not hold: structure can be fixed in our sense whilst varying between KPMs since it is only required to be invariant (when well-defined) between DPMs. This suggest a natural fine-graining of the concept of fixed structure as follows. Structure that is identical in all models that share the same *type* of constitutive structure is *constitutively fixed*. Structure that is identical in all models that share the same *token* of constitute structure but varies between at least some tokens is *contingently fixed*. As indicated in Chapter 5 this distinction between two forms of fixed structure will prove of great relevance to the analysis of time in the context of both simple mechanical theories and relativistic mechanics.

The second special type of structure is the *compete structure* of $\mathcal{T}$. This is structure that is well-defined under all transformations that are possible on the space of DPMs.

Definition 9.11 *Complete Structure*. *The collection of all structures for some theory $\mathcal{T}$, for which the incomplete subspace of Φ is empty:*

$$Inc(\mathcal{S}) = \emptyset.$$

By Definition 9.8 above, the identifying function of the nomic structure is a complete structure for all theories.

In addition to these important special types of structure we can also use our formalism to distinguish a special type of theory: a *symmetry-reduced theory*. This is a theory within which the distinction between dynamically identified models has been 'projected out', such that the space of dynamically possible models consists only of dynamically distinct models.

Definition 9.12 *Symmetry-Reduced Theory*. *The absolute subspace of projection map part of the nomic structure consists solely of the identity element*

$$Abs(\pi_N) = \{\mathbb{1}\}.$$

Finally, it is worth stipulating further conditions on the structures of theory to be sufficient to ensure that the theory $\mathcal{T}$ will be a *non-trivial theory*. Non-trivial theories are guaranteed to have enough structure to be able to represent physically interesting worlds.

Definition 9.13 *Non-Trivial Theory*. *A non-trivial theory is a theory in which the following three conditions hold:*

1. *At least some structures transform in a well-behaved and non-trivial manner under some transformation: $Inc(\mathcal{S}) \neq \Phi/\{\mathbb{1}\}$;*
2. *At least some structures are invariant under more than just the identity map: $Abs(\mathcal{S}) \supset \{\mathbb{1}\}$;*
3. *There are distinct DPMs since the nomic structure is non-fixed structure and thus has non-trivial relative transformations: $Rel(\pi_N) \neq \emptyset$.*

We can consider whether these are reasonable non-triviality conditions by considering the implications of each of their failure in turn. First, consider a theory for which $Inc(\mathcal{S}) = \Phi/\{\mathbb{1}\}$ for all structures $\mathcal{S}$. In such a theory there would be no covariant or invariant structures, only incomplete structures. Such structures would not have the capacity to represent dynamically interesting worlds and thus could collectively only stand in for a trivial theory $\mathcal{T}$.

Similarly, consider the case that $\mathrm{Abs}(\mathcal{S}) = \{\mathbb{1}\}$ and $\mathrm{Inc}(\mathcal{S}) = \emptyset$ for all structures $\mathcal{S}$. We would then have that $\mathrm{Rel}(\mathcal{S})$ would be the entire space, besides the identity, and thus that $\mathcal{S}$ would be isomorphic to Φ. It is not clear how a theory made up of only of such structure would be equipped to represent space and time since there would be no structure at all that is fixed between dynamical models. Finally, if $\mathrm{Rel}(\pi_N) = \emptyset$ then there is only one dynamically distinct model and thus the theory $\mathcal{T}$ would not provide us with a rich enough representational language to make interesting physical statements.

For the time being, we will simply stipulate that we are aways dealing with a non-trivial theory. We will return to the question of non-triviality as a normative principle for choosing the appropriate structural representation of a theory in §10.1.

9.2   The Nomic-AIR Analysis

The analytical power of our framework will be revealed by considering intersections between the AIR regions of a particular structure and the AR regions of the nomic structure. It will be worthwhile for us to briefly summarize the core idea of the 'Nomic-AIR Analysis' in words before we set out the formal details. In what follows we will admit the 'narrow' in narrow symmetries for simplicity but reintroduce the narrow vs broad distinction where necessary.

The most important ideas is that we can consider the situation when transformations have a trivial (A), ill-defined (I), or well-defined (R) pullback onto a particular structure and simultaneously also have a trivial (A) or non-trivial (R) pullback onto the nomic structure. Transformations that pullback onto the nomic structure trivially are symmetry transformations. Transformations that pullback onto the nomic structure non-trivially are dynamical transformations.

Together this gives us six possible combinations which we will define formally shortly. Informally, we can think of these six combinations as picking out *invariant* structure that does not vary under symmetries transformations (AA), *surplus* structure that is covariant under symmetries (AR), *dynamically absolute* structure that does not vary under dynamical transformations (RA), *dynamically relative* structure that does vary under dynamical transformations (RR), and *kinematically incomplete* (AI) and *dynamically incomplete* (RI) structure, respectively. What is more, we can also consider combinations of these combinations, for example, structure that is invariant dynamically relative or surplus dynamically absolute.

Let us consider the identifying function of the nomic structure as represented by the projection $\pi_N : D \to \tilde{D}$ that maps from the space of dynamically possible models (DPMs) to the space of distinct dynamically possible models (DDPMs). Recall that the definition of π_N requires that it is well-defined everywhere on its domain D. This means its pullback $\phi^\star \pi_N$ under an arbitrary element $\phi : D \to D$ of Φ is always well-defined and thus equivalent to a (possibly identical) token of the relevant structure, say π_N'. This implies that $\mathrm{Inc}(\pi_N) = \emptyset$ as per nomic completeness (i.e. Definition 9.9 above). We can then characterize the two remaining nomic AR regions as follows:

Definition 9.14 $Abs(\pi_N)$: **Symmetry transformations.** *The subspace of all $\phi_i \in \Phi$ such that the pullback of π_N under ϕ_i is trivial:* $\phi_i^\star \pi_N = \pi_N' = \pi_N$. *Such trans-*

formations necessarily leave the space $\tilde{D}$ of DDPMs invariant. These are naturally interpreted as symmetry transformations of the theory.

Definition 9.15 *$Rel(\pi_N)$:* **Dynamical transformations.** *The subspace of all $\phi_i \in \Phi$ such that the pullback of π_N under ϕ_i is well-defined and non-trivial: $\phi_i^\star \pi_N = \pi_N' \neq \pi_N$. The transformations in this region induce diffeomorphisms of $\tilde{D}$ that transform between DDPMs. These are naturally interpreted as the dynamical transformations of the theory.*

For non-trivial theories (Definition 9.12) we have that $Rel(\pi_N) \neq \emptyset$ so that there exists non-trivial dynamical transformations and thus at least two DDPMs. For symmetry-reduced theories (Definition 9.11) we have that there are no non-trivial symmetries in the theory: $Abs(\pi_N) = \{\mathbb{1}\}$.[2]

We have now assembled all the necessary ingredients to introduce the centrepiece of our new framework: a classification scheme that compares the AIR classification of a particular structure with that of the nomic structure of the theory. This classification scheme allows us to identify important classes of structure that can exist within a theory by looking at the different ways in which the nomic AR regions of a theory $\mathcal{T}$ can overlap with the AIR regions of a structure $\mathcal{S}$.

There are six possibilities that arise. We will abbreviate as follows: Abs (A), Rel (R) and Inc (I). The first letter refers to the AIR region of the structure in question and the second letter refers to the overlapping nomic AR region. In each case we will provide a selection of the most salient examples of a temporal structure in physical theory that fits into the relevant category together with the relevant forward reference.

Definition 9.16 *Invariant Structure [AA].* $\mathcal{S}$ *is absolute (and therefore invariant) under all symmetries of $\mathcal{T}$:*

$$Abs(\mathcal{S}) \supseteq Abs(\pi_N)$$

Definition 9.17 *Surplus Structure [RA]:* $\mathcal{S}$ *transforms in a well-behaved manner under some of the symmetries of $\mathcal{T}$:*

$$Rel(\mathcal{S}) \cap Abs(\pi_N) \neq \emptyset.$$

Definition 9.18 *Dynamically Absolute Structure [AR]:* $\mathcal{S}$ *does not vary under some transformations between DDPMs and when $\mathcal{S}$ has non-trivial relative transformations, these are always symmetry transformations. Such structure never transforms as one varies between DDPMs:*

---

[2] A special case, discussed briefly above, occurs when the equivalence classes of D are identical manifolds $\mathcal{F}$ so that the projection π_N equips D with a fibre bundle structure. In this case the symmetry transformations $\phi_v \in Abs(\pi_N)$ are the vertical directions in the bundle and the dynamical transformations $\phi_h \in Rel(\pi_N)$ are the horizontal directions. The base space $\tilde{D}$ is then by definition the dynamical possibility space of a symmetry-reduced theory.

  i. $Abs(\mathcal{S}) \cap Rel(\pi_N) \neq \emptyset$
  ii $Rel(\mathcal{S}) \neq \emptyset \to Rel(\mathcal{S}) \subseteq Abs(\pi_N)$.

Definition 9.19 *Dynamically Relative Structure [RR]*: $\mathcal{S}$ *transforms as one varies between some DDPMs:*

$$Rel(\mathcal{S}) \cap Rel(\pi_N) \neq \emptyset.$$

Definition 9.20 *Kinematically Incomplete Structure [IA]*: $\mathcal{S}$ *transforms in an ill-defined way under some of the symmetries of* $\mathcal{T}$:

$$Inc(\mathcal{S}) \cap Abs(\pi_N) \neq \emptyset.$$

Definition 9.21 *Dynamically Incomplete Structure [IR]*: $\mathcal{S}$ *transforms in an ill-defined way as one varies between some DDPMs.:*

$$Inc(\mathcal{S}) \cap Rel(\pi_N) \neq \emptyset.$$

It is important to note that the six different cases that make up our classification scheme are not all mutually exclusive. Rather, the definitions imply the three following theorems which together circumscribe the admissible combinations (all others being in general allowed):

Theorem 9.2 *Invariant structures and surplus structures are mutually exclusive since all invariant structure is not surplus and all surplus structure is not invariant:*

  i. $Abs(\mathcal{S}) \supseteq Abs(\pi_N) \to Rel(\mathcal{S}) \cap Abs(\pi_N) = \emptyset$
  ii. $Rel(\mathcal{S}) \cap Abs(\pi_N) \neq \emptyset \to Abs(\mathcal{S}) \not\supseteq Abs(\pi_N)$

Proof   Assume a structure $\mathcal{S}$ is invariant (Definition 9.16). By no overlap (Definition 9.6) it follows that $Rel(\mathcal{S}) \cap Abs(\pi_N) = \emptyset$ which implies $\mathcal{S}$ must fail condition to be surplus (Definition 9.17). Assume a structure $\mathcal{S}$ is surplus (Definition9.17). By Definitions 9.6 (no overlap) it follows that $Abs(\pi_N)$ cannot be completely contained within (or equivalent to) $Abs(\mathcal{S})$ region which implies structure must fail to be invariant (Definition 9.16)   □

Theorem 9.3 *Invariant structure and incomplete kinematic structure are mutually exclusive since all invariant structure is not incomplete kinematic and all incomplete kinematic structure is not invariant:*

  i. $Abs(\mathcal{S}) \supseteq Abs(\pi_N) \to Inc(\mathcal{S}) \cap Abs(\pi_N) \neq \emptyset$
  ii. $Inc(\mathcal{S}) \cap Abs(\pi_N) \neq \emptyset \to Abs(\mathcal{S}) \not\supseteq Abs(\pi_N)$

Proof follows exactly the same form as proof of Theorem 9.2.

Theorem 9.4 *Dynamically absolute and dynamically relative structure are mutually exclusive since all dynamically absolute structures are not dynamically relative and all dynamically relative structures are not dynamically absolute.*

    i. $AR \to \neg RR$
   ii. $RR \to \neg AR$

Proof Assume a structure $\mathcal{S}$ dynamically absolute (Definition 9.18). Then either $\mathrm{Rel}(\mathcal{S}) \neq \emptyset \to \mathrm{Rel}(\mathcal{S}) \subseteq \mathrm{Abs}(\pi_N) \to \mathrm{Rel}(\mathcal{S}) \cap \mathrm{Rel}(\pi_N) = \emptyset$ or $\mathrm{Rel}(\mathcal{S}) = \emptyset \to \mathrm{Rel}(\mathcal{S}) \cap \mathrm{Rel}(\pi_N) = \emptyset$. In either case the structure fails the condition to be dynamically relative (Definition 9.19). Assume a structure $\mathcal{S}$ is dynamically relative (Definition 9.19). This implies that $\mathrm{Rel}(\mathcal{S}) \neq \emptyset$. However, by no overlap (Definition 9.6) we also have that $\mathrm{Rel}(\mathcal{S}) \not\subseteq \mathrm{Abs}(\pi_N)$. Thus we have that the structure fails the second condition to be dynamically absolute (Definition 9.18) since the antecedent is true and the consequent false. $\qquad\square$

In addition to these three general circumscriptions on the possible combinations of categories of structures, it is worth highlighting two further restrictions that can be derived for fixed structures (Definition 9.10) and complete structures (Definition 9.11). In particular, we can derive the following two further theorems:

Theorem 9.5 *If the structure doesn't change as one varies between all DPMs it is fixed structure (Definition 9.10), then the structure can neither be surplus nor dynamically relative*

$$Rel(\mathcal{S}) = \emptyset \to \neg RA \wedge \neg RR$$

Proof trivially follows from Definitions 9.17 and 9.19 and the properties of the empty set.

Theorem 9.6 *If the structure is well-defined for all DPMs, i.e. is complete structure (Definition 9.11), then it can neither be incomplete kinematic structure nor incomplete dynamical structure.*

$$Inc(\mathcal{S}) = \emptyset \to \neg IA \wedge \neg IR$$

Proof trivially follows from Definitions 9.20 and 9.21 and the properties of the empty set.

Two further, slightly less trivial theorems follow from considering the implications of compound cases.

Theorem 9.7 *Invariant structure that is dynamically absolute is always fixed structure.*

$$AA \wedge AR \to Rel(\mathcal{S}) = \emptyset$$

Proof Combing Definitions 9.16 and 9.17 gives: i) $\text{Abs}(\mathcal{S}) \cap \text{Rel}(\pi_N) \neq \emptyset$; ii) $\text{Rel}(\mathcal{S}) \neq \emptyset \rightarrow \text{Rel}(\mathcal{S}) \subseteq \text{Abs}(\pi_N)$; and iii) $\text{Abs}(\pi_N) \subseteq \text{Abs}(\mathcal{S})$. This implies $\text{Rel}(\mathcal{S}) = \emptyset$ by modus tollens □

Theorem 9.8 *When a structure is both surplus and dynamically absolute then all relative transformations of the structure will be symmetry transformations.*

$$RA \wedge AR \rightarrow Rel(\mathcal{S}) \subseteq Abs(\pi_N)$$

Proof Combing Definitions 9.18 and 9.17 gives: i) $\text{Abs}(\mathcal{S}) \cap \text{Rel}(\pi_N) \neq \emptyset$; ii) $\text{Rel}(\mathcal{S}) \neq \emptyset \rightarrow \text{Rel}(\mathcal{S}) \subseteq \text{Abs}(\pi_N)$; and iii) $\text{Rel}(\mathcal{S}) \cap \text{Abs}(\pi_N) \neq \emptyset$. This implies $\text{Rel}(\mathcal{S}) \subseteq \text{Abs}(\pi_N)$ by *modus ponens*. □

9.3   Global and Local Structures

The final crucial element in our classificatory scheme relates to the behaviour of structures under inhomogeneous transformations; that is, transformations that do not act in the same way across different regions of the event space, given by the manifold, $\mathcal{M}$. When a structure holds a property locally then the Definitions 9.16–9.21 holds for transformations that are inhomogeneous. When a structure holds a property globally then the Definitions 9.16–9.21 holds for transformations that are homogeneous.

We can define homogeneity and inhomogeneity of transformations as follows. Consider a particular transformation $\phi : D \rightarrow D$. We can write out this transformation in terms of the constitutive structures of the two models as: $\phi : \{\mathcal{M}, G_i, T_\alpha\} \rightarrow \{\mathcal{M}', G_i', T_\alpha'\}$. In this context we can consider the pointwise action of ϕ on $\mathcal{M}$ that we write as $\phi_p : \mathcal{M} \rightarrow \mathcal{M}'$ or $\phi(p) = p' \in \mathcal{M}'$. Now, consider a *vector field* ξ on $\mathcal{M}$ which is a map that assigns to every point in $\mathcal{M}$ a vector $\xi(p)$ in the tangent space $\mathcal{M}_p$. We can then use conditions on the *Lie derivative* of ϕ_p with respect to an arbitrary smooth vector field ξ on $\mathcal{M}$ to pick out the cases where the transformations are homogeneous and inhomogeneous.

Intuitively, we can think of the Lie derivative of a map, ϕ_p for $p \in \mathcal{M}$, with respect to a vector field, ξ, as the 'rate of change' of the map relative to a standard of constancy set by the vector field. Thus, the value of $(\mathcal{L}_\xi \phi)|_p$ tell us how quickly the pointwise action of ϕ_p on $\mathcal{M}$ varies as we move across $\mathcal{M}$.[3] A vanishing Lie derivative thus indicates a transformation that has a homogenous action on $\mathcal{M}$ and a non-vanishing Lie derivative indicates a transformation that has an inhomogeneous across action on $\mathcal{M}$:

Definition 9.22 *Homogeneity: A transformation is homogeneous when it has vanishing Lie derivative with respect to an an arbitrary smooth vector field ξ on $\mathcal{M}$:*

---

[3] In fact, for a smooth map on a manifold the Lie derivative reduces to the ordinary directional derivative in the direction picked out by the vector field. Thus we have that $(\mathcal{L}_\xi \phi)|_p = \xi(\phi)|_p$. We have only introduced the Lie derivative here since it will be useful for our later discussions to define our notions of globality/locality such that they can be applied to more general tensorial objects. See Nakahara (2003, §5.3.2) and Malament (2012, §1.6) for full discussion of the Lie derivative.

$$\forall p \in \mathcal{M}, (\mathcal{L}_\xi \phi)|_p = 0.$$

Definition 9.23 *Inhomogeneity: A transformation is inhomogeneous when it has non-vanishing Lie derivative with respect to an an arbitrary smooth vector field ξ on $\mathcal{M}$:*

$$\exists p \in \mathcal{M}, s.t. (\mathcal{L}_\xi \phi)|_p \neq 0.$$

Note that we have not here assumed that ϕ is a diffeomorphism of the event space. As per above, let us distinguish that special case as $\psi = \text{Diff}(\mathcal{M}) \in \{\phi_i\}$. We would then have that $\psi : \{\mathcal{M}, G_i, T_\alpha\} \to \{\mathcal{M}, \psi^\star G_i', \psi^\star T_\alpha'\}$ where $\psi^\star$ is the pullback of the diffeomorphisms onto the geometric and matter structures.

With the definitions of homogeneity and inhomogeneity of transformations in hand, it is straightforward to give a precise sense of what it is for a structure to have a property globally or locally; that is, we can say that a structure is *globally* dynamically absolute or *locally* surplus. The definitions of the globality and locality of a property are:

Definition 9.24 *Globality: S has a property globally when relevant Definitions 9.16–9.21 hold for transformations that are homogenous across $\mathcal{M}$ with respect to arbitrary smooth vector field ξ.*

Definition 9.25 *Locality: S has a property locally when relevant Definitions 9.16–9.21 hold for transformations that are inhomogeneous across $\mathcal{M}$ with respect to arbitrary smooth vector field ξ.*

These distinctions will be of particular relevance when the status of spatiotemporal structure in special relativity and general relativity. They are also key to understanding the difference between temporal structures that are invariant under rigid shifts in time parameter and temporal structures invariant under time reparameterizations. As such, the global vs local distinction will prove highly significant in our later discussion.

9.4 Chapter Summary

In this chapter we have developed the Nomic-AIR framework for the analysis of symmetry and structure in physical theory. This is a powerful and general formal framework that allows for the exhaustive analysis and classification of the transformation properties of nearly any type of structure in nearly any type of theory. The framework is also flexible enough to be compatible with different approaches to the definition of the narrow symmetries of a theory.

The core idea of the framework is to track the behaviour of a given structure induced by transformations on the space of dynamically possible models in comparison to the induced transformations on the relevant nomic structure. In such circumstances, transformations can have a trivial (A), ill-defined (I), or well-defined (R) pullback onto a particular structure and simultaneously also have a trivial (A) or non-trivial (R) pullback onto the nomic structure. Transformations that pullback onto the nomic structure trivially are (narrow) symmetry transformations. Transformations that pullback onto the nomic structure non-trivially are dynamical transformations.

Together this gives us six possible combinations which we will define formally shortly. Informally, we can think of these six combinations as picking out invariant structure that does not vary under symmetries transformations (AA), surplus structure that is covariant under symmetries (AR), dynamically absolute structure that does not vary under dynamical transformations (RA), dynamically relative structure that does vary under dynamical transformations (RR), and kinematically incomplete (AI) and dynamically incomplete (RI) structure, respectively. We can also consider combinations of these combinations and prove a variety of simple theorems that relate the structural classifications.

Finally, the framework also allows for implementation of the global vs local transformation distinction in terms of inhomogeneous transformations. These are transformations that do not act in the same way across different regions of the event space which underlies the dynamically possible models. We can then define what it means for a structure to have a property globally or locally in terms of the relevant structural definition holding for transformations that are inhomogeneous.

Part III

THE PROBLEM OF TIME IN CLASSICAL MECHANICS

Part I presented analysis of key debates in the natural philosophy of time. We applied an informal descriptive framework for the isolation of different temporal structures and their transformation properties in the context of the analysis of time provided in the work of Newton, Leibniz, and Mach. Our discussion focused, in particular, on the two precursor problems of time; that is, of fixing respectively determinate chronordinal and chronometric structure which is invariant and dynamically relative.

Part II provided a comprehensive analysis of symmetry and structure in physical theory and led us to construct a new framework within which we can express the full set of transformation properties of a structure within a theory. By design, this framework allows us to explicitly articulate the informal idea of a spatiotemporal structure transforming in each of the various ways; that is, given a theory we can provide a formal analysis under which spatiotemporal structures can be identified and their transformation properties determined.

One goal of Part III is to bring the ideas of the previous two parts together; that is, we will demonstrate how the new framework can be applied to temporal structure in the explicit formalization of classical mechanical theories. This is not our primary goal, however. The target of our analysis is not the interpretative goal of identifying the structure of a mechanical theory with a stable formal articulation. Rather, our goal is to allow for the re-articulation of a theory such that the relevant structure can be *modified* or *precisified*. This is the context in which we will introduce and solve the problem of time in classical mechanics.

The first of the four chapters that follow will serve to situate the tools of the new framework in the context of a problem of theory re-articulation. We start by offering some general comments regarding structure and heuristics within the new framework. We will then draw upon our earlier illustrative examples of NM Barbour–Bertotti theory in order to provide an explicit example of the modification of spatiotemporal structure under theory re-articulation. The particular example is that of inertial frames and the modification will be the conversion of a structure *from* invariant *to* surplus structure.

The remaining three chapters will then be focused upon the problem of time. We will characterize that problem in terms of the observation that under the standard approach to gauge symmetry, the symmetry properties of *reparameterization invariant theories* with regard to chronometric and chronobservable structure enforce indeterminacy with regard to chronordinal structure. Our solution will be to first demonstrate that the standard approach to gauge symmetry is inadequate to the treatment of reparameterization invariant theories and second formulate a general and fully adequate alternative based upon a more perspicuous analysis of temporal symmetry and structure.

One major goal of our project will be to dispel the putative confusion between symmetry and evolution based upon naïve application of Dirac's theorem. In particular, the idea that in reparameterization invariant theories, 'motion is the unfolding of a gauge transformation' (Henneaux and Teitelboim, 1992, p. 103). We will classify such an understanding as the *pseudo-problem of time* and demonstrate that the supposed equivalence between motion and gauge symmetry rests upon a formal conflation. We will implement a new formal analysis that allows one to explicitly differentiate the generators of evolution and 'gauge' symmetry in reparameterization invariant theories.

Our approach is such that reparameterization invariance is demonstrated to be fully compatible with determinate chronordinal structure. As such, the novel approach to symmetry and evolution in classical mechanics that we will establish will provide a resolution of the problem of time in classical mechanics. We will provide a formal analysis of the symmetry and structure of reparameterization invariant theories that allows for determinate and well-behaved chronometric, chronobservable, and chronordinal temporal structures.

10

Spatiotemporal Structure and Theory Re-Articulation

The framework for the analysis of symmetry and structure in physical theory allows one to identify explicit heuristics for theory rearticulation. One of the most powerful of these heuristics is the surplusing heuristic which is built from the conversion of invariant to surplus structure within a theory. When applied to the pivotal structure of inertial frames within Newtonian Mechanics this heuristic leads directly from Newtonian Mechanics to Barbour–Bertotti theory. In Newtonian mechanics, inertial frames are invariant under the narrow symmetries of the theory given, under the preliminary stance, by rigid Euclidean transformations but variant under the dynamical transformations given by Galilean boosts. Inertial frames are thus dynamically relative invariant structures of the theory. Implementing the surplusing heuristic means converting structure from invariant to surplus and this is exemplified in the case of moving from Newtonian Mechanics to Barbour–Bertotti theory since in the latter inertial frames are dynamically relative surplus structure.

10.1   Structures and Heuristics

10.1.1   The Dynamics of Reason

Theory re-articulation is the formal manipulation of a theory, understood as a collection of constitutive, nomic, and spatiotemporal structures, such that a new theory, with a distinct set of constitutive, nomic and spatiotemporal structures, is derived. One may understand such a new theory as a *reformulation* of the original theory or as a *genuinely new theory*. In our view, not much should hang on this difference, which is best understood as one of degree rather than kind. The important aspect of theory re-articulation for our purposes is not interpretative but pragmatic; that is, we are interested in the *how* question rather than the *what* question. In particular, given a theory which has been presented in terms of a set of structures and subjected to analysis within our new framework, one may seek concrete strategies to rearticulate such a theory such that the transformation properties of a given spatiotemporal structure are modified.

In this regard, a remarkable value of the new framework introduced in the previous chapter is that it allows us to identify explicit heuristics for theory re-articulation. We take our heuristics be part of a *regulative meta-framework for theory construction*; that is, we take our heuristics to be extra-theoretical guides to the modification of the structure of a theory, including in some cases the constitutive structure, that will at a minimum always leave us in a situation where the structure of the old theory can be recovered as a limiting case of the structure of the new, but which will also, in some defeasible sense, serve to guide us in a rational and progressive process of scientific development.

A partial inspiration for this part of our project can be found within the fascinating analysis of the 'dynamics of reason' within the process of physical theory construction provided by Friedman (2001). In particular, we have here been inspired by Friedman's proposals for a framework for the synthesis of the ideas of Kuhn, Kant, and the post-Kantians in the context of theory change. Friedman's project is, of course, rather different to our own, but nevertheless a short overview of his terminology and general outlook will prove highly instructive towards articulating our own ideas. His proposal is based upon three key conceptual moves that we can articulate as follows.

First, drawing upon Kant, Friedman makes the distinction between unrevisable constitutive principles and revisable regulative principles. Whereas, the former are immutable constraints on scientific theorizing, the latter are understood as guides for scientific progress which support but do not constrain. The idea is that via such constraints and guides, science is able to approach but never reach some 'ideal state of completion'. Pursuit of such a state is then a *regulative ideal of reason*, which, importantly, is such that there is no guarantee of actual completion.

Second, Friedman suggests that we adapt Kant's conception of the regulative use of reason such that is applied also in constitutive domain. In particular, we apply such an idea in the context of a Kuhn-inspired conception of theory change as 'paradigm shift' such that we then have regulative ideals which govern scientific reasoning at an inter-paradigm, transhistorical-level universality. The key idea is that although we are never in a position to take our present constitutive principles as universal and fixed, it is perfectly plausible to understand the process by which such principles are constructed as governed by regulative ideals for our reasoning such that our current constitutive principles are 'approximations to more general and adequate constitutive principles that will only be articulated at a later stage' (2001, p. 64).

The third and final step that Friedman enjoins us to make is to then understand the regulative principles themselves as open to contestation and revision within something like a scientific framework for *communicative rationality*. The term originates with Habermas and is taken by Friedman to be a form of rationality that is 'essentially public or intersubjective', and as such is founded on a shared aim towards the establishment of consensus though reasoned argumentation in a context where principles of argument are themselves shared by the parties in dispute. This is understood partially in Kuhnian terms via the idea that there can be an intra-paradigm conception of rationality in terms of a fixed research consensus within a group of scientists. However, in Friedman's reframing, such a consensus operates at the level of meta-framework in which it is the regulative principles of scientific practice that the subject of debate.

Returning to our own project, the correlate of Friedman's regulative principles are our heuristics for theory re-articulation; that is, what we take ourselves to be proposing in formulating the heuristics for transforming theoretical structure are not inflexible principles that constitute an algorithm for articulating new theories that are in some essential sense superior. We take our heuristics to be extremely relevant, and indeed connectable to some of the most significant historical examples of successful theory construction. However, despite this success, we do not take them to be obligatory nor to lead us towards *the truth*. As such they constitute a regulative framework that can guide theory re-articulation but which are open to contestation, and in turn, further development themselves,

We do take our heuristics, however, to lead us towards the construction of new theories from which the structures of the old can always be recovered. This feature, again defensibly, we take to be as close as there is to be a *sine qua non* of theory construction in the context of modern theoretical physics. As Friedman puts things: 'the new conceptual framework or paradigm should contain the previous constitutive framework as an approximate limiting case, holding in precisely defined special conditions' (2001, p. 66). To the extent to which the heuristics provide us with norms for theory re-articulation and extension, these are contextual norms situated within a specific stage of theory constriction seen as a socially and historically embedded enterprise. However, the heuristics by their nature are such that they will lead to theory re-articulation in which the old theory is recoverable from the new. This will hold true even as the heuristics themselves are adapted in light of the new theoretical context which they were themselves crucial in establishing.

Our desire is thus to present our heuristics for the analysis of spatiotemporal structure without the usual attendant metaphysical baggage. This does not mean, however, that the heuristics we present are not closely related to various principles and arguments that have long figured in debates regarding the nature of space and time. In the below we will seek to emphasize these connections by linking each heuristic to at least one relevant historical debates.

10.1.2   Reduction Heuristic

Elimination of surplus structure. This is the most basic and, in a sense, obvious heuristic. Since surplus structure is by definition variant under symmetry transformations, there is a natural theoretical re-articulation strategy under which surplus structure is eliminated in a reformulated theory. Precisely because it is surplus, such structure might be thought to not encode physical differences, and thus be 'otiose'. The elimination of surplus spatiotemporal structure follows a broadly Leibnizian vein of thought as discussed in Chapter 3.

In formal terms, typically the elimination of surplus structure can be achieved by moving to a reduced formulation of the theory in which such structure is quotiented out as has been described extensively in the earlier chapters. It is important to note, however, that reduction is not without its costs. In particular, it is important to recognize that surplus structure can play an important *supporting* or *aiding* role in our theoretical reasoning. Its elimination can, therefore, be problematic in at least some contexts. This feature is particularly significant in the context of non-Abelian gauge theories where the classical auxiliary structure is crucial to understanding anomalies that occur in the quantum theory.

Finally, and most significantly, reduction of non-surplus structure is in general inadvisable. Thus, we can also consider a negative reduction heuristic in terms of a norm against applying reduction to eliminate non-surplus structure. It is in violation of this norm that is, on our account, the route of much confusion in the analysis of the problem of time.

10.1.3   Surplusing Heuristic

Conversion of invariant to surplus structure by expansion of the narrow symmetries of a theory. This is perhaps the most conceptually subtle heuristic and proceeds in three steps. The first step is to identify a structure within the theory which we have good, typically empiricist, reasons to expect to be a surplus structure but which is, based on the preliminary stance, an invariant structure. The second step is to introduce extra variables that parameterize the transformation under which the relevant structure varies. These are intended to be 'dummy variables' that do not alter the physical content of the theory. The third step is to modify the action principle such that the transformations in question are explicitly rendered symmetries and the structure is then converted from invariant to surplus.

The surplusing heuristic is a means by which we can render surplus structure that plays no physical role. It is a true heuristic for theory re-articulation rather than simple reinterpretation since it allows us to formulate different nomic principles. It is a non-reductive heuristic since it does not eliminate structure but rather downgrades the status of structure. We will illustrate this heuristic explicitly for the case of inertial frames in NM and Barbour–Bertotti theory later in this chapter. The more general application of this heuristic will be of particular importance in the context of general relativity as discussed in Volume II of this work. In particular, one can understand the key formal and conceptual move in the famous Kretschmann objection to be essentially an appeal to the general availability of a surplusing heuristic for spacetime coordinate structure in any physical theory.

10.1.4   Dynamical Relativization Heuristic

Dynamical relativization is the conversion of dynamically absolute to dynamically relative structure. Our third heuristic will also be familiar from more traditional discussions of philosophical and physical debates regarding the nature of space and time. The idea is that given a spatiotemporal structure in a theory that is dynamically absolute, one can seek to render such structure dynamical relative. We hope that it is clear that there is an important conceptual link between the dynamical relativization heuristic and the thought of Ernst Mach, as discussed in Chapter 4.

In most cases dynamical relativization will lead to a potential strategy for theory construction rather than re-articulation; that is, in order to make a spatiotemporal structure dynamically relative one will normally need to make fundamental changes to the theory, often including the constitutive structures. There is no guarantee that the dynamical relativization heuristic will lead to a unique path towards theory construction, nor to a theory that is empirically adequate or even internally consistent. Thus any norm for theory construction supported by this heuristic is certainly a defeasible one.

10.1.5   Localization Heuristic

Localization is the conversion of a spatiotemporal structure with global properties to a spatiotemporal structure with local properties; that is, modification of the theory and/or structure such that the transformations on the space of dynamically possible models that the structure is well-behaved under are converted from homogenous to inhomogeneous. Localization of spatiotemporal structure is an extremely important formal tool that will of be central concern in the following chapters.

One approach to localization relates to invariant structure and is achieved via application of a mathematical technique through which the space of kinematically possible models of a theory is artificially enlarged via the addition of compensator fields as per the best matching procedure we described in the previous chapter. Such kinematic enlargement is accompanied by the additional constraints leading, in some cases, back to the same set of dynamical possibilities. The end result, however, is that the transformations under which the laws are invariant is enlarged. Typically this will also mean that the relevant spatiotemporal structure is then invariant under the same larger group of symmetry transformations, and typically the principal difference between the original and new symmetry transformations will be inhomogeneity as parameterized by the compensator fields.

A second related approach that is relevant in the context where we have a fixed structure under a set of homogeneous transformations but where the relevant inhomogeneous transformations are such that they are diffeomorphisms of the space of kinematically possible models $\Psi \in \mathrm{Diff}(K)$ that do not preserve the nomic partition and are thus excluded from $\mathrm{Diff}(D)$. We briefly introduced the general idea of these transformations as $\bar{\psi} \notin \Phi$ in 9.1. Consider the situation where a given structure S is fixed structure under homogenous transformations and does not vary under inhomogeneous transformations but these inhomogeneous transformations are members of $\bar{\psi}$. In such cases we have a clear motivation to reformulate the nomic principle such that inhomogeneous transformations can be included in $\mathrm{Diff}(D)$ and thus the full invariance of the relevant fixed structure is represented dynamically. In the Chapter 11 we will consider the conversion of invariance under rigid shifts in time parameter to invariance under time reparameterizations as the pivotal example of the application of a localization heuristic.

10.1.6   Completion Heuristic

The final heuristic is the most powerful yet also the most novel. To our knowledge the idea of the *completion* of spatiotemporal structure has not previously been explicitly discussed in the physics or philosophy of physics literature. The idea is fairly straightforward to state. Given a theory in which a structure is incomplete one can attempt to either rearticulate or extend the theory such that this structure is rendered complete in the new theory (or new formulation).

The most simple concrete strategy for completion is restricting the class of kinematical models to specifically exclude the possibility of incompleteness. A good exemplar of this strategy, that will feature heavily in our discussions of time in general relativity in Volume II, is topological spacetime structure corresponding to the ability to decompose any given spacetime manifold into a spatial manifold and a temporal manifold,

$M = \Sigma \times \mathbb{R}$. In the covariant formulation of general relativity this topological structure is an incomplete dynamical structure. It is transformed in an ill-defined way—i.e. a way that does not take us to a new token of the type of structure—as one varies between some distinct dynamically possible models.

In more general terms we propose that methodologically the completion by truncation procure as applied to dynamically incomplete structure (IR) should be clearly distinguished from the completion by truncation procedure as applied to kinematically incomplete structure (IA). Whereas the former is always based around removing a subset of the dynamical models, which by definition have empirical content the latter involves simply removing or modifying the surplus structure. The more interesting and potentially heuristically powerful completion strategy is to look to enrich the kinematic and nomic structure such that the structure in question is completed.

10.2 Spatiotemporal Structure in Newtonian Mechanics

10.2.1 Constitutive and Nomic Structure

First, let us recap the constitutive and nomic structure of Newtonian Mechanics (NM) as discussed at length in §6.5. The constitutive structure is the structure that we need to define the space of basic, pre-nomic possibilities and consists of the following elements:

$$\mathcal{K} = \left\{ \mathbb{R}_t \times \mathbb{R}_x^{3N}, g_t, \delta, \gamma : (t_1, t_2) \to T\mathcal{C}, m_I, c_\alpha \right\} . \tag{10.1}$$

where $\mathbb{R}_t \times \mathbb{R}_x^3$ is the product space of the temporal and spatial manifolds, g_t is the temporal metric, δ is the flat Euclidean metric on the spatial manifold $\mathbb{R}^3$, $(t_1, t_2) \in \mathbb{R}_t$, $\mathcal{C} = \mathbb{R}^{3N}$ and $m_I \in \mathbb{R}^+$ and c_α are the particle masses and other coupling constants that enter the potential, V.

The kinematically possible models of NM are then given by the collection of all possible (images of) curves γ in $T\mathcal{C}$ conjoined with a specification of all the particle masses m_I and couplings c_α of the theory.

$$K = \{\gamma : (t_1, t_2) \to T\mathcal{C}, m_I, c_\alpha\} \tag{10.2}$$

where I indexes the number of particles and c_α indexes the number of couplings. We can distinguish between tokens of this constitutive structure and the relevant sets of KPMs by, for instance, considering tokens that differ with regard to the number of particles, $K^a = \{\gamma : (t_1, t_2) \to T\mathcal{C}, m_{I=2}, c_\alpha\}$, $K^b = \{\gamma : (t_1, t_2) \to T\mathcal{C}, m_{I=3}, c_\alpha\}$ or between tokens that differ by the coupling constants that appear in the potential, say a two particle gravitational and pendulum model, $K^c = \{\gamma : (t_1, t_2) \to T\mathcal{C}, m_{I=2}, G_{\text{Newton}}\}$, $K^d = \{\gamma : (t_1, t_2) \to T\mathcal{C}, m_{I=2}, k_{\text{spring}}\}$. Each token then picks out a set of KPMs that differ solely in virtue of the curves γ in $T\mathcal{C}$.

We then defined the nomic structure via the initial value formulation. The partitioning map, $n : K \to D$, is given by the procedure of first performing the Legendre transform of the curves γ, then projecting these onto the space of curves satisfying Hamilton's equations before finally undoing the Legendre transform to specify DPMs as the dynamical curves γ_{cl}. The projection map $\pi_N : D \to \tilde{D}$ from the space of dynamically possible models (DPMs) to the space of distinct dynamically possible models

(DDPMs) can then be picked out based on our analysis of the canonical conserved charges. In particular, recall that we considered the charges:

$$P_i = \sum_I p_i^I \qquad L_i = \sum_I \epsilon_{ij}{}^k q_I^j p_k^I \qquad C^i = \sum_I \left(m_I q^i - t_N \delta^{ij} p_j^I \right), \qquad (10.3)$$

Together with the Hamiltonian H, these charges generate the vector fields $(v_P^i, v_L^i, v_C^i, v_H)$ which generate projective representations of Barg$(3,1)$, which are ordinary representations of Galilean group Gal(3). Significantly, the full group of transformations of Gal(3) are *not* narrow symmetries since the time-dependent charges C^i generate transformations which are not variational symmetries. These are the Galilean boots and require a special treatment.

Adopting our preliminary stance on narrow symmetries implies that DPMs related by boosts should be taken to be dynamically distinct. Physically this is well justified by the privileged empirical role that the conserved charge associated with boosts plays in the theory. By contrast, considering the sub-group of transformations generated by the time-independent charges P_i and L_i correspond to ISO(3) transformations of the spatial coordinate system

$$x^i \to R^i{}_j(\theta)x^j + a^i. \qquad (10.4)$$

that are variational symmetries. The conserved charges correspondingly play no privileged empirical role and no dynamical information is lost in fixing a value of these charges.

The first stage of our analysis of NM is thus complete, we can define the projection map π_N by fixing a value of the charges P_i and L_i which is equivalent to specifying as dynamically identical DPMs related by time-independent spatial shifts (Leibniz shifts) and rotations. We thus have that $Abs(\pi_N)$ is made up of the ISO(3) transformations and $Rel(\pi_N)$ is made up of Galilean boots and transformations of the initial data generated by the Hamiltonian H.

10.2.2 Spatiotemporal Structures

We are now in a position to consider the key spatiotemporal structures of Newtonian particle mechanics. These can be broken down into temporal structure, spatial structure, and inertial structure. With regard to the temporal structure we are particularly interested in three types of temporal structure. These are chronordinal, chronometric, and chronobservable. Chronordinal structures are structures that provide a representation of time-ordering s of sets of events. Chronometric structures are structures that provide a representation of temporal distance relations between sets of events. Chronobservable structures are structures that function as empirically accessible representations of some aspect of time. They will to sets of temporal quantities that are observables by the lights of the theory.

The chronordinal structures of NM are fairly straightforward in the formulation we are considering. In particular, there is a total temporal-ordering $>$ inherited from the natural ordering on the real line $\mathbb{R}$. This is an undirected ordering structure. We can also, conventionally, pick out a future, $+$, and past, $-$, directed time orientation given by the vectors $t_\pm^a = (\pm 1, 0, 0, 0)$ on the space-time manifold $\mathbb{R}_t \times \mathbb{R}_x^3$. We are free to choose one or the other direction by convention. Thus, the directed chronordinal structure has a different status to the undirected chronordinal structure. As already noted we will not, for the most part, venture into the analysis of the subtle issues in the interpretation of symmetries and directed chronordinal structure in this book.

The chronometric structure of NM is significant in terms of its contrast with the spatial metric structure. In particular, spatially, the theory is equipped with a family of privileged *Cartesian* coordinate charts x^i on the spatial manifold $\mathbb{R}^3$ such that the Euclidean metric takes the form $\delta = \text{diag}(1, 1, 1)$. These are defined by the isometries of δ up to the Euclidean transformations

$$x^i \to R^i_{\ j}(\theta)x^j + a^i \,, \tag{10.5}$$

where $R(\theta)$ is an element of the rotation group $SO(3)$ in 3D. The analogous chronometric structure is given by the privileged temporal coordinate t_N. This is such that Newton's second law takes the form

$$\frac{\mathrm{d}^2 q^i_I}{\mathrm{d}t^2_N} = -\frac{\partial V}{\partial q^i_I} \,. \tag{10.6}$$

Significantly the privileged temporal coordinate is only defined up to time translation

$$t_N \to t_N + a_0 \,. \tag{10.7}$$

As such, a *Newtonian time*, T_N, is given by a pairing of t_N *and* a specific value of the offset a_0. Plausibly, the chronometric structure of time in NM *as Newton interpreted it*, and as discussed in Chapter 2, should be given by T_N rather than t_N.[1] We will return to this point regarding chronometric structure shortly.

The key chronobservable structure is provided by the set of clock observables τ satisfying

$$\{\tau, H\} = f \,, \tag{10.8}$$

for some non-zero and (by convention) positive function f.[2] An *ephemeris* clock observable τ_E obeys the special case of (10.8) given by:

$$\{\tau, H\} = 1 \,, \tag{10.9}$$

Whereas both sets of clock observables function as empirically accessible representations of some aspect of time within NM, only the ephemeris clock observables will

---

[1] It is worth noting here that our formulation Newtonian time is still *less* underdetermined than the Newtonian time function, defined as an affine parameter on geodesic trajectories, in the discussion due to Friedman (1983, p. 114). This is because our Newtonian time is underdetermined up to an offset, but Friedman's Newtonian time function is underdetermined up to an offset *and* a multiplicative constant.

[2] Clock observables as we have defined them here are subject of a general result which implies that for Hamiltonians which are bounded from below, the associated Hamilton vector field, X_τ will be incomplete. This result, which is the classical analogue of Pauli's theorem, indicates global issues in the definition of classical time observables but does not prohibit their rigorous deployment in local physics. In particular, since the direction in which the relevant vector field is incomplete is tangential to the dynamics, clock observables can still provide a complete parameterization of physical motions. See Roberts (2014*b*) for further details and discussion.

'march in step' with the privileged temporal coordinate. However, even this observable will not match the Newtonian time, T_N, in terms of the value that it reads, since the offset a_0 is completely dynamically underdetermined.

Finally we can consider the inertial structure. Our formulation above includes a privileged set of inertial frames. These are given by the frames that are co-moving with a coordinate system on the space-time manifold $\mathbb{R}_t \times \mathbb{R}_x^3$ in which Newton's second law takes the form (10.6), when expressed in a Cartesian coordinate chart and Newtonian time. When expressed in the privileged charts, the inertial frames are defined up to the Galilean transformations

$$x^i \to R^i{}_j(\theta)x^j + a^i + v^i t_N \,, \tag{10.10}$$

$$t_N \to t_N + a_0 \,, \tag{10.11}$$

where v^i are the Galilean boost parameters. Note that the inertial frames are given dynamical definitions, unlike the privileged Cartesian coordinates but like the privileged temporal coordinate.

10.2.3 Nomic-AIR Analysis

Let us now consider each spatiotemporal structure in turn and apply our AIR classification scheme.

First we have the (undirected) chronordinal structure provided by the temporal order $>$. This structure is fixed for all curves in TC and is independent of space of masses or coupling constants. Thus, $\mathrm{Abs}(>) = \Phi$. This means that the time-ordering structure is *fixed structure* which does not vary either between dynamically identical models (i.e. invariant structure) nor between dynamically distinct models (dynamically absolute structure). We know from the (rather trivial) Theorem 9.5 above that this means that the time-ordering structure of NM is neither surplus nor dynamically relative. More significantly, we can consider the local vs global status of the ordering structure as fixed structure. In NM, as we have formulated it, inhomogeneous transformations have been excluded from Φ thus we automatically have that ordering structure is globally fixed structure, both temporally and spatially.

The next structure we considered was the spatial structure given by the Cartesian coordinate charts x^i on the spatial manifold $\mathbb{R}^3$. Above we noted that the sub-group of transformations generated by the time-independent charges P_i and L_i correspond to ISO(3) transformations of the spatial coordinate system and are variational symmetries. This means that the Cartesian coordinate charts vary under the narrow symmetries of the formalism and are thus surplus structure. The Cartesian coordinate charts do not, however, vary under transformations that move us between dynamical models. In particular, we can consider dynamical transformations generated by the Hamiltonian H, which corresponds to DPMs with distinct initial data surfaces, or generated by changing the Galilean boost parameters, v^i, which corresponds to DPMs defined in a distinct inertial frame. Such transformations, which make up $Rel(\pi_N)$, do not change the Cartesian coordinate charts x^i on the spatial manifold $\mathbb{R}^3$, and thus this spatial structure is dynamically absolute. x^i is thus dynamically absolute surplus structure.

Next we can consider the chronometric structures, the privileged temporal coordinate, t_N, and the Newtonian time, T_N, given by a pairing of t_N and a specific value of the offset a_0.

Let us focus on t_N first. The only element of Φ that acts on t_N is the time translation $t_N \rightarrow t_N + a_0$. However, by definition t_N is invariant under such a translation so we immediately have that $\text{Abs}(t_N) = \Phi$ and thus that the Newtonian time is fixed structure. Furthermore, recall from the discussion following Definition 9.10 that we differentiated two forms of fixed structure. Structure that is identical in all models that share the same type of constitutive structure is constitutively fixed. Structure that is identical in all models that share the same token of constitutive structure but varies between at least some tokens is contingently fixed. It will prove significant in our later comparative analysis that the privileged temporal coordinate is constitutively fixed since there evidently is no basis for it to vary between different tokens of the Newtonian constitutive structure.

Let us then consider the Newtonian time, T_N. In contrast to t_N we know that changing the offset $t_N \rightarrow t_N + a_0$ corresponds to moving to a different Newtonian time $T_N \rightarrow T_N'$. Thus this structure varies under the relevant narrow symmetry transformation and is surplus structure. Furthermore, since such a transformation is dynamically irrelevant, we know that the structure is also dynamically absolute. Thus the Newtonian time, T_N, is a dynamically absolute surplus structure. Following Theorem 9.8 above, this means that all relative transformations of the structure will be symmetry transformations. Which is obviously matches intuitions in this case also.

Significant for what follows is that both of the chronometric structures t_N and T_N have their respective status (i.e. dynamically absolute invariant and dynamically absolute surplus) temporally and spatially globally. This was guaranteed since inhomogeneous transformations have been excluded from Φ in our formulation.

Finally we have the chronobservable structures given by the clock observables structure and the more specific ephemeris clock observables structures. Clock observables satisfies $\{\tau, H\} = f$ will in general commute with the absolute charges of the theory $\{\tau, k_\alpha\} = 0$. Changes of the absolute charges leave both the nomic structure and the clock observable structure invariant, and we have $\text{Abs}(\pi_N) = \text{Abs}(\tau)$. Thus a clock observable is an invariant structure in NM. Clock observables need not, however, have vanishing Poisson bracket with all the non-trivial HJ constants of the theory. Thus a clock observable can co-vary under elements of $\text{Rel}(\pi_N)$ and we thus have that the general class of clock observables in NM are dynamically relative invariant structure. The special case of ephemeris clock observables for which $f = 1$ will have the same structure categorization with the satisfaction of the relevant condition varying between dynamically distinct models in terms of the initial conditions required for satisfaction. It is worth noting that in the relevant regard clock observables in NM have the same structural status as proper time in general relativity. Once more, the clock observables are automatically dynamically relative invariant structure *spatially and temporally globally*. And this is of course the important contrast to proper time.

10.3 Spatiotemporal Structure in Barbour–Bertotti Theory

10.3.1 Constitutive and Nomic Structure

We restrict attention the partially reduced BB theory presented in Section 8.5.3 because the nomic structure π_N is possible to represent explicitly in the Hamiltonian formalism for this theory. This gives us an explicit means of characterizing the space of DPMs and DDPMs for the theory. Moreover, it also leads to a simple comparison with NM because the constitutive structures are shared between both theories.

The full specification of the space of kinematical structures is identical to that we defined for NM:

$$\mathcal{K} = \left\{ \mathbb{R}_t \times \mathbb{R}_x^{3N}, g_t, \delta, \gamma : (t_1, t_2) \to T\mathcal{C}, m_I, c_\alpha \right\} . \tag{10.12}$$

The kinematically possible models of (partially reduced) Barbour–Bertotti theory are then given by the collection of all possible (images of) curves γ in $T\mathcal{C}$ conjoined with a specification of all the particle masses m_I and couplings c_α of the theory.

The nomic structure of partially reduced BB theory is then given by:

Partitioning map $n : K \to D$: the nomic partition can be defined restricting to integral curves of the vector field $v_{H^{\text{red}}_{\text{BB}}}$ that lie on *any* of the gauge-fixed surfaces defined by the functions a^i.

Projection Map $\pi_N : D \to \tilde{D}$ from the space of DPMs to the space of DDPMs is then a projection onto the space given (equivalently) by: i) considering *families* of integral curves related by different value of a^i which correspond to different gauge-fixed surfaces; or ii) considering the projection of the integral curves of $v_{H^{\text{red}}_{\text{BB}}}$ onto the reduced phase space of Dirac observables

Recall that we can define $\text{Abs}(\pi_N)$ as the space of transformations on D that leave the nomic structure π_N invariant. The space of DPMs is therefore the space of all integral curves of $v_{H^{\text{red}}_{\text{BB}}}$ that lie on any of the gauge-fixed surfaces defined by the functions a^i and the space of DDPMs is the space of *families* of integral curves related by different value of a^i.

This space can be explicitly constructed (whenever $V(q)$ and a^i are bounded and continuous) in terms of the Dirac observables (8.110) and (8.111), which commute with P_i^{tot}. The space of DDPMs is then isomorphic to the space of the projection of the integral curves of $v_{H^{\text{red}}_{\text{BB}}}$ onto the reduced phase space of Dirac observables.

The space $\text{Abs}(\pi_N)$ is therefore the space of maps generated by the Hamilton vector fields of P_i^{tot}. We thus have

$$\text{Abs}(\pi_N) = \exp\left\{ \mathcal{L}_{a^i v_{P_i^{\text{tot}}}} \right\}\big|_{P_i^{\text{tot}} = 0}, \tag{10.13}$$

for the arbitrary phase space functions a^i. $\text{Rel}(\pi_N)$ is is then made up solely of transformations of the initial data generated by the Hamiltonian H.

10.3.2 Spatiotemporal Structure

We can then proceed to consider the spatiotemporal structures. Clearly since the kinematic structures are identical we can pick out *almost* the same spatiotemporal structures.

With regard to temporal structures, there is again a chronordinal structure given by the temporal order $>$ inherited from $\mathbb{R}_t$ and past and future orientation vectors $t^a_\pm = (\pm 1, 0, 0, 0)$ on the space-time $\mathbb{R}_t \otimes \mathbb{R}^3_x$.

We also again have a privileged notion of duration dt_N, giving our chronometric structure, which is such that Newton's law holds in centre-of-mass coordinates and a family of preferred temporal frames related by

$$t_N \to t_N + a_0 \tag{10.14}$$

that result from this privileged notion of duration.

Finally, there are dynamically defined clock variables, giving chronobservable structure, which are τ satisfying

$$\{\tau, H^{\mathrm{red}}_{\mathrm{BB}}\} = 1 \,, \tag{10.15}$$

and a canonically privileged Newtonian clock variable τ_E that additionally satisfies

$$\{\tau_E, Q\} = \{\tau_E, P\} = 0 \tag{10.16}$$

for the maximal independent set of constants of motion (Q, P). We will see below that such constants should be functions of the Dirac observables for τ_E to be an invariant structure of the BB theory. Note that $H^{\mathrm{red}}_{\mathrm{BB}}$ and (Q, P) differ from standard NM.

The spatial structures of shift-eliminated BB theory are *nearly* identical to those of NM. There is still a family of fixed Cartesian coordinates, x^i, related by time-independent ISO(3) transformations

$$x^i \to R^i{}_j(\theta) x^j + a^i \,. \tag{10.17}$$

However, the inertial frames have an importantly different categorization within the Nomic-AIR scheme.

10.3.3   Nomic-AIR Analysis

Let us analyse the structures that have a different nomic AIR analysis in BB theory compared to NM. All other structures have the same transformation properties.

First is the total linear momentum of the system P^{tot}_i. In NM, the only symmetry transformations are the time-independent Euclidean transformations ISO(3). P^{tot}_i is invariant under these. However, P^{tot}_i varies under some dynamical transformations of NM because any initial value of P^{tot}_i is allowed in NM. Thus, the total linear momentum is a dynamically relative invariant structure of NM.

By contrast, in BB theory, while P^{tot}_i is invariant under all symmetry transformations, it does *not* vary between DDPMs because all DPMs of the theory are required to satisfy the constraint $P^{\mathrm{tot}}_i = 0$. The total linear momentum is therefore a dynamically absolute invariant structure, and thus a fixed structure, in BB theory. This of course is something we have engineered into the theory deliberately through our best matching and then partial reduction procedure. It is *because* we interpreted the total linear momentum of NM as not empirically well motivated as dynamically relative that we followed the procedure to construct the Barbour–Bertotti formalism in which the structure becomes fixed and thus plays no dynamical role.

The most significant structures which are different in the BB theory as compared with NM are the inertial frames, x^i, related to the centre-of-mass frame by Galilean boosts:

$$x^i = q^i_{\text{cm}} + v^i t_N \,. \tag{10.18}$$

In NM, these structures are invariant under the narrow symmetries, that is, ISO(3) transformations, but variant under at least some of the dynamical transformations, that is, the Galilean boosts themselves. Thus the inertial structure is dynamically relative invariant structures in NM. In BB theory, however, the frames x^i vary under at least some of the narrow symmetry transformations of the theory, that is, Galilean boosts, and vary under at least some dynamical transformations, that is, transformations of the initial data generated by the Hamiltonian H, since two distinct DPMs can have different values of q^i_{cm}. The inertial frames in BB theory are thus dynamically relative surplus structure. The move from NM to BB theory thus illustrates our surplusing heuristic as promised.

10.4   Chapter Summary

Our framework for the analysis of symmetry and structure in physical theory allows one to identify explicit heuristics for theory re-articulation. One of the most powerful of these heuristics is the surplusing heuristic which is built from the conversion of invariant to surplus structure within a theory. When applied to the pivotal structure of inertial frames within NM this heuristic leads directly from Newtonian theory to Barbour–Bertotti theory. In NM inertial frames are invariant under the narrow symmetries of the theory given, under the preliminary stance, by rigid Euclidean transformations but variant under the dynamical transformations given by Galilean boosts. Inertial frames are thus dynamically relative invariant structure of the theory. Implementing the surplusing heuristic means converting structure from invariant to surplus and this is exemplified in the case of moving from NM to Barbour–Bertotti theory since in the latter inertial frames are dynamically relative surplus structure.

11
Local Temporal Symmetry

A theory is reparameterization invariant when its equations of motion are invariant under the temporal relabelling symmetry given by locally rescaling a monotonic and smooth time function. This understanding of reparameterization invariance is more general than that provided by study of the action since the properties of the equations of motion generated by a variational principle depend only on the local terms, and not the boundary terms, of that variation principle. The treatment presented here outlines the connections between reparameterization invariance, various conditions on the Lagrangian and Hamiltonian functions, and the relevant generalized Bianchi identities. In the next crucial stage of the analysis a novel first-order geometric analysis of reparameterization invariance is provided in terms of the study of tangential transformations. This provides a basis for the unambiguous differentiation between the objects responsible for generating evolution and those that generate degeneracy along tangential directions.

11.1 Time Reparameterization Invariance

In the theories considered thus far we have been focused on the consideration of the global transformation properties of temporal structure. This was guaranteed since inhomogeneous transformations have been excluded from the spaces of transformations in the simple mechanical theories that we have considered. The more general case is, of course, that in which we include inhomogeneous in the space and thus consider the behaviour of temporal structure under inhomogeneous and homogeneous transformations.

The importance of this extension is seen straightforwardly by correspondence between invariance of a temporal structure under inhomogeneous transformations and the status of such a structure as locally invariant. Most vividly, we can consider a mechanical theory in which chronometric structure is a local invariant and thus the structure which gives temporal distances does not vary between dynamically possible models related by inhomogeneous narrow temporal symmetry transformations. Such structure is a direct analogue of a proper time structure in relativistic theory. In more fully understanding such structure in the context of canonical theories we are therefore laying the grounds for a canonical treatment of time in general relativity. This is, of course, where we encounter the classical mechanical manifestation of the problem of time. The solution of this problem will be the major preoccupation of the next three chapters.

Before we proceed towards that subject it is important to emphasize two facets of the local temporal symmetry that we will consider. The first and most basic is that it is a temporally rather than spatially local symmetry. The reparameterization transformations we will consider are inhomogeneous in time but not in space. This is important since the full set of transformations that are inhomogeneous in both time and space bring with them additional formal and conceptual problems that require a specialized analysis. The relevance of such refoliations transformations to our analysis will be outlined in the final chapter as a foreshadowing one of the major projects of Volume II of this work.

The second point of significance is that the class of theories in which inhomogeneous temporal transformations are narrow symmetries is not a narrow subset of mechanical theories since such an invariance can be engineered into virtually any mechanical theory by application of our localization heuristic. Such theories need not be seen, however, as artificial constructs. Rather, one could understand reparameterization invariant mechanics as the general case, and mechanical theories with a fixed time parameterisation as the artificial constructs. In any case, the important point is that for all its mathematical complexities, reparameterization invariance is a strong candidate for a fundamental symmetry and its relevance is thus a general one.

A theory is said to be *reparameterization invariant* when its equations of motion are invariant under the symmetry

$$t \to \bar{t}(t), \tag{11.1}$$

where t is a time parameter on the domain, I, of γ and $\bar{t}$ is some *monotonic* smooth function $\bar{t} : I \to \mathbb{R}$ on this domain such that

$$f \equiv \frac{d\bar{t}}{dt} > 0. \tag{11.2}$$

It will be important for our considerations below that this definition is given in terms of an invariance of the equations of motion rather than the properties of an action. That is because the properties of the equations of motion generated by a variational principle depend only on the *local terms*, and not the boundary terms, of that variation principle. In other words, a theory will be reparameterization invariant if the *local form* of the action; that is,

$$\mathrm{d}t\, L\left(q^i, \frac{dq^i}{dt}\right) = \mathrm{d}\bar{t}\, L\left(q^i, \frac{dq^i}{d\bar{t}}\right) \tag{11.3}$$

is invariant under (11.1). Note that we have already restricted ourselves to second-order theories with no explicit t-dependence in L. Importantly, the local condition (11.3) is *not* equivalent to invariance of the action. This is because, in general,

$$\int_{t_1}^{t_2} \mathrm{d}t\, L\left(q^i, \frac{dq^i}{dt}\right) \neq \int_{\bar{t}_1}^{\bar{t}_2} \mathrm{d}\bar{t}\, L\left(q^i, \frac{dq^i}{d\bar{t}}\right), \tag{11.4}$$

even if (11.3) holds. This claim may seem strange at first, but its truth is obvious once one realizes that the quantity $\mathrm{d}t\, L$ is not, in general, a constant along a DPM.

While the functional form of the integrands of both sides of the expression above are identical, their regions of integration, (t_1, t_2) versus $(\bar{t}_1, \bar{t}_2)$, are not identical unless $f(t_1) = f(t_2) = 0$. Evaluating these integrals over different endpoints will thus, in general, lead to a different result—a free particle is a notable exception. Reparameterizations are, thus, not strict variational symmetries unless one imposes the condition $f(t_1) = f(t_2) = 0$. This point will be central to our analysis and interpretation of reparameterization symmetry below.

Before giving our first-order analysis, let us derive some immediate consequences of the invariance (11.3). A sufficient condition for (11.3) is that the Lagrangian $L(q^i, \dot{q}^i)$ be homogeneous of degree 1 in the velocities $\dot{q}^i$; that is, by definition the Lagrangian being homogeneous of degree 1 in the velocities $\dot{q}^i$ means that:

$$L(q^i, \dot{q}^i) \to L(q^i, f^{-1}\dot{q}^i) = f^{-1}L(q^i, \dot{q}^i) \tag{11.5}$$

We then have that $\mathrm{d}t \to \frac{\mathrm{d}\bar{t}}{f}$ and $L(q^i, \dot{q}^i) \to L(q^i, f\frac{dq^i}{d\bar{t}}) = fL(q^i, \frac{dq^i}{d\bar{t}})$ so that $\mathrm{d}t\, L$ is invariant. As noted by Dirac (1950, 1964), application of Euler's homogenous function theorem implies that the condition (11.5) is equivalent to the condition:

$$\dot{q}^i \frac{\partial L}{\partial \dot{q}^i} - L = 0. \tag{11.6}$$

This is equivalent to the vanishing of the Hamiltonian function in the second-order formalism. Significantly, this condition is valid *off-shell*. The vanishing of the classical Hamiltonian is often considered the origin of the frozen-formalism problem. In addition to the vanishing of H, if we differentiate (11.6) with respect to $\dot{q}^i$ we obtain

$$\dot{q}^i W_{ij} = 0, \tag{11.7}$$

Which implies says that the velocities are in the kernel of the Hessian W_{ij}. This suggests that reparameterization invariance is associated with some underdetermination in the equations of motion. We will see in our first-order analysis that this under-determination is associated with a particular null vector of ω_L and involves the freedom to arbitrarily choose the time parameter along γ. We will also see, however, that there is an additional null vector of ω_L, due to the vanishing of H, which is *not* associated with any underdetermination in the equations of motion but instead defines the classical solutions. We will interpret the second null vector as the generator of evolution.

Let us now consider the narrower class of reparameterization symmetries defined in terms of the action. Here we are mostly following the treatment of Logan (1977, §8). The first step is to consider the two *Zermelo conditions*. This second condition is given by (11.6). The first is simply the time independence of the Lagrangian:

$$\frac{\partial L}{\partial t} = 0 \tag{11.8}$$

Together the first and second Zermelo conditions, (11.8) and (11.6), can be proved to be necessary and sufficient for the *reparameterization invariance of the action*:

$$S = \int_{t_1}^{t_2} \mathrm{d}t \, L\left(q^i(t), \frac{\mathrm{d}q^i(t)}{\mathrm{d}t}, t\right) = \int_{\tilde{t}_1}^{\tilde{t}_2} \mathrm{d}\tilde{t} \, L\left(\tilde{q}^i(\tilde{t}), \frac{\mathrm{d}\tilde{q}^i(\tilde{t})}{\mathrm{d}\tilde{t}}, \tilde{t}\right) \qquad (11.9)$$

where $\tilde{q}(\tilde{t}) \equiv q \cdot h^{-1}(\tilde{t})$ and the equivalence holds for arbitrary transformations $\tilde{t} = h(t)$ where $\frac{\mathrm{d}h}{\mathrm{d}t} > 0$.

In the context of reparameterization invariance of the actions, it is interesting to note that the generalized Bianchi identities associated with the invariance (11.9) are already directly implied by the Zermelo conditions; that is, it is straightforward to use the two Zermelo conditions to derive the further *Weierstrass condition* which takes the form:[1]

$$\dot{q}^i \alpha_i = 0 \qquad (11.10)$$

where α_i is the Euler–Lagrange derivative (Logan, 1977, p. 156). The Weierstrass condition is the generalized Bianchi identity that results from application of Noether's second theorem to a parameter invariant action (Logan, 1977, pp. 158–61).[2]

There is an important sense in which Noether's second theorem is trivial in this case: the Weierstrass condition follows from the reparameterization invariance of the action without the need to consider boundary conditions (or the adjoints of differential operators) as per the full derivation of generalized Bianchi identities as required in Noether's second theorem. This is perhaps the simplest formal argument that 'gauge identity' associated with temporal symmetries are of a very different kind to those found in the more general class of theories with irregular nomic structure.

Before proceeding, let us note that homogeneity of degree 1 in velocities is not a necessary condition for reparameterization invariance in local sense. Although homogeneity of degree 1 in velocity is necessary and sufficient for the specific form of temporal symmetry encoded in reparameterization invariance as expressed in Equation (11.9), it is neither a necessary nor sufficient condition for reparameterization invariance in the most general sense as per our definition (11.3). Most prominently the ADM action in $3 + 1$ general relativity is not homogeneous of degree 1 in velocity but is reparameterization invariant because the lapse and shift variables transform like velocities under temporal diffeomorphisms even though they are not, strictly speaking, velocities themselves. In order for H to vanish in this more general case, the quantities, N^α, in question must be such that $\frac{\partial L}{\partial N^\alpha} = 0$. The lapse and shift are examples of such variables in GR. In GR and theories like it, this condition arises as an on-shell condition so that H is seen to vanish only when this on-shell condition is satisfied. This issue will be the major topic of Volume II and we will return to it in brief in the final chapter of this volume.

11.2 Dynamical Redundancy and Evolution

The first step in our first-order geometric analysis is to return to the variation (8.44) and consider the tangential variations that we previously excluded from our treatment.

---

[1] Geometrically, this condition take the form $\mathfrak{L}_u S = 0$ once you realize that the first Zermelo condition reduces $u \propto X$ for time reparameterizations. The Weierstrass condition thus follows rather trivially from the geometric perspective.

[2] It may be of historical interest to note that Noether mentions this connection in her original paper. See Kosmann-Schwarzbach (2010) and Neuenschwander (2017).

Recall that the variation was decomposed as:

$$\delta S[\gamma; u] \equiv \mathcal{L}_u S[\gamma] = - \int_{t_1}^{t_2} \left(\iota_u \left(\iota_X \omega_L + \mathrm{d}H \right) \mathrm{d}t + H \mathcal{L}_u \mathrm{d}t \right) + \iota_u \theta_L \Big|_{t_1}^{t_2} . \tag{11.11}$$

This equation involved expressing the variational derivative of the action $S[\gamma]$ in terms of infinitesimal variations generated by a vector field $u \in T\Gamma$, where $\Gamma = TC$ is the velocity phase space equipped with local coordinates (q^i, v^i). Above we expressed such variations as $\delta q^i \to \mathcal{L}_u q^i$ and split these variations into those that lie tangent to γ, and therefore could correspond to variations of t and are parallel to X, and those that are transverse to γ, and therefore satisfy $\mathcal{L}_u t = 0$ and are spanned by the Y_i vectors above. For a general u, the variation of S_1 can be written in terms of the Lie drag of S by the vector field u. We also defined the exact two-form $\omega_L = \mathrm{d}\theta_L$ where

$$\theta_L = \frac{\partial L}{\partial v^i} \mathrm{d}q^i . \tag{11.12}$$

We would like to study the case where the action $S[\gamma]$ has symmetries that lie along an arbitrary trial curve γ with tangent vector X. This is the situation where the action $S[\gamma]$ is reparameterization invariant. Using $u = fX$ for some arbitrary positive function $f \in \mathbb{R}^+$ (note that X is not required to solve (8.45)), we find that the skew symmetry of ω_L (since it is a two-form) implies

$$\delta S_1[\gamma; fX] = - \int_\gamma \left(\mathcal{L}_X H f \mathrm{d}t + H \mathrm{d}f \right) + f \iota_X \theta_L \Big|_{t_1}^{t_2} = 0 . \tag{11.13}$$

but now for some *arbitrary* tangent vector field X and positive function f. Because $X = \frac{\mathrm{d}}{\mathrm{d}t}$, the variation $u = fX$ can be interpreted as the pullback by γ of a diffeomorphism on the domain of γ; that is, a temporal diffeomorphism or reparameterization.[3] The vanishing of the new term arising from $\mathcal{L}_u \mathrm{d}t = \mathrm{d}f \neq 0$ immediately implies

$$H = 0 . \tag{11.14}$$

Note that the vanishing of the Hamiltonian is an *off-shell* relation implied by the reparameterization invariance of S and holds along any trajectory in phase space. The theory, however, can of course be supplemented with additional initial value constraints of the form (8.53) at least for u_i transverse to X. We will see how to implement these below.

According to (11.13), ω_L is not required to have null directions off-shell as is the case when u_i is transverse to X. Instead, ω_L need only be a two-form. The vanishing of H and the Hamilton equations (8.45) resulting from transverse variations of S, however, do imply that the classical solutions $X = X_{\mathrm{ev}}$ themselves must be null directions of ω_L. Since ω_L is fixed in terms of the Lagrangian of the theory, it is a straightforward (though perhaps laborious) mathematical exercise to find its kernel. When $H = 0$ and if there are no additional symmetries-over-histories of S, then one can easily identify

---

[3] Recall that, in our abuse of notation, $\mathrm{d}t$ is the differential induced by the embedding of γ, and therefore varies as γ is varied.

this direction X_{ev} as the tangent to the classical solutions. One can then use this null direction to construct a *Hamiltonian constraint* function $\mathcal{H}$ on the canonical extension of the image of the Legendre transform with symplectic two-form ω_e as the solution to

$$\iota_{X_{\mathrm{ev}}}\omega_e = \mathrm{d}\mathcal{H} \tag{11.15}$$

and then obtain solutions to Hamilton's equations (8.45) by restricting this flow to $\mathcal{H} = 0$. Note that, while (11.15) looks formally like Hamilton's equations, its method of solution is different owing to the fact that $\mathcal{H} = 0$. In this case, it becomes an equation that says that X_{ev} is defined in terms of the kernel of ω_e (in a way we will make more precise below).

It is significant to note that Equation (11.15) only fixes $\mathcal{H}$ up to a constant $\mathcal{H} \to \mathcal{H}+E$. Such underdetermination is common to both the transverse variations considered earlier and the tangential variations we are considering now. However, in the transverse case, the relevant constant was fixed using the relations (8.53), which resulted from the off-shell vanishing of the boundary variation. Now, by contrast, the relevant boundary variation must vanish for *any trial curve* X since the variation is along the trial curve and the action is reparameterization invariant along any such curve.

For non-trivial θ_L, the vanishing of the boundary variation for any X is therefore only guaranteed if

$$f(t_1) = f(t_2) = 0\,. \tag{11.16}$$

Let us now demonstrate explicitly that the relevant boundary term produces no new interesting initial value constraints even along a classical solution. All we need to do is use the vanishing of H to rewrite the boundary variation as

$$f\iota_X\theta_L\Big|_{t_1}^{t_2} = f\iota_X L\mathrm{d}t\Big|_{t_1}^{t_2} = fL\Big|_{t_1}^{t_2}\,. \tag{11.17}$$

For a non-trivial theory, we cannot require L to remain constant along a curve. We thus see that no interesting canonical constraint can possibly arise from the vanishing of the boundary variation. The vanishing of f at the endpoints says that this infinitesimal diffeomorphism induced by u must preserve the boundary. We thus get no new constraint from the boundary variation to fix the constant offset of $\mathcal{H}$. The classical solutions are therefore given by the *full set* of null vectors of ω_L.

The fact that one gets no new constraint from the vanishing of the boundary variation follows from the nature of the symmetry transformation $u = fX$ and is essentially contained in the distinction between transverse and tangential variations. Recall that the distinction between transverse and tangential variations was rooted in the orientation of the transformation relative to the vector field X. Being transverse to X of course implies being tangential to the initial data surfaces and being tangential to X implies being transverse to the initial data surfaces. This means that transverse variations are tangential to initial data surfaces and thus 'reshuffle' the initial data. Such transformations can can therefore be used to define equivalence classes of initial data sets for solving (8.45).

By contrast, tangential variations, like reparameterizations, have a very different action: they change the parameterization along a particular solution, and therefore define different instants along a trajectory. The direct implication is then that *reparameterizations do not define equivalence classes of initial data* since they act transversely to these initial data surfaces. Note that this is not simply a choice: one must impose $f(t_1) = f(t_2) = 0$ in order for the reparameterizations generated by $u = fX$ to be genuine narrow symmetries-over-histories of the theory. It is these endpoint conditions that leave the constant offset of $\mathcal{H}$ arbitrary. This also implies the intuitively obvious statement that reparameterizations are clearly not narrow (or broad) symmetries-at-an-instant.

For those reparameterization invariant theories where the Lagrangian is homogeneous of degree 1 in the velocities, we have $v^i \frac{\partial L}{\partial v^i} = L$. Differentiating this with respect to v^j gives

$$v^j W_{ij} = 0 \,. \tag{11.18}$$

Thus, such theories will always have the velocity vector v^i in the kernel of W_{ij}.

More generally, by inspecting the general expansion of the u vectors, (8.67), in terms of components u^I that are independent at a time, we find that since there exists a $u = fX$ for a general reparameterization invariant theory, then there must be some linear combination such that $\bar{T}^i = \sum_I \lambda^{I\alpha} T_{(I)}{}^i{}_\alpha = v^i$ for some velocity phase space functions $\lambda^{I\alpha}$. This tells us that we can generally expect the velocity vector v^i to be in the kernel of W_{ij}. In terms of this particular linear combination, the null direction $\bar{T}^i = v^i$ implies two linearly independent null directions

$$\bar{u}^0 = v^i \frac{\partial}{\partial q^i} + \dot{v}^i \frac{\partial}{\partial v^i} \qquad\qquad \bar{u}^1 = v^i \frac{\partial}{\partial v^i} \,. \tag{11.19}$$

The first of these is simply X (by construction) while the second reflects the homogeneity of degree 1 of the Lagrangian in terms of the velocities. It is the latter that reflects the reparameterization invariance of the theory. Significantly, such narrow symmetries-over-histories *do not lead to initial value constraints* because the velocities are already varied freely on the boundary; that is, $\iota_{\bar{u}^1} \theta_L = 0$ automatically.

We can use (11.17) to isolate the true generator of evolution (i.e. the tangential component of u) from the transverse components. More specifically, if we define the change of basis

$$\bar{u}^I_\alpha = \sum_{J,\beta} \lambda^{I\alpha\beta}_J u^J_\beta \tag{11.20}$$

such that

$$\iota_{\bar{u}^I_\alpha} \theta_L = \delta^I_0 \delta^1_\alpha \,, \tag{11.21}$$

then we can simply define $\bar{u}^0_1$ as the generator of evolution. That such a non-trivial condition exists for the generator of evolution is essential to our argument that evolution can be clearly separated from symmetry in reparameterization invariant theories.

The evolution generator $\bar{u}^0_1$ has an interesting geometric interpretation in the image of the Legendre transform. If we assume that all degeneracies due to the transverse components have been eliminated (and restricting to the image of the Legendre transform eliminates the degeneracy due to $\bar{u}^1_1$), then $d\theta_L$ has only a one-dimensional kernel spanned by $\bar{u}_1^0$ with $\iota_{\bar{u}_1^0}\theta_L = 1$. This means that $(\mathrm{d}\theta_L)^{\wedge(N-1)}$ is a contact form on this space. Moreover, the normalization $\iota_{\bar{u}_1^0}\theta_L = 1$ implies that $\bar{u}_1^0$ is simply the Reeb field on this contact space. The dynamical evolution can then be interpreted as the Reeb flow in the image of the Legendre transform. Finally, the condition $\iota_{\bar{u}_1^0}\theta_L = 1$ can be used to identify a conservation law for the Reeb flow. In this case, this simply expresses the usual Hamiltonian constraint. The Reeb flow can be calculated in at least two different ways: (1) symplectify the contact manifold, write the Reeb vector as a Hamiltonian vector field, and restrict this Hamiltonian vector field to the original contact manifold; (2) compute u_0^1 on TC directly using the expansion (8.67) and (8.71) and, if necessary, project this flow onto phase space. We will return to these constructions in the context of an explicit model in §13.1.

Identification of this criterion, which to our knowledge has never been achieved before, provides an explicit refutation of the vivid yet fundamentally mistaken claim of Henneaux and Teitelboim (1992) that in the context of reparameterization invariant theories 'motion is the unfolding of a gauge transformation' (p. 103). The putative confusion of dynamical redundancy with evolution that results from the Hamiltonian of reparameterization invariant theories being a first-class constraint is what we have earlier called the *pseudo-problem of time*. Our approach dissolves the pseudo problem in terms of an unambiguous identification of the symmetry and evolution generators.

11.3 Temporal Leibniz Shifts

Let us now briefly consider the case of time reparameterizations which are homogenous in space *and* time, that is, the 'temporal Leibniz shifts' which are given by homogenous tangential variations. These transformations are an important test case for our analysis since their formal treatment is not subject to dispute. Furthermore, since the transformations in question have a straightforward connection to conserved charges via Noether's first theorem this is also a good test case for our treatment of boundary terms.

In the case of a homogenous variation tangent to X, we have $u = a_0 X$ for some constant a_0. We can then use the variation of S in the form (11.11). However, we now have that a new term,

$$H\mathfrak{L}_u \mathrm{d}t = H\mathrm{d}a_0 = 0\,, \tag{11.22}$$

is vanishing since the time reparameterization is simply a global shift by a_0.

The significance of the globally of the transformation is therefore that we *do not* get a condition of the form $H = 0$. Global as opposed to local time translations do not lead to a vanishing Hamiltonian. This is precisely as we would expect since invariance under homogenous transformations should not generate canonical constraints.

We then have that since the Hamiltonian is not zero for the case of global time reparameterizations, Hamiltonian's equations no longer imply that u is a null direction of ω_L. However, the term

$$\iota_X \iota_u \omega_L = a_0 \iota_X \iota_X \omega_L = 0\,, \tag{11.23}$$

still vanishes automatically for all off-shell trajectories. The vanishing of the integrand of (11.11) then implies

$$\mathfrak{L}_X H = 0 \,. \tag{11.24}$$

This combined with Hamilton's equations implies, through the symplectic Noether theorem, that the Hamiltonian is a constant of motion.

As before, the boundary terms cannot be set to zero for a non-trivial theory unless $u(t_1) = u(t_2) = 0$. We therefore find no initial value constraint and, therefore, no constraint on the value of the (conserved) total energy. This reproduces Noether's first theorem for global symmetries along the dynamical trajectory (i.e. time translations).

Note that the argument for the non-vanishing of E in Noether's first theorem is identical to the argument that general global reparameterizations do not imply independent initial value constraints, and that the off-set of the generator of evolution is not fixed (although the Hamiltonian itself is still vanishing). Thus our formalism recovers the standard results and intuitions regarding temporal Leibniz shifts.

11.4   Chapter Summary

A theory is reparameterization invariant when its equations of motion are invariant under the temporal relabelling symmetry given by locally rescaling a monotonic and smooth time function. This understanding of reparameterization invariance is more general than that provided by study of the action since the properties of the equations of motion generated by a variational principle depend only on the local terms, and not the boundary terms, of that variation principle. In our treatment we have outlined the connections between reparameterization invariance, various conditions on the Lagrangian and Hamiltonian functions, and the relevant generalized Bianchi identities. We have also noted that there is an important sense in which Noether's second theorem is trivial in this case. We then introduced a novel first-order geometric analysis of reparameterization invariance in terms of the study of tangential transformations. This has allowed us to unambiguously differentiate between the objects responsible for generating evolution and those that generate degeneracy along tangential directions.

12

Reparameterization Invariant Dynamics

For the subset of Jacobi theories where the metric is conformally flat there is always a dynamically privileged ephemeris time parameter that can be defined along a classical curve in terms of the fixed total energy of the system. There are available two importantly different interpretations of the modal status of the total energy constant. Under the first, total energy is interpreted as a constant of nature. The value of the total energy is thus the same for all dynamically possible models with the same token of the constitutive structure. Total energy is therefore a contingently fixed structure of the theory. Under the second, total energy is interpreted as a constant of motion. The value of the total energy is then a free function of the initial data and thus can vary between dynamically possible models with the same token of the constitutive structure.

12.1   Jacobi Actions as Geodesic Principles on Configuration Space

Let us define the class of *Jacobi Theories* as theories in which the nomic structure is expressed via a geodesic principle on configuration space, $\mathcal{C}$. Such a definition requires the existence of a connection, A, over $\mathcal{C}$, that can be used to define the geodesic and thus in turn a tangent bundle $T\mathcal{C}$ over configuration space together with a notion of horizontality on this bundle.

The simplest way to implement a Jacobi action is to define a metric g on $\mathcal{C}$; that is, a positive definite symmetric (and therefore invertible) bilinear map $g : T\mathcal{C} \times T\mathcal{C} \to \mathbb{R}^+$. The metric g then defines a unique horizontal lift, $A \equiv \Gamma$, through the requirement that it covariantly conserve g along an arbitrary direction in $\mathcal{C}$. The geodesics are then defined as those curves whose tangents are parallel transported by Γ.

The constitutive structures of a Jacobi theory then consist of a spatial manifold Σ, a 'temporal' manifold $I \subset \mathbb{R}$, and matter fields, q^i, which can be used in various ways to construct the configuration space $\mathcal{C}$. From this, we can define the tangent bundle $T\mathcal{C}$ and the metric $g : T\mathcal{C} \times T\mathcal{C} \to \mathbb{R}$. Finally, we define the trial curves $\gamma : I \to \mathcal{C}$.

The nomic structure can be implemented by a variational principle with an action of the form

$$S_{\mathrm{Jac}}[\gamma; q_1, q_2] = \int_{q_1}^{q_2} \mathrm{d}t \sqrt{g_{ab}\dot{q}^a \dot{q}^b}\,, \qquad (12.1)$$

where t is an arbitrary parameter along γ and $\dot{q} \in \mathcal{C}$. A geodesic is then the curve of minimal length S with endpoints (q_1, q_2).

It is straightforward to see that S_{Jac} is reparameterization invariant owing to the fact that it is homogeneous of degree 1 in the velocities. This action principle can be used to generate a set of local equations in terms of the Euler–Lagrange equations. These reproduce the general form of the geodesic equation:

$$\ddot{q}^a + \Gamma^a{}_{bc}\dot{q}^b\dot{q}^c = \kappa u^a \tag{12.2}$$

in terms of the metric compatible connection

$$\Gamma^a{}_{bc} = \frac{1}{2}g^{ad}\left(\partial_b g_{dc} + \partial_c g_{db} - \partial_d g_{bc}\right) \tag{12.3}$$

and $\kappa = \frac{\mathrm{d}}{\mathrm{d}t}\left(\log L\right)$, where $L = \sqrt{g_{ab}\dot{q}^a\dot{q}^b}$.[1]

In most cases it will prove insightful to define the kinetic energy in terms of the flat metric $\tilde{g}$ whose metric-compatible connection $\tilde{\Gamma}$ (computed by inserting $\tilde{g}$ into (12.3)) has zero curvature. We then have that the conformal mode ϕ of g such that $g = e^\phi \tilde{g}$ can be parameterized by the potential energy function $V = E - \frac{1}{2}e^\phi$ of the theory while the bilinear structure of the corresponding flat metric $\tilde{g}$ can be used to define a notion of kinetic energy through

$$T = \frac{1}{2}\tilde{g}(\dot{q}^a, \dot{q}^b)\,, \tag{12.4}$$

with $\dot{q}^a \in T\mathcal{C}$. The potential energy and total energy E are then given by solving the relation

$$g = 2(E - V)\tilde{g}\,. \tag{12.5}$$

The constant factor E, which we will later interpret as the total energy, has been inserted in order to ensure that e^ϕ is non-negative along all geodesics for a given potential function V.

The potential energy function $V : \mathcal{C} \to \mathbb{R}$ may also contain additional coupling constants k_i, which represent the coefficients of a power series expansion of $V(q^i)$. The zeroth term of this expansion can be absorbed into the total energy E. Unless $V = 0$, changing E will in general lead to different metrics g and therefore different geodesics. We can then label DPMs by E and consider the two different nomic principles which: (1) fix the value of E and (2) allow E to take any value. We will return to this crucial distinction in what follows.

For a conformally flat metric the Jacobi action principle becomes:

$$S_{\text{Jac1}}[\gamma; q_1, q_2] = \int_{q_1}^{q_2} \mathrm{d}t\sqrt{2(E - V)T}\,. \tag{12.6}$$

The Euler–Lagrange equations in terms of these quantities are easily computed. Using the fact that the flat metric $\tilde{g}$ is time-independent, we can write these equations in the familiar Newtonian form

---

[1] Note that for an affine parameter, $\kappa = 0$ and we obtain the more familiar form of the geodesic equation: $\dot{q}^a{}_{;b}\dot{q}^b = 0$.

$$\frac{\mathrm{d}^2 q_a}{\mathrm{d}\tau^2} = -\frac{\partial V}{\partial q^a},$$

(12.7)

where $q_a = \tilde{g}_{ab} q^b$ and we importantly have

$$\mathrm{d}\tau = \sqrt{\frac{T}{E - V}}\,\mathrm{d}t.$$

(12.8)

The convenient change of variables from an arbitrary time parameter t to τ suggests the existence of a dynamically privileged time parameter τ. Using $\frac{\mathrm{d}\tau}{\mathrm{d}t} = 1$ we find that in this time parameterization the quantity $T + V$ is conserved:

$$T + V = E.$$

(12.9)

For this expression to hold, the velocities in T must be computed using τ derivatives. The existence of this time parameter implies that, provided the Euler–Lagrange equations are integrable, there will always be a choice of time parameter where the total energy, $T + V$, is a conserved quantity. Moreover, it is easy to see that the definition of τ is itself reparameterization invariant. This means that the preferred parameter can be computed using only the geometric properties of the geodesics.[2]

12.2   Total Energy as a Constant of Motion

This sub-section will introduce one of the core aspects of the approach to the problem of time advocated in this book: the reinterpretation of total energy as a constant of motion. The starting point for this crucial move is to differentiate two different approaches to the formulation of the nomic structure of a Jacobi action principle.

Within a *Boundary Value Problem* (BVP) approach we consider histories 'as-a-whole' in a manner analogous to the idea of symmetries-over-histories that we discussed in Part II. DPMs are given by γ that correspond to the geodesics of $(\mathcal{C}, g)$. In a BVP approach one fixes the endpoints (q_1, q_2) of the image of γ, which both live in $\mathcal{C}$, and identifies a DPM as a curve of minimum length with endpoints (q_1, q_2). Within an *Initial Value Problem* (IVP) approach we consider an 'at-an-instant' nomic principle and understand a curve as being inferred given some initial data given 'at-an-instant'. DPMs are again given by γ that correspond to the geodesics of $(\mathcal{C}, g)$. In an IVP approach one fixes initial data in a particular parameterization $(q_1, \dot{q}_1)$ and then identifies the DPMs as the solution to the geodesic equation corresponding to the initial data in question. No endpoint need be specified, and thus we take the complete DPM to correspond to the maximal extension of the geodesic on $(\mathcal{C}, g)$ starting at q_1 and with initial velocity $\dot{q}_1$.

---

[2] Note that the preferred time τ in which $V + T$ is conserved is manifestly *not* an affine parameter for the geodesics of g. Instead of requiring $V + T = E$, an affine parameterization would require that the Lagrangian $L = 1$. This leads to the uninformative condition $2T = \frac{1}{V - E}$. A more illustrative way to understand an affine parameterization is to recognize that such a parameterization is given in terms of the on-shell action, which is the arc length along the geodesic. But the on-shell action is simply the Hamilton–Jacobi function $S(q_1, q_2)$ treated as a function of the endpoints of the action. While this is certainly an interesting quantity geometrically, it is cumbersome to compute explicitly for non-trivial theories.

The problem now is to construct the space of DPMs for the BVP and IVP. This is a standard geometric problem since it is equivalent to finding the space of geodesics on some differential manifold. We will assume that g has been chosen to be non-degenerate everywhere in $\mathcal{C}$. In the BVP formulation existence of solutions is guaranteed for every choice of distinct initial and final endpoints.[3] Under the assumption of uniqueness, the space of DPMs is given by the geodesics labelled by the space of distinct pairs of points (q_1, q_2) in $\mathcal{C}$. In the IVP formulation, the space of DPMs is given by curves chosen to be the maximal extension of the solution to the geodesic equation for the initial data $(q_1, \dot{q}_1)$. For non-degenerate g, the geodesic equation is usually locally integrable; that is, the maximal extension of the solution exists and is unique.

In cases where the solutions fail to be integrable at some point q_s in the evolution, the manifold $(\mathcal{C}, g)$ will be *geodesically incomplete* and q_s will be a singular point of $(\mathcal{C}, g)$. The maximal extensions starting with initial data $(q_1, \dot{q}_1)$ can nevertheless still be found unless q_1 itself is a singular point. We assume that such choices can always be avoided. This is *not* a very restrictive assumption since singular points can usually be readily identified once the geometry of $(\mathcal{C}, g)$ has been specified and they almost always represent physically pathological states of the theory. One example of a singular point is the total collision in the Newtonian N-body problem with zero energy and zero angular momentum.

Let us now compare the action of the reparameterizations in BVP and IVP. In both cases, the reparameterizations act at *every point* t along a solution by rescaling the norm $|\dot{q}(t)|$ for all $t \in I$. However, since this action leaves all points $q(t), \forall t \in I$ invariant, the BVP in terms of (q_1, q_2) is invariant under reparameterization. In the IVP, which requires a specification of the norm, $|\dot{q}_1|$ is transformed under reparameterization. In this way, we see that the initial value of $|\dot{q}_1|$ is made redundant by the reparameterization symmetry while the BVP is unaffected by this symmetry. Nevertheless, we will now see that there is a sense in which $|\dot{q}_1|$ can be thought to convey physically relevant information once we take into account information contained in the DPMs themselves.

To see how this is possible, we will restrict to Jacobi theories where the metric g is conformally flat. Then, the reparameterization invariant definition

$$d\tau = \sqrt{\frac{\frac{1}{2}\tilde{g}_{ab}\dot{q}^a\dot{q}^b}{E - V(q)}}\, dt \qquad (12.10)$$

tells us that there is always a privileged *ephemeris time* parameter τ such that

$$\tfrac{1}{2}\tilde{g}_{ab}q'^a q'^b + V(q) = E\,, \qquad (12.11)$$

where primes indicate differentiation with respect to τ.

---

[3] Uniqueness will depend on the metric and topological properties of the manifold $(\mathcal{C}, g)$. In general, non-unique solution will only exist when there are symmetries present in the geometry of $(\mathcal{C}, g)$. These can be studied on a case-by-case basis. We will simply assume for simplicity that the solutions are unique or that there is a simple way (perhaps using some physical principle) of choosing a unique representative for each solution. Note that this is a non-trivial assumption. But our intention is to focus on general aspects of reparameterization symmetry so ignoring global issues regarding the solution space, which usually depend on the individual nature of the problem in consideration, shouldn't affect the features we want to illustrate.

The Equation (12.11) for ephemeris time is central to our analysis and can be read in two importantly different ways. In the first reading, it can be seen as an equation that fixes the dynamically privileged ephemeris time parameter τ along a classical curve in terms of the fixed total energy of the system E; that is, knowing that a particular dynamical curve is a geodesic tells you that there will be some preferred time parameter τ where the quantity $T + V$ is conserved, and E is that fixed quantity. This is how the ephemeris time has been traditionally interpreted in the context of Jacobi actions.[4] The potential for conventionality within the specification of the value of the total energy in Jacobi theory is noted, however, by Sklar (2004), although the implications for ephemeris time are not drawn.

The alternative reading, which we shall advocate, is to first note that (12.11) can be rewritten as

$$|q'| = 2(E - V).\tag{12.12}$$

In this form, we see that the conservation of energy identity is relating the norm of the velocity in a particular parameterization to the total energy E. Thus, we can also read (12.11) as a way of fixing E in terms of the norm of the initial velocity. What makes this procedure well-defined is the knowledge that $|q'|$ is the norm of the velocity in the particular parameterization where the quantity $T + V$ is preserved along the dynamical evolution. Thus, what we are doing is *not* defining the energy in a particular privileged time parameterization, but rather defining the time parameterization in a particular *contingent* value of the total energy.

The crucial point is that the norm of the initial velocity vector can be used to fix the total energy E provided we specify that this norm is computed in the dynamically preferred ephemeris time parameterization. Since this parameterization is fixed by the dynamical condition $T + V = E$ and not by some externally specified condition, the whole procedure is reparameterization invariant. What is necessary for this procedure to work is that the total energy E be thought of as a freely specifiable parameter given as a function of the norm of the initial velocity in a dynamically privileged ephemeris time parameterization. We will further elucidate this procedure and how it can be implemented in practical terms on phase space in §13.2.

We now have available two different readings of the conservation of energy equation (12.11). These in turn lead to two different interpretations of the *modal status* of total energy constant:

> **Energy as a Constant of Nature.** E is the same for all DPMs with the same token of the constitutive structure. Total energy is therefore a contingently fixed structure of the theory. The ephemeris time is found by rescaling $|q'|$ appropriately to satisfy (12.12)). The initial velocity $|\dot{q}_1|$ is seen as redundant and E is fixed to a single value. The space of DPMs for BVP is $\mathcal{C} \times \mathcal{C}$ and for IVP is $T\mathcal{C}/\mathbb{R}$.

> **Energy as a Constant of Motion.** E is a free function of the initial data and thus can vary between DPMs with the same token of the constitutive structure. DPMs are parameterized by the free value of the total energy E. For the BVP this means that the space of DPMs is augmented to $\mathcal{C} \times \mathcal{C} \times \mathbb{R}$, which is parameterized

---

[4] See in particular Barbour and Bertotti 1982; Barbour and Foster 2008; Barbour 2009*b*; Pooley 2001; Pooley 2004.

by the triple (q_1, q_2, E). For the IVP, Equation (12.12) can be used to fix the energy E for arbitrary values of $|\dot{q}|$ using the knowledge that $t = \tau$. Thus, the space of DPMs is simply $T\mathcal{C}$; that is, the space of *unrestricted* initial data on the tangent bundle.[5]

The significance of this distinction is best understood by simultaneously looking back to the comparison with the Newtonian formulation of mechanics and looking forward towards quantization and the problem of time.

In the first regard, notice that the total energy as a constant of motion approach leads to a IVP formulation of the space of DPMs which is isomorphic to the space of DPMs in NM. The two spaces do not, however, have the same *representational capacities*, in particular with regard to the temporal structures that live within them. We will return to this comparison in §13.3.3. The comparison between Jacobi action principle under the energy as a constant of nature interpretation and NM is a more stark one. This is because *the stipulation of a unique energy value selects a space of DPMs that does not admit Galilean boosts as a broad symmetry*. Here the contrast is with NM in general terms irrespective whether one wishes to treat models related by Galilean boosts to be dynamically identical and thus a narrow symmetry. Under the energy as a constant of nature approach Galilean boosts fail to be endomorphisms on the space of DPMs and thus cannot be interpreted as symmetries in either the broad or narrow sense.

This is, of course, not to say that in general terms Jacobi actions under the energy as a constant of nature approach are inconsistent with applying a best-matching procedure and reformulating a theory with a Jacobi action such that Galilean boosts kinematically eliminated along the lines discussed in §8.5.1. The energy appearing in this context is best interpreted as the energy in a centre of mass frame, which is manifestly Galilean invariant. However, the capacity to explicitly represent individual models related by boosts within the space of DPMs clearly does depend on the ability to consider different values of the total energy for the same token of the constitute structure. As such, the energy as a constant of motion approach offers greater representational capacities than the energy as a constant of nature approach precisely because it leads to a space of DPMs which is isomorphic to the space of DPMs in NM. We take this to be a strong argument in favour of the energy as a constant of motion approach.

12.3   Generalized Hamilton–Jacobi Formalism

In this section we will provide an analysis of the integrals of motion of a totally constrained system in terms of a generalized Hamilton–Jacobi formalism. Our treatment follows that provided in Gryb and Thebault (2015).[6] We begin by presenting a generalized Hamilton–Jacobi formalism for constrained classical systems. This approach is closely related to those found in the literature (Rovelli, 2001; Rovelli, 2004; Souriau,

---

[5] Here and below we are of course ignoring the parameterization of the space of DPMs by the absolute charges as discussed at length in §10.2.

[6] Note this is the preprint archive version rather than the published version. There are only minor difference between the two versions.

1997). This formalism will be applied towards a concise rendering of the influential *Complete Observables* formalism in the following section.

Consider a totally constrained Hamiltonian theory on a phase space Γ, coordinatized in some chart by (q, p), defined by the first-class system of r constraints $C_\alpha(q, p) \approx 0$, for $\alpha = 1, \ldots r$. The canonical action reads

$$S = \int \left[p \cdot \dot{q} - \lambda^\alpha C_\alpha(q, p) \right] \mathrm{d}t. \tag{12.13}$$

For the moment, we use an abstract index-free notation for phase space variables (q, p) so that the dot product represents an abstract inner product on Γ. Thus, our considerations will generally apply to the infinite-dimensional case. Integration is over the arbitrary parameter t and over-dots represent t-derivatives. The symplectic term requires Γ to be equipped with the symplectic potential $\theta = p \wedge \mathrm{d}q$ which defines a symplectic two-form $\omega = \mathrm{d}\theta$ (where exterior derivatives and wedge products are defined on Γ). In addition, we will assume that Γ is equipped with a canonical Poisson structure such that $\{q, p\} = \mathbb{1}$ in some chart and all other Poisson brackets are zero.

Our goal is to find a canonical transformation that parameterizes the flow of the constraints C_α locally on Γ and allows us, in particular, to restrict this flow to the constraint surface defined by $C_\alpha \approx 0$. To do this, consider the modified action:

$$S_{\mathrm{e}} = \int \left[p \cdot \dot{q} + \dot{\phi}^\alpha \cdot \mathcal{E}_\alpha - \lambda^\alpha \left(\mathcal{E}_\alpha - C_\alpha(q, p) \right) \right] \mathrm{d}t \tag{12.14}$$

defined on the *extended* phase space $\Gamma(q, p) \to \Gamma_{\mathrm{e}}(q, p; \phi^\alpha, \mathcal{E}_\alpha)$ coordinatized in some finite patch by the additional variables $(\phi^\alpha, \mathcal{E}_\alpha)$. The symplectic potential on Γ_{e} is now

$$\theta_{\mathrm{e}} = p \wedge \mathrm{d}q + \mathcal{E}_\alpha \wedge \mathrm{d}\phi^\alpha. \tag{12.15}$$

We additionally assume the Poisson structure $\{\phi^\alpha, \mathcal{E}_\beta\} = \delta^\alpha_\beta$ so that ϕ^α and $\mathcal{E}_\alpha$ are canonically conjugate. Eventually, we will use ϕ^α to locally parameterize a canonical transformation along the orbits of C_α. The momenta $\mathcal{E}_\alpha$ are easily seen to be constants of motion, since their equations of motion imply

$$\dot{\mathcal{E}}_\alpha = \{\mathcal{E}_\alpha, H\} = 0. \tag{12.16}$$

For the special initial condition $\mathcal{E} = 0$, we see that the extended theory defined by S_{e} is classically equivalent to the original theory defined by S.

In order for the constraints $\mathcal{E}_\alpha - C_\alpha(q, p) \approx 0$ to form a first-class surface on the extended phase space Γ_{e}, the functions $C(q, p)$ must be Abelian (i.e. $\{C_\alpha, C_\beta\} = 0$). For simplicity, we will assume that the $C(q, p)$s have already been *Abelianized*, which is always possible locally on Γ due to Darboux's theorem (Abraham and Marsden, 1978, p. 175).

We are now in a position to define a canonical transformation that parameterizes the flow of the constraints. This can be achieved by requiring that the new coordinates $(P, Q; \Phi^\alpha, E_\alpha)$ have zero flow under the transformed C_α, so that they are analogous to the 'initial data' of standard Hamilton–Jacobi theory. For the old coordinates (q, p), we additionally require that the momenta $\mathcal{E}_\alpha$ are constrained to be equal to the $C_\alpha(q, p)$.

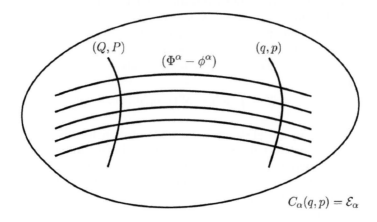

Fig. 12.1: The canonical transformation gives the general coordinates (q, p) lying along the integral curves of the constraints, $C_\alpha(q, p) = \mathcal{E}_\alpha$, in terms of the 'initial data' (Q, P) and the parameters ϕ^α.

The new symplectic potential $\theta'_e = P \wedge dQ + E_\alpha \wedge d\Phi^\alpha$ must differ from the old one by an exact form dS. This ensures that the symplectic two-form $\omega_e = d\theta_e = d\theta'_e$ is invariant. Implementing the above requirements, we obtain

$$dS(q, Q; \phi^\alpha, \Phi_\alpha) = \theta_e - \theta'_e$$
$$= p \wedge dq + \mathcal{E}_\alpha \wedge d\phi^\alpha - P \wedge dQ - E_\alpha \wedge d\Phi^\alpha$$
$$= p \wedge dq + C_\alpha(p, q) \wedge d\phi^\alpha - P \wedge dQ, \qquad (12.17)$$

where in the last line we set $\mathcal{E}_\alpha = C_\alpha(q, p)$ and $E_\alpha = 0$. The reason for setting $E_\alpha = 0$ is because E_α represents the generator of flow on the transformed coordinates (Q, P), which we want to vanish. This is completely analogous to requiring that the transformed Hamiltonian vanish in standard Hamilton–Jacobi theory.

We see from this that $S(q, Q; \phi^\alpha, \Phi_\alpha)$ is a type-1 generating functional for a canonical transformation taking the lower-case coordinates to upper-case ones subject to our requirements. Figure 12.1 shows how to interpret it in terms of a flow along the integral curves of $C_\alpha(q, p)$.[7]

It is convenient to convert S to a type-2 generating functional $F(q, P; \phi^\alpha, \Phi_\alpha)$ in the (q, p) coordinates by adding the boundary term $Q \wedge P$. Thus,

$$dF(q, P; \phi^\alpha, \Phi_\alpha) = p \wedge dq + C_\alpha(q, p) \wedge d\phi^\alpha + Q \wedge dP. \qquad (12.18)$$

From the above, we find the equations defining the canonical transformation we are looking for:

---

[7] This figure has been previously used in the published article Gryb and Thébault (2016*a*).

$$\frac{\partial F}{\partial q} = p \qquad\qquad \frac{\partial F}{\partial \phi^\alpha} = C_\alpha(q, p) \qquad\qquad (12.19a)$$

$$\frac{\partial F}{\partial P} = Q \qquad\qquad \frac{\partial F}{\partial \Phi_\alpha} = 0 . \qquad\qquad (12.19b)$$

The first equation in (12.19a) is just the definition of the momenta p. Combined with the second equation in (12.19a), we get the generalized Hamilton–Jacobi relations

$$\frac{\partial F}{\partial \phi^\alpha} = C_\alpha \left(q, \frac{\partial F}{\partial q} \right) . \qquad\qquad (12.20)$$

The second equation of (12.19b) simply says that since (Q, P) are preserved along the flow of the transformed constraints there is no parameter Φ^α to parameterize their flow. The first equation of (12.19b) becomes the Hamilton–Jacobi equation of motion.

We can reduce these equations using the following separation Ansatz for F

$$F(q, P; \phi^\alpha, \mathcal{E}_\alpha) = W(q, P) + \mathcal{E}_\alpha \phi^\alpha , \qquad\qquad (12.21)$$

where we have slightly abused notation using the same symbol for the separations constants $\mathcal{E}_\alpha$ as the canonical coordinates $\mathcal{E}^\alpha$. This abuse is forgivable since they are equal on-shell. Our Ansatz converts the Hamilton–Jacobi relations (12.20) to the reduced set of equations

$$\mathcal{E}_\alpha = C_\alpha \left(q, \frac{\partial W}{\partial q} \right) . \qquad\qquad (12.22)$$

Inserting this back into (12.21), we obtain

$$F(q, P, \phi^\alpha) = W(q, P) + \phi^\alpha C_\alpha \left(q, \frac{\partial W}{\partial q} \right) , \qquad\qquad (12.23)$$

which gives the generator of the canonical transformation we are looking for parameterized, as advertised, by ϕ^α.

12.4 Integrals of Motion and Complete Observables

The generalized Hamilton–Jacobi formalism that we have just developed allows an explicit and formally insightful characterization of the popular partial and complete observables approach to the classical mechanical problem of time pioneered by Rovelli (2002, 2004, 2007) and formalized by Dittrich (2006, 2007).[8] This approach has important similarities and differences to the positive approach developed in the following chapter and will benefit from a concise and perspicuous presentation that makes clear the deep connection between integrals of motion and complete observables.

---

[8] For detailed overview see Thiemann 2007; Tambornino 2012. Important developments of the approach include Gambini and Porto 2001; Gambini, Porto, Pullin and Torterolo 2009 and Bojowald, Höhn and Tsobanjan 2011*a*; Bojowald, Höhn and Tsobanjan 2011*b*; Höhn 2019. Critical responses include Kuchař 1991; Kuchař 1992; Kuchař 1999; Dittrich, Höhn, Koslowski and Nelson 2017. For a review of the various notions of observable, that includes discussion of the limitations of the partial and complete observables approach, see Anderson 2014; Anderson 2017. For philosophical analysis of the ontological implications of the partial and complete observables approach an excellent extended discussion can be found in Rickles (2007, pp. 161–171). For a further overview see Thébault (2021*b*), which contains further references and discussion.

First, consider the final form of the Hamilton–Jacobi equation of motion that we derived in the previous section.

$$Q = \left. \frac{\partial F}{\partial P} \right|_{C_\alpha \left(q, \frac{\partial W}{\partial q} \right) = \mathcal{E}_\alpha}, \qquad (12.24)$$

This equation should be read as an equation for q in terms of the 'initial data' (or reference section), (Q, P), the constants of motion, $\mathcal{E}_\alpha$, and the parameters ϕ^α. In this form, it represents an integral of motion for the $q(\phi^\alpha)$ in terms of some initial data.

An alternative formal expression can be given for the integrals of motion of the system in terms of the flow parameters ϕ^α by directly integrating Hamilton's equations for the extended action in (12.14). These are the flow equations of the Hamilton vector field of the Hamiltonian. For the extended variables, they are

$$\dot{\phi}^\alpha = \{\phi^\alpha, H\} = \lambda^\alpha \qquad\qquad \dot{\mathcal{E}}_\alpha = \{\mathcal{E}_\alpha, H\} = 0. \qquad (12.25)$$

Using these, we can rewrite the flow equations for (q, p) as

$$\dot{q} = \{q, H\} = \lambda^\alpha \{q, C_\alpha\} = \dot{\phi}^\alpha \{q, C_\alpha\} \qquad (12.26)$$

$$\dot{p} = \{p, H\} = \lambda^\alpha \{p, C_\alpha\} = \dot{\phi}^\alpha \{p, C_\alpha\}, \qquad (12.27)$$

or

$$\delta q = \delta\phi^\alpha \{q, C_\alpha\} \qquad\qquad \delta p = \delta\phi^\alpha \{p, C_\alpha\}. \qquad (12.28)$$

We can exponentiate the Lie flow defined by these expressions as:

$$q(\phi^\alpha) = \exp\left[\phi^\alpha \mathcal{L}_{\mathcal{X}_{C_\alpha}}\right] \cdot Q = \sum_{n=0}^{n=\infty} \frac{(\phi^\alpha)^n}{n!} \{C_\alpha, Q\}_n \qquad (12.29a)$$

$$p(\phi^\alpha) = \exp\left[\phi^\alpha \mathcal{L}_{\mathcal{X}_{C_\alpha}}\right] \cdot P = \sum_{n=0}^{n=\infty} \frac{(\phi^\alpha)^n}{n!} \{C_\alpha, P\}_n. \qquad (12.29b)$$

Where $\{C_\alpha, Q\}_0 := Q$ and $\{C_\alpha, Q\}_{n+1} = \{C_\alpha, \{C_\alpha, Q\}_n\}$. Here Q and P are required to satisfy the initial value constraint $C_\alpha(Q, P) = \mathcal{E}_\alpha$. This form of the integrals of motion is equivalent to solving (12.24), since both result from the solution of the same variational principle. The form (12.29) will later allow us to relate our formalism to the complete and partial observables approach.

There are then two options[9] for the construction of observables:

i) Retain the parameters ϕ^α and solve (12.24) (or equivalently (12.29)) for the integrals of motion, q, as a functions of ϕ^α and some reference section as specified by (Q, P). This will give different representations of the observables in different gauges labelled by the ϕ^α.

---

[9] For explicit discussion of these two options see Rovelli (2004, pp. 113–114) and Gryb and Thébault (2011, §3.2).

ii) Solve (12.29) (or equivalently (12.24)) to get an explicit expression for the integrals of motion, q, as functions of ϕ^α, Q, and P (as per i). Invert these expressions for the independent parameters, ϕ^α, in terms of some subset of the integrals of motion for q^α and the (Q, P). Then, reinsert this result into the remaining equations for the q's in terms of the q^α and the full set of Q's and P's. Observables are then specified as families of functions $F(Q, P, q^\alpha)_{|q^\alpha = \kappa^\alpha}$ where κ^α are a set of real numbers that pick out one member of each family of observables.

Option ii corresponds precisely to an implementation of the partial and complete observable approach with Q and P the partial observables, and $F(Q, P, q^\beta)_{|q^\beta = \kappa^\beta}$ the complete observables. The conditions under which this strategy makes sense will be discussed in the context of a simple model in the following section.

12.4.1   Explicit Model

The finite dimensional particle model we will study has a total Hamiltonian given by:

$$H(\vec{q}_i, \vec{p}_i) = N\mathcal{H}(\vec{q}_i, \vec{p}_i) + \vec{\lambda} \cdot \vec{\mathcal{P}}(\vec{p}_i), \qquad (12.30)$$

which is defined on the phase space $\Gamma(\vec{q}_i, \vec{p}_i)$, where i ranges over the number of particles n. In this free 'Jacobi' theory, evolution of the particle positions, $\vec{q}_i$, and momenta, $\vec{p}_i$, is generated by the Hamiltonian constraint

$$\mathcal{H}(\vec{q}_i, \vec{p}_i) = \sum_i \frac{\vec{p}_i^2}{2m_i} - E \approx 0, \qquad (12.31)$$

which characterizes a tangential variational symmetry, as we've just explained. The 'Gauss-like' constraint,

$$\vec{\mathcal{P}} = \sum_i \vec{p}_i \approx 0, \qquad (12.32)$$

implements a gauging of the spatial translational invariance of the free particle system. The lapse function N and the vector $\vec{\lambda}$ act as Lagrange multipliers enforcing the vanishing of these constraints. The vanishing of the total linear momentum implies that the position of the centre of mass, $\vec{q}_{\rm cm} = \frac{1}{\sum_i m_i} \sum_i \vec{q}_i$, is pure gauge.

Following the general procedure outlined in §12.3, we define the extended theory

$$S_{\rm e} = \int dt \left[\sum_i \dot{\vec{q}}_i \cdot \vec{p}_i + \dot{\vec{\sigma}} \cdot \vec{\Upsilon} - \dot{\tau}\mathcal{E} - N\left(\mathcal{E} - \mathcal{H}\right) - \vec{\lambda} \cdot \left(\vec{\Upsilon} - \vec{\mathcal{P}}\right) \right], \qquad (12.33)$$

where the extended variables $(\tau, \vec{\sigma})$ are arbitrary labels parametrizing the time and centre of mass of the system respectively. The energy, $\mathcal{E}$, can be thought of as a redefinition of the zero of the total energy of the system $E \to E + \mathcal{E}$. The other conjugate momentum variable, $\vec{\Upsilon}$, is the total linear moment of the system. We have conventionally added a minus sign to the $\dot{\tau}\mathcal{E}$ term to ensure the usual relations between time and energy. The extended theory is physically indistinguishable from the original in the case of the $(\tau, \mathcal{E})$ extension, and when $\vec{\Upsilon} = 0$ in the case of the $(\vec{\sigma}, \vec{\Upsilon})$ extension.

After making the appropriate identifications, the Ansatz (12.23) for F, takes the form

$$F(\vec{q}_i, \vec{P}_i; \vec{\sigma}, \tau) = W(\vec{q}_i, \vec{P}_i) + \vec{\mathcal{P}}\left(\vec{q}_i, \frac{\partial W}{\partial \vec{q}_i}\right) \cdot \vec{\sigma} - \mathcal{H}\left(\vec{q}_i, \frac{\partial W}{\partial \vec{q}_i}\right) \tau, \tag{12.34}$$

where $\mathcal{H}$ and $\vec{\mathcal{P}}$ obey the reduced Hamilton–Jacobi relations

$$\mathcal{E} = \mathcal{H}\left(\vec{q}_i, \frac{\partial W}{\partial \vec{q}_i}\right) \qquad\qquad \vec{\Upsilon} = \vec{\mathcal{P}}\left(\vec{q}_i, \frac{\partial W}{\partial \vec{q}_i}\right). \tag{12.35}$$

Using the definitions (12.31) and (12.32) and the additional Ansatz for W

$$W(\vec{q}_i, \vec{P}_i) = \sum_i \vec{q}_i \cdot \vec{P}_i, \tag{12.36}$$

we get

$$F(\vec{q}_i, \vec{P}_i; \vec{\sigma}, \tau) = \sum_i \vec{q}_i \cdot \vec{P}_i - \left(\sum_i \frac{\vec{P}_i^2}{2m_i} - E\right)\tau + \vec{\sigma} \cdot \sum_i \vec{P}_i. \tag{12.37}$$

The Hamilton–Jacobi equation of motion, (12.24), for this system is then

$$\frac{\partial F}{\partial \vec{P}_i} = (\vec{q}_i + \vec{\sigma}) - \frac{\vec{P}_i \tau}{m_i} = \vec{Q}_i, \tag{12.38}$$

which can easily be inverted for $\vec{q}_i$

$$\vec{q}_i = \left(\vec{Q}_i - \vec{\sigma}\right) + \frac{\vec{P}_i \tau}{m_i}. \tag{12.39}$$

The $\vec{P}_i$'s in this equation must obey the constraints

$$\mathcal{E} = \sum_i \frac{\vec{P}_i^2}{2m_i} - E \qquad\qquad \vec{\mathcal{P}} = \sum_i \vec{P}_i = 0 \tag{12.40}$$

in order for the reduced Hamilton–Jacobi equations (12.35) to be satisfied.

Now, consider the symmetry associated with the coordinate σ. For this simple symmetry, there is a natural way to parameterize the reduced phase space. If we sum (12.39) over the mass weighted particle indices and divide by the total mass, $m_{\text{tot}} = \sum_i m_i$, we obtain (upon using the second constraint of (12.40))

$$\vec{\sigma} = \vec{Q}_{\text{cm}} - \vec{q}_{\text{cm}}, \tag{12.41}$$

which is just the change in the centre of mass of the system. Reinserting this back into the integral of motion (12.39) gives

$$\vec{q}_i^{\text{cm}} = \vec{Q}_i^{\text{cm}} + \frac{\vec{P}_i}{m_i}\tau, \tag{12.42}$$

where the 'cm' refers to the use of centre of mass coordinates. Because of the redundancy implied by the translational symmetry, the centre of mass coordinates over-parameterize the reduced phase space. We then have the choice between options i and ii detailed above. In this case, i) corresponds to treating (12.42) as an over-complete set of relational observables. These are easily verified to be Dirac observables for the system.

Option ii, on the other hand, leads us to treat the coordinates $\vec{q}_i$ as explicit functions of the initial data $(\vec{Q}_i, \vec{P}_i)$ and the change in centre of mass $\vec{\sigma}$ (and the time, τ). This is equivalent to treating the change in centre of mass as a partial observable that parameterizes the different representations of the coordinates $\vec{q}_i$ in different gauges. Although this can be done for this system, it is very unnatural because it parameterizes things that are not physically measurable—the explicit coordinates, $\vec{q}_i$—in terms of something else that is not physically measurable—the change in centre of mass, $\vec{\sigma}$. For this example of transverse variational symmetries, the perspective of Thiemann (2007), in which partial observables are non-measurable quantities, is more appropriate than the perspective Rovelli (2002), where the partial observables are understood as measurable. Moreover, since it is only the Dirac observables that have operational significance, one could simply follow the more direct definition via option i and avoid the partial and complete observables entirely.

Contrast this with what happens when we apply the same options to the reparameterization symmetry. Option i is just equivalent to a standard gauge fixing, which we have the ability to do arbitrarily. For example, the integral of motion in terms of centre of mass coordinates is equivalent to using a gauge fixing $\vec{\sigma} = 0$. The analogous condition $\tau = 0$ indeed gives us a parameterization of the reduced phase space. Then, the 'integral of motion' reduces to the trivial statement

$$\vec{q}_i^{\,\mathrm{cm}} = \vec{Q}_i^{\,\mathrm{cm}} \,. \tag{12.43}$$

The only non-trivial equations in this gauge are, perhaps unsurprisingly, the initial value constraints (12.40). Clearly this option does not provide us with any dynamical information.

The only way to obtain a notion of evolution for this system is to consider (12.42) as a genuine evolution equation for the system We can identify the physically relevant observables of the system as those corresponding to the *entire* set of configuration space variables $\vec{q}_i^{\,\mathrm{cm}}$ after removing the centre of mass. These observables evolve according to (12.42), tracing out curves labelled by the arbitrary parameter τ, which is of course itself not an observable. Rather, τ is an independent parameter, and, as such, can be specified independently of quantities which are deemed measurable within the theory. The curves defined by (12.42) are reparameterization invariant even if the equation makes reference to the unphysical labelling parameter.

One might, however, wish to give a 'parameter-free' expression for the relative variation of the observables. It is in that context that the complete and partial observables program—option ii—offers a unique strength. Consider our formalism in one-dimension. We can choose the partial observables to be the centre of mass coordinates, q_i^{cm}, meaning the q_i^{cm} defined via (12.42), play the role of the 'flow equations'. A natural choice of clock variables is the centre of mass coordinate of one of the particles, say q_1^{cm}. We can invert (12.42) for particle 1 to obtain

$$\tau = \frac{m_1}{P_1} \left(Q_1^{\mathrm{cm}} - q_1^{\mathrm{cm}} \right) \,. \tag{12.44}$$

The first initial value constraint of (12.40) gives P_1 in terms of the remaining P_is. Using this, we can de-parameterize the evolution of the remaining q_as (where $a = 2 \ldots n$) in terms of q_1

$$q_a^{\text{cm}}(Q_i^{\text{cm}}, P_i, q_1^{\text{cm}}) = Q_a^{\text{cm}} - \frac{m_1}{m_a} P_a \frac{(Q_1^{\text{cm}} - q_1^{\text{cm}})}{\left[2m_1 \left(E + \mathcal{E} - \sum_a \frac{P_i^2}{m_i}\right)\right]^{1/2}} . \tag{12.45}$$

For any $q_1^{\text{cm}} = \kappa \in \mathbb{R}$ this expression defines a 'complete observable', which will also be a Dirac observable. Following this approach one can use the complete and partial observables program to de-parameterize the evolution purely in terms of observable quantities. This evolution is, however, fundamentally controlled by (12.42) and is *always* well-defined, even when a particular deparameterization breaks down. Thus, on our view, even if one wishes to use parameter-free 'complete observable' expressions, one is still required to retain the full 'partial observables' representation given by (12.42). This indicates that the Rovelli perspective, in which partial observables are measurable quantities, is more appropriate than the Thiemann perspective, where the partial observables are understood as non-measurable. Moreover, it indicates that the strategy represents an implicit departure from the standard Dirac analysis. This idea will be developed in a more rigours and complete direction in the following chapter.

12.5   Chapter Summary

Jacobi theories are theories in which the nomic structure is expressed via a geodesic principle on configuration space. Jacobi-type actions are such that they will automatically be reparameterization invariant. For the subset of Jacobi theories where the metric is conformally flat there is always a dynamically privileged ephemeris time parameter that can be defined along a classical curve in terms of the fixed total energy of the system. There are available two importantly different interpretations of the modal status of the total energy constant. Under the first, we interpret total energy as a constant of nature. The value of the total energy is thus the same for all dynamically possible models with the same token of the constitutive structure. Total energy is therefore a contingently fixed structure of the theory. Under the second, we interpret total energy as a constant of motion. The value of the total energy is then a free function of the initial data and thus can vary between dynamically possible models with the same token of the constitutive structure. Dynamically possible models are parameterized by the free value of the total energy. For the initial value problem approach to formulating the dynamics of Jacobi theories we then have that the space of dynamically possible models is simply the space of unrestricted initial data on the tangent bundle. Further insight into the dynamics of reparameterization invariant systems can be gained by formulating a generalized Hamilton–Jacobi formalism for totally constrained systems. In particular, such an approach allows explicit decomposition of integrals of motion in terms of variables that parameterize the flow of constraints. This framework then provides an explicit and formally insightful characterization of the popular partial and complete observables approach which can be understood as complementary, yet in a certain specific sense more limited, than our own analysis.

13
Temporal Structure Regained

The pseudo-problem of time rests on a putative confusion of dynamical redundancy with evolution that results from the Hamiltonian of reparameterization invariant theories being a first-class constraint. Such confusion has led to paradoxical claims within the literature such as 'motion is the unfolding of a gauge transformation'. Application of the diagnostic framework for isolating dynamical redundancy developed in previous chapters provides a general solution of the problem of time in classical mechanics via an unambiguous identification of the mathematical objects associated with narrow symmetry and evolution in reparameterization-invariant theories. In particular, based upon such an identification the pseudo-problem of the supposed equivalence between dynamics and gauge transformations can be avoided and the actual problem can be solved; that is, a framework in which the symmetry properties of reparameterization invariant theories with regard to chronometric and chronobservable structure can be provided that is compatible with such structure being determinate.

13.1 Differentiating Dynamical Redundancy and Evolution

We will now apply the first-order formalism for tangential variations developed in Chapter 11 to the Jacobi actions considered in Chapter 12. This will provide an entirely explicit identification of the mathematical objects responsible for the distinct functions of generating dynamics and generating reparameterization symmetries within this class of theories.

We start from the basic feature that the Jacobi action is invariant under infinitesimal diffeomorphisms of the temporal manifold I and their pullbacks by γ. Using the coordinate t on I we can write such diffeomorphisms as

$$t \rightarrow t + \dot{\epsilon}\,, \tag{13.1}$$

where ϵ is the group parameter of the diffeomorphism $t \rightarrow \epsilon(t)$ for $\dot{\epsilon} > 0$. This transformation can be pulled back onto TC by γ. In coordinates, this can be done by taking the first term of a Taylor expansion of $q^a(t + \dot{\epsilon})$ about t. We obtain:

$$q^a \rightarrow q^a + \dot{\epsilon}\dot{q}^a\,. \tag{13.2}$$

This transformation can be written as the action of a first-order differential operator T^a whose general expansion (8.14) can be matched with the expression above. Using the expansion coefficients of (8.14) and adapting to the notation of this section tells us that $T_{(0)}^a = 0$ and $T_{(1)}^a = v^a$. We can then read off the non-zero, independent 'u' vectors on velocity phase space using (8.67). This leads to

$$u^1 = \dot{q}^a \frac{\partial}{\partial q^a} + \dot{v}^a \frac{\partial}{\partial v^a} = X \qquad\qquad u^2 = v^a \frac{\partial}{\partial v^a}\,. \qquad (13.3)$$

We find that u^1 is a variation along the trial curve X, and is therefore a tangential variation. The fact that u^1 is a tangential variation can also be obtained by writing:

$$\theta_{L_J} = \frac{\partial L_J}{\partial v^a} \mathrm{d}q^a = \hat{v}_a \mathrm{d}q^a\,, \qquad (13.4)$$

where we have defined the covectors $v_a = g_{ab}v^b$, the norm $|v| = \sqrt{g_{ab}v^a v^b}$ defined by the metric g, and the unit vectors $\hat{v}^a = v^a/|v|$. We can then verify directly that the condition for a tangential variation

$$\frac{1}{L_J} \iota_{u^1} \theta_{L_J} = 1\,, \qquad (13.5)$$

holds on the surface $\dot{q}^a = v^a$. We will see below that the condition $\dot{q}^a = v^a$ is an on-shell condition for this first-order system, which is necessary for providing the equivalence between first-ß and second-order formalisms.

According to the arguments of §11.1, when there is a tangential symmetry the Hamiltonian, $H = v^a \frac{\partial L_J}{\partial v^a} - L_J = 0$, vanishes exactly. This can be straightforwardly confirmed for Jacobi theory by direct computation using $L_J = |v|$. The tangential variation u^1 is not required to be in the kernel of the (pre)-symplectic two-form $\omega_{L_J} = \mathrm{d}\theta_{L_J}$ *except* on-shell, which is the remaining part of Hamilton's equations when $H = 0$.

We can therefore find the solutions of Jacobi theory by computing the kernel of ω_{L_J} and removing the vectors that are linearly independent of whatever transverse vectors that also lie in the kernel. To do this we compute the (pre)-symplectic two-form:

$$\omega_{L_J} = \left(\hat{v}^c \partial_a g_{bc} - \frac{1}{2} \hat{v}_b \hat{v}^c \hat{v}^d \partial_a g_{cd}\right) \mathrm{d}q^a \wedge \mathrm{d}q^b + \frac{1}{|v|} \left(g_{ab} - \hat{v}_a \hat{v}_b\right) \mathrm{d}v^a \wedge \mathrm{d}q^b\,. \qquad (13.6)$$

Note that the coefficient of the second term is the Hessian

$$W_{ab}^J = \frac{\delta^2 L_J}{\delta v^a \delta v^b}$$

$$= \frac{1}{|v|} \left(g_{ab} - \hat{v}_a \hat{v}_b\right) \qquad (13.7)$$

of the Legendre transform which is proportional to the orthogonal decomposition of the metric g along the directions $\hat{v}^a$. As such, we obtain the usual Weierstrass condition

$$v^a W_{ab} = 0\,. \qquad (13.8)$$

In our picture this condition results from the fact that $u^2 = v^a \frac{\partial}{\partial v^a}$ is a transverse symmetry generator and therefore must be in the off-shell kernel of ω_{L_J} according to the arguments of §11.2. This can be verified by explicitly computing

$$\iota_{u^2} \omega_L = v^a \left(g_{ab} - \hat{v}^a \hat{v}^b\right) \mathrm{d}q^b = v^a W_{ab} \mathrm{d}q^b = 0\,, \qquad (13.9)$$

which vanishes for all $\mathrm{d}q^b$. Note that u^2 does *not* generate an initial value constraint because it acts trivially on the symplectic potential. It does, however, generate a genuine off-shell symmetry on $T\mathcal{C}$. The action of u^2 is to rescale the norm $|v|$. This is precisely what happens when one changes the time parameter along a curve as discussed in §11.2. We therefore interpret u^2 as the generator of infinitesimal instantaneous reparameterizations.

The remaining part of the kernel of ω_{L_J} is u^1. This can be used to generate Hamilton's equations. The interior product $\iota_{u^1}\omega_{L_J}$ has two terms that can be computed and set to zero separately. The first is the term proportional to dv^a:

$$\omega_{L_J}\left(\tfrac{\partial}{\partial v^a}, u^1\right) = \frac{\dot{q}^a}{|v|}\left(g_{ab} - \hat{v}_a\hat{v}_b\right) = \dot{q}^a W_{ab} = 0\,. \tag{13.10}$$

Since the kernel of W_{ab} is spanned by v^a, this equation tells us that $\dot{q}^a = v^a$, which is Hamilton's first equation for this system. The second term is proportional to dq^a and leads to the non-trivial condition

$$\omega_{L_J}\left(\tfrac{\partial}{\partial q^a}, u^1\right) = \frac{1}{|v|}\left(\Gamma_{abc}v^b v^c - \frac{1}{2}\left(v^b\hat{v}^c\hat{v}^d\partial_b g_{cd}\right)\hat{v}_a + \dot{v}^b\left(g_{ab} - \hat{v}_a\hat{v}_b\right)\right) = 0 \tag{13.11}$$

after using $\dot{q}^a = v^a$ and applying several simplifications.[1]

The prescription to generate well-defined evolution equations thus involves removing the dynamical redundancy associated with the action of u^2. This can be achieved via two equivalent procedures: reduction and gauge fixing. In the reduction procedure we first take the quotient of $T\mathcal{C}$ by the action of u^2. This eliminates $|v|$ and we obtain the quotient space $T\mathcal{C}/\mathbb{R}^+$, which is the standard contact bundle over $\mathcal{C}$. Evolution on this contact bundle is generated by the projection of u^1 onto this space. The condition $\frac{1}{L_J}\iota_{u^1}\theta_{L_J} = 1$ tells us that this flow is proportional to the flow of the Reeb vector field $R = \frac{1}{|v|}u^1$ on this contact manifold. The evolution is a Reeb flow and there is an associated conserved quantity we will interpret as the generalized energy. This entire construction is manifestly reparameterization invariant. In the gauge fixing procedure we treat the original tangent space $T\mathcal{C}$ as a principal fibre bundle over $T\mathcal{C}/\mathbb{R}^+$. We can then define a connection on this space that could be used to define a section, or gauge fixing, of the evolution directly in $T\mathcal{C}$. This corresponds to some explicit prescription for selecting the value of $|v|$ along a curve.

Our framework has thus delivered an unambiguous means to distinguish between dynamical redundancy and evolution through the distinction between u^1 and u^2. Dynamical redundancy is entirely due to the transverse vector field u^2 which generates *off-shell* narrow symmetries and can capture the *full set* of time reparameterization symmetries of Jacobi theory. By contrast, the tangential vector u^1 *does not* contribute to any underdetermination in the equations of motion but rather *defines* the equations of motion themselves.

The explicit differentiation between dynamical redundancy and evolution is the crucial result of our formalism. Such a result does not appear in any previous treatments in the literature. The result provides a complete refutation of any supposed

---

[1] Note that this is equivalent to the geodesic equation (12.2). To show this one needs to expand the definition of κ in (12.2) and use $\dot{q}^a = v^a$ to prove that $\frac{1}{2}v^b\hat{v}^c\hat{v}^d\partial_b g_{cd} = |v|\kappa - \dot{v}^a\hat{v}_a$. In this form, however, we see that in order to solve for the acceleration $\dot{v}^a$ in terms of the velocities v^a and configurations q^a one would need to invert the matrix $W_{ab} = (g_{ab} - \hat{v}_a\hat{v}_b)/|v|$, which is degenerate. This obstruction becomes manifest in our first-order geometric approach.

equivalence between motion and gauge transformations as per the pseudo-problem of time. Furthermore, it is precisely because our formalism allows one to explicitly distinguish between symmetry and evolution in the classical mechanical theory that secure foundations for extension of our ideas into the quantum domain have been set. The first steps towards such an extension will be taken in the next section where we extend our analysis to the canonical formalism. We will return to the relation between our ideas and quantization in outline form in Chapter 14 as a prospectus towards the extended discussion provided in Volume II.

13.2 A Dynamical View of Hamiltonian Constraints

To perform the Legendre transform we define the momenta for the system. Explicit calculation shows that they are the unit co-vectors $\hat{v}_a$:

$$p_a = \frac{\partial L_J}{\partial v^a} = \hat{v}_a \, . \tag{13.12}$$

In this form, it is obvious that the momenta obey the so-called *Hamiltonian constraint*

$$\mathcal{H}_J \equiv g^{ab} p_a p_b - 1 = 0 \, , \tag{13.13}$$

where g^{ab} is the inverse of g_{ab}. A more systematic way to derive this identity is to re-write the condition $\iota_R \omega_{L_J} = 1$, which implies the generalized conservation of energy for the system, on phase space. For Jacobi theory, since $p_a = \hat{v}_a$ we obtain the implicit (underdetermined) equation $v^a = |v| g^{ab} p_b$. Inserting this into the identity for the Reeb flow gives

$$\iota_R \theta_{L_J} = 1 \qquad \Rightarrow \qquad \frac{1}{|v|} v^a p_a = p_a p_b g^{ab} = 1 \, . \tag{13.14}$$

Within our approach, the sole role of the Hamiltonian 'constraint' $\mathcal{H}_J = 0$ is to define the image of the Legendre transform reflecting its non-injective nature. The Legendre transform induces a quotient by the action of the transverse vector u^2 by explicitly removing the variable $|v|$. Thus the cotangent space itself is simply the standard contact bundle $T^*\mathcal{C} = T\mathcal{C}/\mathbb{R}^+$ over $\mathcal{C}$. Thus the representation space it affords is already manifestly reparameterization invariant.

The dynamics of this system is generated by the Reeb vector R associated to the contact 1-form obtained by pulling back the symplectic potential θ_{L_J} by the Legendre transform. Since we can calculate $R = u^1/L_J$ directly from the degeneracies of ω_{L_J}, this can be done without ever having to use the flow of $\mathcal{H}_J$ on the auxiliary phase space obtained from the canonical symplectification of $T^*\mathcal{C}$ as is done in the standard Dirac procedure.

Significantly, since u^1 is tangential, it places no nomic restriction on the boundary data. This allows us to consider *unrestricted* initial data on phase space and treat $\mathcal{H}_J = |p| - 1 = 0$ as an equation for fixing the scale of $|p|$. Thus far we have been working in units where $|p|$ is set to 1. If we introduce the dimensionful quantity m, the Hamiltonian constraint reads $\mathcal{H}_J = |p| - m = 0$, and we can see the choice of unrestricted initial p_a as a way of fixing the value of m.

We can provide a more concrete illustration of this new way of thinking about the Hamiltonian constraint by specializing to the conformally flat case where:

$$\mathcal{H}'_J = T(q,p) + V(q) - E = 0\,, \tag{13.15}$$

where now $T(q,p) = \frac{1}{2}\tilde{g}^{ab}(q)p_a p_b$. Significantly, we treat $\mathcal{H}'_J = 0$ as a *family* of surfaces embedded in an unrestricted phase space labelled by different values of E. Formally, one can construct this space by forming a tensor product, $\Gamma = T^*\mathcal{C} \times \mathcal{E} = T\mathcal{C}/\mathbb{R}^+ \times \mathbb{R}$, of the original contact bundle with a manifold $\mathcal{E} = \mathbb{R}$ representing the values of the energy E.[2]

The picture on the phase space Γ is therefore the following. The dynamical evolution is restricted to constant-E hypersurfaces defined by $\mathcal{H}_J = 0$. This constraint is then the function that generates the dynamical evolution, via the canonical symplectic form on the unrestricted phase space, for *unrestricted* momenta p_a. Only if one chooses to restrict to a particular constant-E surface does one find constraints among the initial values of the momenta. The phase space itself is invariant under instantaneous reparameterizations, which are only non-trivial on $T\mathcal{C}$, where they are generated by u^2. We thus see that the attitude towards the energy as a constant of motion or constant of nature feeds directly into the treatment of the Hamiltonian constraint. However, this is entirely independent of the treatment of reparameterizations.

To complete the picture let us consider the geometric relationship between Γ and $T\mathcal{C}$. The Legendre transform is non-injective and thus each constant-E hypersurface in Γ maps to a one-parameter family of surfaces, labelled by $|v|$, on $T\mathcal{C}$ that are transverse to u^2. For each constant-E surface there is a preferred surface transverse to u^2. This is the surface defined in §12.1 by (12.12) with $|q'| \to |v|$, where the quantity $T + V$ is manifestly preserved.[3] There exists a bijection between the dynamically privileged surface in $T\mathcal{C}$ and the corresponding constant-E hypersurface on Γ where E is equal to the constant value of $T + V$. Using such a bijection for all values of E, we can define a bijection between the entire unrestricted phase space Γ and $T\mathcal{C}$. Thus, the set of *unrestricted* initial momenta maps to a set of *unrestricted* initial velocities. The value of the norm of these velocities must then be interpreted, as in §12.1, as the particular value belonging to the dynamically privileged time parameter τ such that the quantity $T + V$ is equal to the constant value E.

We will explain the sense in which we understand such surfaces to be dynamically privileged in §13.3, and highlight why this does not introduce any external temporal structure into the theory. For now we simply note that the dynamics can always be expressed in an arbitrary parameterization, by using u^2 to generate an infinitesimal reparameterization at all values of τ.

---

[2] Note that this is precisely the canonical symplectification of $T^*\mathcal{C}$ foliated by constant-E hypersurfaces. Of course, in this formalism one can always *choose* to restrict to a single constant E-surface, but this is not required by the consistency of the variational principle. On the contact manifold $T^*\mathcal{C}$, the different values of E label a family of Reeb vectors R_E and one obtains a similar structure.

[3] In the usual Dirac picture, this is the gauge obtained by evolving the total Hamiltonian of the system using the lapse $N = 1$.

13.3   The Structure of Jacobi Theories

13.3.1   Constitutive and Nomic Structure

Recall that constitutive structure is the structure that we need to define the space of basic, pre-nomic possibilities: the space of *kinematically possible models* (KPMs). For Jacobi theories, following our discussion of NM in §10.2, we can specify the total space of *kinematical structures* via the following elements:

$$\mathcal{K} = \left\{ \mathbb{R}^3_x, \delta, \gamma : I \to \mathcal{C}, m_I, c_\alpha, E \right\} . \tag{13.16}$$

where $\mathbb{R}^3_x$ is the spatial manifold, δ is the flat Euclidean metric on the spatial manifold, $\mathcal{C} = \mathbb{R}^{3N}$, $\gamma : I \to \mathcal{C}$ are geodesic curves given by an embedding of a one-dimensional interval, I, into the configuration space $\mathcal{C}$ with $I \subset \mathbb{R}$, $m_I \in \mathbb{R}^+$ and c_α are the particle masses and other coupling constants that enter the potential, V, and E is the total energy. The indices I and α run over the number particles and couplings respectively. Note that in contrast to the structure of a Newtonian theory we do not specify a temporal metric.

The construction of the space of KPMs then crucially depends upon an *interpretational choice* between treating the total energy as a constant of nature and as a constant of motion. Adopting the constant of nature interpretation means that the total energy is treated like an additional coupling constant. KPMs that differ solely with regard to the value of the total energy correspond to different token level specifications of the constitutive structure and are therefore constitutively distinct. The kinematically possible models of a Jacobi theory under the energy-as-constant-of-nature interpretation are specified via the collection of all possible images of curves γ in $\mathcal{C}$ conjoined with a specification of all the particle masses m_I and couplings c_α of the theory, and the value of the total energy.

$$K = \{\gamma : I \to \mathcal{C}, m_I, c_\alpha, E\} \tag{13.17}$$

Adopting the constant of motion interpretation means that the total energy is a free function of the initial data and thus can vary between KPMs which are *constitutively identical* in that they are identical constitutively at the token level. The energy is thus treated exactly like an additional constant of motion. Two constitutively identical KPMs can differ solely with regard to the value of the total energy. The kinematically possible models of a Jacobi theory under the energy-as-constant-of-motion interpretation are specified via the collection of all possible images of curves γ in $\mathcal{C}$ conjoined with a specification of all the particle masses m_I and couplings c_α of the theory.

$$K = \{\gamma : I \to \mathcal{C}, m_I, c_\alpha\} \tag{13.18}$$

This specification is identical to that provided in the context of NM in §10.2.

We then specify the nomic structure via the partitioning map, $n : K \to D$ which maps from the space of KPMs to the space of DPMs. Following the discussion of §12.2 and focusing on the IVP approach, DPMs are given by the curves γ that correspond to the geodesics of $(\mathcal{C}, g)$. We fix initial data in a particular parameterization $(q_1, \dot{q}_1)$ and then identify the DPMs as the solution to the geodesic equation corresponding to the initial data in question. A complete DPM corresponds to the maximal extension of the geodesic on $(\mathcal{C}, g)$ starting at q_1 and with initial velocity $\dot{q}_1$. Under suitable uniqueness and existence assumptions, the space of DPMs is then given by the curves that are the maximal extensions of the solutions to the geodesic equation for the initial data $(q_1, \dot{q}_1)$.

The crucial next step towards the application of the Nomic-AIR formalism is the characterization of the subspaces of dynamical transformations, $\text{Rel}(\pi_N)$, and narrow symmetries, $\text{Abs}(\pi_N)$. These subspaces are mutually exclusive (and complete) subspaces of the full set of diffeomorphisms Φ of the space of DPMs D; that is, $\Phi = \text{Diff}(D)$. To define these spaces, we need to specify the projection map $\pi_N : D \to \tilde{D}$ from D to the space $\tilde{D}$ of *distinct dynamically possible models* (DDPMs). In terms of this map, the narrow symmetry transformations, $\text{Abs}(\pi_N)$, are those transformations that are *invariant* under the projection while the dynamical transformations, $\text{Rel}(\pi_N)$, are those that are *variant* under the projection:

$$\text{Abs}(\pi_N) = \{A : D \to D | \pi_N(A(d)) = \pi_N(d), \forall d \in D\} \qquad (13.19)$$

$$\text{Rel}(\pi_N) = \{R : D \to D | \pi_N(R(d)) \neq \pi_N(d), \forall d \in D\} = \neg \text{Abs}(\pi_N). \qquad (13.20)$$

The analysis of DDPMs via absolute charges associated with the time-independent Euclidean transformations is identical to that for NM and implications for π_N will be identical. By contrast, following from our earlier discussion, the analysis of the Galilean boosts will be different depending on whether we adopt the energy as a constant of motion or constant of nature interpretation. In the first, the analysis will be identical to that of NM. Thus, excluding time symmetries, π_N takes the same form. By contrast, under the energy as a constant of nature interpretation, Galilean boosts are no longer endomorphisms of the space of DPMs, meaning that they are necessarily precluded from Φ and therefore cannot feature in π_N. This issue is not our primary focus here, however, and we will neglect a full characterization of the relevant transformations in order to devote our analysis fully to questions of temporal symmetries.

Our particular focus is of course on the specific characterization of the action of the reparameterization transformations. We will start by considering the theory under the constant of nature interpretation with energy E fixed kinematically, we will then consider the theory under the constant of motion interpretation with varying energy E. In all cases, we will be considering the IVP formulation of Jacobi theory as this is the one that is most relevant for describing local observations.

We will work in the tangent space $T\mathcal{C}$, where the kinematical possibilities are represented as integral curves of smooth vector fields X. The space D is given by the space of integral curves γ_{TC} of the vector u^1:

$$D = \{\gamma_{TC} | \dot{\gamma}_{TC} = u^1(\gamma_{TC})\}. \qquad (13.21)$$

This space can be parameterized by a constant-time hypersurface τ that is everywhere transverse to u^1. For convenience, we will now drop the TC subscript on γ below as it should be clear that we are referring to curves on tangent space and not configuration space.

The projection map π_N then identifies all integral curves γ of u^1 that are related by narrow symmetry transformations. As we argued in §13.1, these should be all

γ's related by transformations generated by the kernel of the Hessian. In this case, that kernel is spanned by the vector u^2. We then find that the equivalence relation $e : D \to D$ defining π_N is given by

$$e(\gamma) = \exp\{\epsilon u^2\}(\gamma)\,, \forall \gamma \in D\,, \tag{13.22}$$

where $\epsilon(t)$ is a time-dependent parameter for the group generated by u^2. Using the equivalence relation (13.22) to define π_N, it is then clear that the projection π_N is invariant under the flow of u^2 and thus that $\mathrm{Abs}(\pi_N)$ is simply given by

$$\mathrm{Abs}(\pi_N) = \{A(\gamma) = \exp\{\epsilon u^2\}(\gamma)\,, \forall \gamma \in D \text{ and } \forall \epsilon : I \to \mathbb{R}\}\,. \tag{13.23}$$

Since u^2 spans the kernel of the Hessian, $\mathrm{Rel}(\pi_N)$ will simply be given by the flow of the span of all the vectors in the tangent space to τ that are linearly independent of u^2. If we call these vectors z^I then

$$\mathrm{Rel}(\pi_N) = \{R(\gamma) = \exp\{\mathrm{Span}(z^I)\}(\gamma))\,, \forall \gamma \in D\}\,. \tag{13.24}$$

We have thus provided explicit representations of π_N and $\mathrm{Abs}(\pi_N)$ in terms of the action of the vectors u^1 and u^2, which we've computed above. Since we've identified u^2 as the generator of infinitesimal reparameterizations, our analysis implies that reparameterizations are the narrow symmetry transformations of Jacobi theory as expected. The space $\mathrm{Rel}(\pi_N)$ of dynamical transformations is given in terms of the vectors z^I. Since these are required to be transverse to u^1, an explicit representation of z^I is equivalent to solving the initial value problem for Jacobi theory. In our considerations below, we will focus on the infinitesimal representations of $\mathrm{Rel}(\pi_N)$ and $\mathrm{Abs}(\pi_N)$ for simplicity since these are simply given in terms of the action of vector fields on TC.

To treat the theory with a freely specifiable energy E, and thus apply the energy as a constant of motion approach, we need to extend the tangent space $TC \to TC \times \mathbb{R}$ to include a specification of the energy $E \in \mathbb{R}$. The space of DPMs is then the space of curves $\gamma : \Sigma_t \to TC \times \mathbb{R}$, where the embedding into $\mathbb{R}$ simply assigns energy E to the system. Importantly, the vector fields u^1 and u^2 have zero Lie drag on E: $\mathfrak{L}_{u^1} E = \mathfrak{L}_{u^2} E = 0$. Thus, $\mathrm{Abs}(\pi_N)$ takes the same form as in (13.23) but with γ defined on its extended image. The only significant difference in $\mathrm{Rel}(\pi_N)$ is the fact that the space transverse to u^1 has an additional vector $\frac{\partial}{\partial E}$ spanning the extended direction. This vector generates solutions with different energies.

13.3.2 Gauge Fixings

Before providing an analysis of the temporal structure of Jacobi theory under the two interpretations of energy. It will prove highly instructive to explicitly introduce gauge fixing structures into our formalism. Above we saw that the momenta are unit covectors $p_a = \hat{v}_a$ on tangent space. This means that phase space functions are invariant under the flow of u^2, which affects only $|v|$. It also means that the Legendre transform is non-invertible since any state related by the flow of u^2 projects to the same state on phase space. We can nevertheless define an invertible map from phase space to a *gauge-fixed* surface on TC, which corresponds to a surface that is transverse to u^2. A simple way to represent surfaces of this kind is through a gauge fixing condition of the form

$$G = |v| - f(q, v) = 0 \,. \tag{13.25}$$

The surface generated by G is transverse to u^2 provided $\mathfrak{L}_{u^2} G \neq 0$, or $v^a \frac{\partial f}{\partial v^a} \neq |v|$ (otherwise u^2 would be in the tangent space of the gauge-fixed surface). Thus, the function f is arbitrary up to this restriction. Different choices of f correspond to different ways of parametrizing the gauge-invariant evolution on tangent space. We will see below that this choice will pick out different clock choices on phase space.

The Jacobi equations of motion (13.11) are non-invertible because of the degeneracy of $W_{ab} = g_{ab} - \hat{v}_a \hat{v}_b$. Note, however, that these equations of motion can be rewritten as

$$\dot{v}_a = \frac{\mathrm{d} \ln |v|}{\mathrm{d}t} v_a - \Gamma_{abc} v^b v^c \,, \tag{13.26}$$

where the first term contains the $\dot{v}^a$ dependence leading to the degeneracy. Restricting to the surface $G = 0$ restricts $|v|$ to some independently specifiable function $f(q, v)$. This removes the $\dot{v}^a$ dependence of the RHS of (13.26) and leads to well-posed equations for X.

Two useful choices of f are as follows. The choice $f_{\text{affine}} = 1$ reduces (13.26) to

$$\dot{v}^a + \Gamma^a{}_{bc} v^b v^c = 0 \,, \tag{13.27}$$

which is the geodesic equation with an affine parameterization. In the conformally flat case, the choice

$$f_{\text{eff}} = 2(E - V) \tag{13.28}$$

leads to

$$\tilde{g}_{ab} \dot{v}^b = -\frac{\partial V}{\partial q^a} \,, \tag{13.29}$$

which reduces to Newton's second law in its standard form (where the masses have been absorbed into $\tilde{g}_{ab}$). It's clear then that gauge fixing the flow of u^2 results in a fixing of the parameterization of the evolution generated by X. This illustrates once more that u^2 generates local reparameterizations. It will also prove instructive for our characterization of the temporal structures of Jacobi theory in the following section.

13.3.3 Temporal Structure

Recall from Chapter 10 that we introduced three categories of temporal structure: chronordinal, chronometric, and chronobservable. Let us consider each in turn in the context of Jacobi theory, highlighting the comparison with our analysis of the temporal structure of Newtonian theory as we go.

We commenced our discussion of Jacobi theories by pointing to the fact that the common defining feature that links such theories is that their nomic structure implements some form of geodesic principle on configuration space. By definition a geodesic is a curve, $\gamma : I \to \mathcal{C}$, that is given by an embedding of a one-dimensional interval, I, into the configuration space $\mathcal{C}$. Since we require $I \subset \mathbb{R}$ and because points in I label different instantaneous configurations in $\mathcal{C}$, the domain I of γ is naturally interpreted as a temporal manifold.

We will now show that it is possible to interpret the symmetry transformations generated by u^2 as the pullback of γ by a temporal diffeomorphism on I. To do this, note that these infinitesimal transformations act on γ as

$$\delta_{\epsilon u^2}(\gamma) = \epsilon(t)v^a \,, \tag{13.30}$$

where we have written γ in terms of the coordinates $\gamma = (q^a(t), v^a(t))$. This is a time-dependent rescaling of the velocities. But that is exactly the effect of applying a time reparameterization $t \to h(t)$ on I and then applying γ when we identify $\epsilon(t) = \dot{h}(t)$. We can thus speak of temporal diffeomorphisms on I interchangeably with the transformations generated by u^2 with the understanding that these maps should be passed through γ when appropriate.

It is now possible to define a chronordinal total time-ordering structure $>$ on DPMs using the topological structure of I inherited from $\mathbb{R}$. This marks a key structural difference with the Partial and Complete Observables approach discussed in §12.4 where monotonicity is not guaranteed and thus temporal-ordering structure will in general both only be given by a partial ordering and vary between DPMs. By contrast, on our approach such structure is invariant under both symmetry transformations and dynamical transformations of the theory given the identifications we have made above because the topological structure on I is diffeomorphism invariant. We thus have that $\mathrm{Abs}(>) = \Phi$ which means that the time-ordering structure is *fixed structure*. Since the structure does not vary even between distinct tokens of the Jacobi constitutive structure, for example two-body and three-body systems, the ordering structure is constitutively fixed.

Thus far the structural categorization is the same as the chronordinal structure of NM. A significant difference is with regard to the local vs global status of the ordering structure as fixed structure. In NM, inhomogeneous transformations on I (composed with γ) were excluded from Φ; thus the temporal-ordering structure was globally fixed structure. In Jacobi theories we have included inhomogeneous temporal transformations and thus the ordering structure is temporally locally fixed.

A further significance of this distinction can be made clear by returning to the idea of a subset of diffeomorphisms of K which are *not* diffeomorphisms of D. These transformations were briefly introduced in §9.1 as transformations $\bar{\psi} \notin \Phi$ which are diffeomorphisms of K that do not preserve the nomic partition; that is, $\bar{\psi}^\star n \neq n$. Interestingly, not only do we find that inhomogeneous temporal transformations play this role in NM, but that these transformations are such that they preserve the time-ordering structure; that is, inhomogeneous temporal transformations in Newtonian theory push forward onto ordering structure such that $\bar{\psi}^\star(>_a) = >_a \,, \forall a$.

This analysis allows us to understand the transition from NM to Jacobi theory in terms of the localization heuristic discussed in §10.1.5. In particular, one can observe that in the Newtonian theory we have a temporal structure that is fixed structure under a set of homogeneous transformations and does not vary under the inhomogeneous transformations either, but where the inhomogeneous transformations are members of the $\bar{\psi}$ set. This motivates a reformulation of the nomic principle such that inhomogeneous transformations can be included in $\mathrm{Diff}(D)$ and thus that the full invariance properties of the relevant fixed structure be represented dynamically. This is precisely what is achieved by rewriting a Newtonian action in the Jacobi form.

The next type of structure we will consider highlights a more direct contrast with NM. This is chronometric structure given by the privileged temporal coordinate t_N. Focusing on the Jacobi theory under the energy-as-a-constant-of-nature interpretation, the privileged temporal coordinate defines a preferred parameterization of the domain I of $\gamma : I \to \mathcal{C}$. We can represent this as some monotonic function $t_N : I \to \mathbb{R}$ on I. Diffeomorphisms on I thus reparameterize t_N. Such transformations can only preserve t_N when $\epsilon = 0$. Thus, t_N is not preserved by the symmetry transformations of Jacobi theory: it is kinematically incomplete structure as per our Definition 9.20. By contrast, the z^I vectors are tangent to the constant time hypersurfaces and therefore involve a reshuffling of the theory's initial data. Thus, t_N is invariant under dynamical transformations and is dynamically absolute structure.

The treatment of the theory with energy as a constant of motion is almost identical because $\mathrm{Abs}(\pi_N)$ is still generated by the u^2 vector. t_N is therefore also incomplete under symmetry transformations in the extended theory. The extra direction $\frac{\partial}{\partial E}$ in the generators of $\mathrm{Rel}(\pi_N)$ still has no effect on the parameterization of I. Thus, t_N is again dynamically absolute. This highlights the fact that extending the system by including arbitrary energies does not induce a preferred parameterization into Jacobi theory: the AIR analysis of t_N is completely unchanged.

That the chronometric structure of the theory given by the privileged temporal coordinate parameter is incomplete is hardly surprising since this is not a temporal structure that plays a physical role in the theory. Rather, what we find is that the chronometric structures of the theory that are crucial are those provided by chronobservable structures. As we shall see shortly, understanding how a reparameterization invariant theory can have observable structures that are chronometric crucially relies upon the analysis of gauge fixing structures introduced in §13.3.2

To get an explicit representation of the key chronobservable structures, let us first recall that the Hamiltonian constraint expresses the fact that the momenta are unit vectors in terms of the metric g:

$$\mathcal{H} = p_a p_b g^{ab} - 1 = 0 \, . \tag{13.31}$$

The extended Hamiltonian $H = N\mathcal{H}$ then leads to the following equations of motion

$$\dot{q}^a = \{q^a, N\mathcal{H}\} = 2N g^{ab} p_b \, , \tag{13.32}$$

$$\dot{p}_a = \{p_a, N\mathcal{H}\} = N p_a p_b \partial_c g^{bc} \, . \tag{13.33}$$

By inverting the first equation and inserting it into the second, we obtain

$$\dot{v}^a = \frac{\mathrm{d} \ln N}{\mathrm{d}t} v^a - \Gamma^a{}_{bc} v^b v^c \, , \tag{13.34}$$

where $v^a = \dot{q}^a$. We clearly have a correspondence between the canonical formalism and equation (13.26) when $N = c|v|$, for some constant c. Thus, gauge fixing u^2 selects a lapse up to a global normalization. More precisely, choosing $|v|$ fixes the *combination*

$N\mathcal{H}$ since N in our derivations above was chosen for a specific normalization of $\mathcal{H} \approx 0$. This, in turn, selects a privileged clock variable τ_f which is such that

$$\{\tau_f, N\mathcal{H}\} = 1. \tag{13.35}$$

For example, the choice $|v| = f_{\text{eff}}$ leads to the *ephemeris clock*

$$\left\{\tau_{\text{eff}}, \frac{1}{2}\tilde{g}_{ab}v^a v^b + V(q) - E\right\} = 1 \tag{13.36}$$

whereas the choice $|v| = f_{\text{affine}}$ leads to the *affine clock*

$$\left\{\tau_{\text{affine}}, p_a p_b g^{ab} - 1\right\} = 1. \tag{13.37}$$

Because the clock observables are phase space functions, they are automatically invariant under the narrow symmetries associated with reparameterization transformations, which only have an action on $\mathcal{TC}$. As with the Newtonian description, clock observables will also in general commute with the absolute charges of the theory $\{\tau, k_\alpha\} = 0$. Changes of the absolute charges leave both the nomic structure and the clock observable structure invariant, and we have $\text{Abs}(\pi_N) = \text{Abs}(\tau_f)$. Thus a clock observable in Jacobi theory is an invariant structure like in NM.

Clock observables need not, however, have vanishing Poisson bracket with all the non-trivial Hamilton–Jacobi constants of the theory. The condition (13.35) only fixes τ_f up to some initial condition. Thus a clock observable can covary under elements of $\text{Rel}(\pi_N)$ and thus clock observables in Jacobi theories are dynamically relative invariant structure. This is the same status as in NM. The crucial difference is that in Jacobi theories this status is temporally local and we can thus understand the transition from Newtonian to Jacobi mechanics in terms of a form of localization heuristic. Furthermore, we also have a precise sense in which Jacobian clock observables are more closely analogous to proper time in relativistic theories than their Newtonian counterparts.

13.3.4   Ephemeris Time and Inertial Clocks

A physically compelling feature of the particular choice of ephemeris time is that it can be shown to 'march in step' with the time marked out by an inertial clock. Recall from §4.3 that inertial clocks were introduced in the late-nineteenth century with the idea of transferring reference to absolute time in Newton's mechanics to material bodies moving such that they can play an equivalent functional role.

The significant difference, however, is that inertial clocks require the unreasonable idealization of a lack of gravitational coupling for all times and thus an artificial division between the clock and the universe. In contrast, ephemeris time is a time built from dynamical coupling between all the degrees of freedom. It is precisely a measure of the change of things, to recall Mach's iconic phrase, and thus provides the determinate temporal metric structure that satisfies both relational and absolutist desiderata.

The proof of equivalence proceeds as follows. First consider a Jacobi theory defined by a conformally flat metric g. Suppose also that there is a single free degree of freedom q^0 that is completely dynamically decoupled from the rest of the system coordinatized by $\{q^a\}_{a=1}^N$ so that the configuration space decomposes as $\mathcal{C} = \mathcal{C}_{\text{iso}} \times \mathcal{C}_{\text{sub}}$. This condition can be implemented by requiring that the potential V be only a function of the q^a; that is, that the potential for q^0 and the interaction potential between subsystems be zero. Geometrically, this is equivalent to requiring that the Lie drag of V along the q^0 direction be zero. This in turn is simply equivalent to requiring that

$$V(q^0, q^a) = V(q^a) \tag{13.38}$$

in the particular coordinates (q^0, q^a). We further need to require a similar condition on the Lie flow of the metric g along the q^0 direction. Again, this condition is easiest to implement in coordinates by requiring that g_{00} be a constant on $\mathcal{C}$ and that $g_{0a} = 0$.[4]

Next, note that the dynamical isolation conditions imply that (12.12) can be written as:

$$g_{00}\dot{q}^0\dot{q}^0 + g_{ab}\dot{q}^a\dot{q}^b = 2(E - V(q^a)). \tag{13.39}$$

A simple rearrangement tells us that:

$$g_{ab}\dot{q}^a\dot{q}^b + 2V = 2E - g_{00}\dot{q}^0\dot{q}^0. \tag{13.40}$$

Since the RHS is independent of q^a and the LHS is independent of q^0, these two expressions must be both equal to a new constant E'. We therefore obtain the two independent equations:

$$\tfrac{1}{2}g_{00}\dot{q}^0\dot{q}^0 = E - E' \tag{13.41}$$

$$\tfrac{1}{2}g_{ab}\dot{q}^a\dot{q}^b = E'. \tag{13.42}$$

This system can be solved for the ephemeris time τ of the whole system by solving (13.41) and (13.42) separately at the cost of introducing a separation constant, E', into the theory. We can solve (13.41) by noting that g_{00} is a constant. Thus,

$$\dot{q}^0 = \frac{2}{g_{00}}(E - E'), \tag{13.43}$$

which can be integrated to give $q^0 = c\tau + q_0^0$, where $c = \frac{2}{g_{00}}(E - E')$ is a constant fixed by E', E, and g_{00}. Note here that, in particle mechanics, g_{00} is simply the mass of the isolated particle.

We thus have that q^0 is proportional to the ephemeris time, τ, of the whole system up to a constant offset. Moreover, the constant of proportionality is fixed by the energies of the isolated particle and subsystem through the solution of (13.43). The second equation, (13.42), can simply be read as defining τ to be the ephemeris time of the subsystem $\{q^a\}_{a=1}^N$ with energy E' according to the general definition (12.12).

---

[4] This can easily be seen by expanding the Lie drag of g along a vector pointing in the q^0-direction using ADM-style coordinates.

Thus, q^0/c is, up to a constant offset, both the ephemeris time for the total system and the ephemeris time for the isolated system q^a with fixed energy E'.

A consequence of the above which is of particular significance to our analysis is that unless you are sure that you have captured the motions of *all* degrees of freedom in the universe, no matter how dynamically isolated they are, then you will never know the true value of the total energy E. Thus, from a practical perspective, it is better to assume that the energy can take any value. This provides motivation for the energy as a constant of motion viewpoint discussed in §12.2.

13.4   Chapter Summary

In this chapter we have dispelled the pseudo-problem of time by correcting the putative confusion of dynamical redundancy with evolution that results from the Hamiltonian of reparameterization invariant theories being a first-class constraint. Such confusion has led to paradoxical claims within the literature such as 'motion is the unfolding of a gauge transformation'. Application of our diagnostic framework for isolating dynamical redundancy has led to a general solution of the problem of time in classical mechanics via an unambiguous identification of the mathematical objects associated with narrow symmetry and evolution in reparameterization invariant theories. In particular, based upon such an identification the pseudo-problem of the supposed equivalence between dynamics and gauge transformations can be avoided and the actual problem can be solved; that is, we can provide a framework in which the symmetry properties of reparameterization invariant theories with regard to chronometric and chronobservable structure are compatible with such structure being determinate.

This chapter has provided a detailed illustration of these ideas for the case of Jacobi theories in which the action generates a geodesic principle. In addition, we have articulated two alternative interpretations of these theories in terms of the *modal status* of energy; that is, whether the total energy is a constant nature with a fixed value for any token of the constitutive structure or a constant of motion whose value acts as an additional parameter of initial data. With these ideas in mind we have applied the Nomic-AIR framework towards a characterization of the temporal structure of Jacobi theories. In structural terms we have shown that: i) chronordinal structure is fixed structure as in Newtonian theory; ii) the chronometric structure given by the privileged time parameter is kinematical incomplete structure; iii) chronobservable structures can be defined such that they are invariant under the relevant narrow symmetries and thus provide gauge invariant means of chronometry; and iv) the temporal structure of the theory has the same form irrespective of whether energy is treated as a constant of nature or constant of motion.

14

Conclusion and Prospectus

This chapter will provide a short summary of the key steps in our solution to the problem of time in classical mechanics described in the present Volume I. We will then provide a prospectus to the ideas and arguments relating to the full problem of time in classical and quantum gravity that is the subject of Volume II of this work.

14.1   The Problem of Time in Classical Mechanics

The problem of time in classical mechanics is that the symmetry properties of reparameterization invariant theories with regard to chronometric and chronobservable structure enforce indeterminacy with regard to such structure under the standard approach to gauge symmetry. Our solution of this problem involves three formal and conceptual moves.

First, we have demonstrated that the standard approach to gauge symmetry is inadequate to the treatment of reparameterization invariant theories. In particular, through the crucial distinction between transverse and tangential variations, we have established that reparameterizations do not define equivalence classes of initial data since they act transversely to initial data surfaces. Reparameterizations are not narrow symmetries-at-an-instant and do not lead to dynamical redundancy.

Second, we have then applied our first-order formalism to unambiguously differentiate between the objects responsible for generating evolution and those that generate degeneracy along tangential directions. Within our formalism, evolution can clearly separated from symmetry in reparameterization invariant theories and the complex web of relationships between irregular Lagrangians, Noether's second theorem, and the existence of first-class Hamiltonian constraints is unentangled. The basic formal conflation behind the idea that motion is the unfolding of a gauge transformation resulting from the Hamiltonian of reparameterization invariant theories being a first-class constraint is thus totally dispelled.

Third, and finally, we have applied the nomic-AIR formalism to the case of Jacobi theories to demonstrate that in the case of such theories reparameterization invariance is fully compatible with determinate chronordinal and chronometric structure. In particular, we have isolated the a total time-ordering structure, as a determinate structure that is invariant under narrow symmetry transformations and dynamical transformations of the theory, together with chronometric structure given by the class of clock observables, which are dynamically relative invariant structure.

14.2   Quantization and Redundancy

The framework for the analysis of theories with regular and irregular Lagrangians according to the unconstrained and constrained Hamiltonian analysis maps directly onto a division in terms of approaches to quantization; that is, in the case of theories with regular Lagrangians, the unconstrained Hamiltonian analysis can be used as a platform for standard canonical quantization techniques. One takes the symplectic structure of phase space to generalize naturally to an operator algebra on a physical Hilbert space and the well-established algorithm for first quantization can be applied.[1]

In the case of theories with irregular Lagrangians one expects, in general, an ill-posed initial value problem and, moreover, the presence of constraints of the phase space blocks a direct canonical quantization. However, it was towards the problem of quantization that the constraint Hamiltonian formalism was in fact constructed. In particular, in the original, rather informal, Dirac (1964) treatment, the algorithm for characterizing the full set of first-class constraints is understood as an intermediary step towards the quantum theory.

The basic, rather schematic idea, is to conceive of the construction of the quantum analogue of a theory with an irregular Lagrangian in terms of two further steps. First, we apply a standard canonical quantization of the *unconstrained* phase space with again the symplectic structure generalized to an operator algebra over a Hilbert space. However, in this case we identify the relevant Hilbert space as a *kinematical* Hilbert space. In a second step, represent the first-class constraints in terms of the operator algebra and then, having resolved suitable operator ordering ambiguities, enforce these constraints as operators annihilating the wavefunction. We thus define the physical Hilbert space via an expression of the form

$$C_\alpha(\hat{q}, \hat{p})\psi_{\text{phys}} = 0. \tag{14.1}$$

where $C_\alpha(\hat{q}, \hat{p})$ are the first-class constraints, given a suitable choice of operator ordering, and ψ_{phys} are states on a further *physical* Hilbert space. Crucially, this equation is understood to *project* quantum states from the the kinematical Hilbert space to the physical Hilbert space, in which the redundancy associated with the classical first-class constraints has been effectively reduced out at the quantum level. One then understands the quantum theory of the physical Hilbert space as implementing the non-redundant dynamics of the 'true' degrees of freedom of the quantum theory in question.

Remarkably, the largely heuristic Dirac quantization algorithm closely parallels modern, fully rigorous approaches. In particular, one can apply the Refined Algebraic Quantization (RAQ) (Giulini and Marolf, 1999*b*) scheme under the conditions that: i) the constraints are self-adjoint operators with an algebra which closes with structure constants; ii) the symmetry group is a finite dimensional locally compact group with respect to a suitable topology. Under this scheme the action of the constraints on the

---

[1] Strictly speaking a more rigorous algebraic quantization method is required to insure these brackets (as well as the structures below) are well defined. Such formal issues are tangential to our current project and so can reasonably be neglected. See Isham and Kakas 1984*a*; Isham and Kakas 1984*b*; Henneaux and Teitelboim 1992; Giulini and Marolf 1999*b*; Giulini and Marolf 1999*a*; Thiemann 2007 for more details.

kinematical Hilbert space is *averaged* over the relevant group symmetry manifold. The group averaging in effect superselects the zero eigenstate of the constraint operator and we recover the results of Dirac's original constraint quantization scheme. Most significantly for our current discussion, the group averaging approach to quantization defines something called a rigging map which genuinely acts to project from kinematic to physical Hilbert space states. What is more, the rigging map also induces an inner product the physical Hilbert space. Most signfiantly, the rigging map is a projection which removes unphysical states, but it also has a further role in 'quotienting out' physically identical states that lie upon a 'quantum gauge orbit' (Corichi, 2008).

The implications for the problem of time should hopefully we clear to the reader. In the case of a transverse variation, the classical constraints do genuinely unambiguously lead to the generators of 'gauge' transformations and there certainly is associated dynamical redundancy. In such cases the constraint quantization algorithm serves as a means to reduce out redundant degrees of freedom at the quantum level. As such, its implementation is founded on the same basis as the classical reduction algorithm; that is, the necessity of a well-posed initial value problem.

By contrast, in the case of tangental variations, as associated with reparameterization invariance, the connection between the first-class constraints, redundancy, and the well-posedness of the initial value problem is rather different. As we have argued at great length, one needs to be extremely careful to differentiate between the objects responsible for generating evolution and those that generate degeneracy along tangential directions. Moreover, the Hamiltonian constraints associated with reparameterization invariance certainly should not be understood as generating gauge transformations, and enforcing such an interpretation is not necessary for a well-posed initial value problem.

The crucial point is that whereas, classically, conflation between the differing treatments needed for tangential and transverse variations leads to a puzzling and somewhat problematic situation in terms of the analysis of evolution, the empirical content of the theory is not directly affected. In the quantum domain, however, following the standard Dirac–Henneaux–Teitelboim route, leads to a quantum theory in which the wavefunction is restricted to zero energy eigenstates. This is the quantum problem of time. In avoiding such a result, our rigorous means for differentiation between the objects responsible for generating evolution and degeneracy along tangential directions in the classical theory can only be the first step. What is needed in addition is a new approach to quantization which implements our approach to the symmetry and structure of reparameterization invariant theories at the quantum level. However, before we outline the core ideas behind our novel quantization scheme, we must first consider the complications implied by the extension from classical mechanical theories to gravitational theory. This is to generalize from the case of time reparameterizations to that of time refoliations.

14.3   Gravitation and Refoliation

The symmetry properties of general relativity are remarkable for their generality. In the covariant formulation of the theory the standard way in which to express the symmetry is in terms of invariance of the nomic structure given by the Einstein equations under metric transformations induced by diffeomorphisms of the underlying spacetime manifold. An alternative, and in our view more perspicuous, approach to the symmetries of the covariant theory is to consider the Einstein–Hilbert action in the context of the framework of dynamical redundancy outlined above. Such an approach would then focus on the isolation of initial value constraints and arrive at an identification of the symmetries-at-a-time of the theory. Finally, in the context of the canonical formulation of general relativity (Dirac, 1958; Arnowitt, Deser and Misner, 1960) in order to characterize the symmetry of the theory we are required to consider the complications implied by the constrained Hamiltonian treatment of theories invariant under time reparameterizations. It is thus no surprise that here we find a problem of time arising from the entangling of symmetry and evolution. The problem in this context is, however, rather more acute than that we have considered and resolved in the context of simple reparameterization invariant models, such as Jacobi theories. Let us briefly set out the technical reasons why.

The formalism of canonical general relativity can be concisely expressed as follows.[2] Consider a four-manifold M with topology $M \cong \mathbb{R} \times \sigma$ where σ is a three-dimensional manifold which is compact without boundary but otherwise has an arbitrary differentiable topology. Consider the diffeomorphism $X : \mathbb{R} \times \sigma \to M; (t, x) \to X(t, x)$. Define a *foliation* of M into hypersurfaces $\Sigma_t := X_t(\sigma)$ where $t \in \mathbb{R}$ and $X_t : \sigma \to M$ is an *embedding* defined by $X_t(x) := X(t, x)$ for the coordinates x^a on σ. Restrict to a foliation of a spacetime, M, into spacelike hypersurfaces, Σ_t and thus spacelike embeddings. In this context, the Einstein–Hilbert action can be cast in canonical form in terms of the Arnowitt–Deser–Misner (ADM) action:

$$S = \frac{1}{\kappa} \int_{\mathbb{R}} dt \int_{\sigma} d^3x \{ \dot{q}_{ab} p^{ab} - [N^a \mathcal{H}_a + |N|\mathcal{H}] \}. \tag{14.2}$$

Here $\kappa = 16\pi G$ (where G is Newton's constant and we assume units where $c = 1$), q_{ab} is a Riemannian metric tensor field on Σ, and p^{ab} its canonical momenta defined by the usual Legendre transformation. N and N^a are multipliers called the lapse and shift. $\mathcal{H}_a$ and $\mathcal{H}$ are the momentum and Hamiltonian constraint functions of the form:

$$\mathcal{H}_a := -2q_{ac} D_b p^{bc}, \tag{14.3}$$

$$\mathcal{H} := \frac{\kappa}{\sqrt{det(q)}} [q_{ac} q_{bd} - \frac{1}{2} q_{ab} q_{cd}] p^{ab} p^{cd} - \sqrt{det(q)} \frac{R}{\kappa}. \tag{14.4}$$

where R is the Ricci scalar and D is a covariant differential operator. We can define smeared versions of the constraints with respect to the multipliers as:

$$H(N) = \int_{\Sigma} d^3x N \mathcal{H} \tag{14.5}$$

$$H(N^a) = \int_{\Sigma} d^3x N^a \mathcal{H}_a \tag{14.6}$$

The constraints then obey relations of the form:

---

[2] Here for the most part we are following the treatment of Thiemann (2007)

$$\{\vec{H}(\vec{N}), \vec{H}(\vec{N}')\} = -\kappa \vec{H}(\mathfrak{L}_{\vec{N}}\vec{N}'), \tag{14.7}$$

$$\{\vec{H}(\vec{N}), H(N)\} = -\kappa H(\mathfrak{L}_{\vec{N}}N), \tag{14.8}$$

$$\{H(N), H(N')\} = -\kappa \vec{H}(F(N, N', q)), \tag{14.9}$$

where $\mathfrak{L}$ is the Lie derivative of operator and $F(N, N', q) = q^{ab}(NN'_{,b} - N'N_{,b})$. The presence of structure functions on the right-hand side of Equation (14.9) prevents closure as an algebra and means that the associated set of transformations on phase space are a groupoid rather than a group.

Just as the single Hamiltonian constraint of a Jacobi-type theory is associated with spatially global and temporally local reparameterizations, the infinite family of Hamiltonian constraints, $\vec{H}(\vec{N})$, of canonical general relativity is associated with spatially and temporally local reparameterizations, or refoliations. As in the simple case, the manner of this association is a subtle issue. In particular, once more we have that the symmetry in questions is defined at the level of histories, whereas the constraint function is defined at the level of phase space data. A slightly more detailed and yet concise expression of the problem can be stated as follows.

On the one hand, we have that the representation theorem due to Teitelboim (1973) allows us to connect the Hamiltonian constraints to the normal deformations of three-dimensional hypersurfaces embedded within four geometries in terms of the specific form of the structure functions in the algebra (14.9). However, on the other hand, the Teitelboim representation theorem is not sufficient to demonstrate that the role of the Hamiltonian constraints is *all and only* that of effecting a refoliation. In particular, as argued in Gryb and Thébault (2016b), at the level of phase space points, we do not even have a unique state-by-state representation of refoliations.

The crucial point here is that one needs to specify the embedding of the phase space data into spacetime in order to define a refoliation. Without this embedding information, it is impossible to construct the explicit refoliation map between two histories on phase space. In the context of the Teitelboim theorem, the same issue can be expressed in terms of the requirement for a spacetime metric in order to define the relevant defomations. This metric can may be defined either explicitly as spacetime geometric data or implicitly via reconstruction of the embedding from canonical data, but without it there simply is no comprehensible sense in which the statement 'Hamiltonian constraints generate normal deformations or refoliations' can be made. Neither concept is well defined at the level of phase space points.

In more general term, the relationship between the four-dimensional diffeomophism group and the transformations generated by the constraints of canonical general relativity is a subtle one. When the vacuum equations of motion and the constraints are satisfied, the groupoid of transformations generated by the constraints (14.7)–(14.9) can be mapped onto a subgroup of the spacetime diffeomorphism group consisting of those diffeomorphisms that preserve the spacelike nature of the embedding and are connected to the identity.[3] However, there does not exist a general algorithm for uniquely distinguishing the dynamics from the dynamical redundancy of the theory

---

[3] Most explicitly, we can construct the so-called *Bergman–Komar group* $BK(M)$ which represents the specific form of four-dimensional field dependent infinitesimal diffeomorphisms that preserve the embedding and thus can be represented in phase space by weighted sums of the constraints, the lapse

within a generic phase space representation. Mainstream approaches rely upon application of the Dirac approach in which the constraints are directly interpreted as gauge generation. If, as in the case of reparameterizations and the associated global Hamiltonian constraints, we wish to treat refoliations and the associated local Hamiltonian constraints as *sui generis*, then new formal techniques, such as a generalization of the first-order formalism, will be required.

14.4   The Two Faces of Classical Gravity

It is in the specific context of the distinction between reparameterization invariance and refoliation invariance and the problem of time that the next crucial ingredient in our approach to the problem of time in classical and quantum gravity comes in. In particular, our approach is based upon a reformulation of classical canonical gravity in which the symmetries of the ADM theory are *traded* for a different set of symmetries such that a dual formalism is derived wherein reparameterization invariance has replaced refoliation invariance. As per our discussion above, this is equivalent to moving from a theory with an infinite set of local Hamiltonian constraints to a theory with a single non-local Hamiltonian constraint. Let us briefly sketch the conceptual and formal moves involved in this reformulation.

General relativity is a theory with very little 'background' spatiotemporal structure in particular spatiotemporal coordinate structure. However, general relativity still privileges length scales; that is, models of the theory related by conformal re-scalings are on standard approaches understood to be physically distinct. Interestingly, the same type of argument that enjoins us to eliminate coordinate dependence in physical theory also motivates us to eliminate scale dependence. In particular, just as we have no empirical access to absolute coordinate structures—only relative ones, we may argue that we have no empirical access to absolute scale structures—only relative ones.[4]

This leads us directly to the question of whether local scale invariance can be implemented within a physically viable formulation of general relativity? The answer here depends upon how we represent scale invariance. If we are looking for a reformulation of general relativity which implements four-dimensional conformally rescallings as invariances of the theory then the answer is unclear—such a formalism has not been isolated despite valiant attempts.[5] By contrast, in the context of three-dimensional conformal rescalings, we have available the Shape Dynamics formalism in which the subset of such three-dimensional conformal transformations that preserve global spatial volume are implemented as invariances of the theory. The steps towards the derivation of this theory from the ADM formalism are as follows.[6]

Following the discussion in Gryb and Thébault (2016*b*), the first step in our reasoning relies upon the notion of a *hidden symmetry*. The idea is to identify, for a particular

---

and shift, the structure functions from (14.9), and a set of further arbitrary multipliers that depend on the form of the diffemorprisms (Bergmann and Komar, 1972; Pons, Salisbury and Sundermeyer, 2010; Pitts, 2014).

[4] The following paragraphs of general discussion regarding Shape Dynamics are reproduced from Gryb and Thébault (2016*b*) with permission of the publisher.

[5] For discussion of a four-dimensionally scale-invariant theory of gravity see Weyl 1922; Mannheim 2012; Hooft 2010.

[6] For introductions see Mercati 2018; Gomes and Koslowski 2013; Barbour 2012.

theory, a pairing between a symmetry that is present in the theory and one that is absent such that the two symmetries satisfy a particular formal requirement. Specifically, the constraint surfaces corresponding to the symmetries need to be 'orthogonal', on phase space.[7] If this is the case, the elements of the formalism can be modified without changing the physical predictions of the theory in such a way that the first symmetry becomes explicitly realized. Remarkably, it has been proved for the case of canonical gravity that given certain reasonable simplicity assumptions, the unique set of hidden symmetries can be identified as three-dimensional conformal transformations which preserve the total volume of space (in the case of spatially closed topologies) (Gomes, 2013). Intuitively, one can think of these as transformation that *redistribute* scale from one region to another in a way that is very similar to what happens to the 2D surface of a balloon when the balloon is squashed or deformed. As we noted, hidden symmetries can only be identified when there exists another symmetry which has the required formal relationship such that the two sets can be understood as *dual* to each other. For the case in hand, the relevant dual to the scale symmetry is *almost all* of the foliation symmetry. This means that if we *symmetry trade* such that the hidden volume preserving scale transformation symmetry becomes manifest, we simultaneously switch to a theory with merely global, rather than local, time relabelling symmetry. This corresponds to a single Hamiltonian constraint.

Specifically, what was proven in Gomes, Gryb and Koslowski (2011) is that there exits two theories on the phase space of General Relativity that are physically equivalent but have different symmetries: one is the standard ADM theory, which is foliation invariant, and the other is Shape Dynamics, which is invariant under (volume-preserving) conformal transformations. The physical equivalence of the formalisms is expressed by the fact that there is a special gauge choice in both theories where the dynamical trajectories on phase space are identical, given some valid initial data.[8]

Once symmetry trading is completed, the theory we get is more naturally amenable to the framework we have developed since we no longer have to deal with the problem of interpreting refoliations and the infinite family of local Hamiltonian constraints. Rather, it has a neatly divided set of constraints: volume-preserving conformal transformations and spatial diffeomorphism symmetries can be classified as gauge symmetries and treated as per the standard Dirac approach; and a single non-local Hamiltonian constraint that is associated with spatially global time reparameterizations. The local problem of time in general relativity is thus reduced to the global problem of time in shape dynamics. Significantly, this is achieved by moving to a *dual* formalism of gravity, in terms of shaped dynamics and not via a restriction to a subclass of models.[9]

[7] For the general theory behind symmetry trading, see (Gomes and Koslowski, 2012).

[8] It is possible, however, for these theories to differ if, for whatever reason, there are global obstructions to imposing the special gauges in both frameworks. These possibilities have been explored, for instance, in the case of black holes Gomes (2014).

[9] It is worth noting in this context that the starting point for the construction of shape dynamics is globally hyperbolic models compatible with constant mean curvature foliations. This is to be compared with the standard ADM formalism which is constituted by globally hyperbolic models without the constant mean curvature restriction. The more physically substantive restriction to the global hyperbolic spacetimes that admit foliation into space-like hyper surfaces is thus *already* applied within any canonical formalism, including loop quantum gravity. In the canonical context at least,

14.5 Relational Quantization and the New Copernican Principle

Relational quantization is a novel approach to the quantization of theories invariant under global time reparameterizations which is such that chronordinal structure is retained in the quantum theory and the evolution equations are of the Schrödinger rather than the Wheeler–DeWitt type. Here we will set out the key general motivation for relational quantization based upon what we will call the *New Copernican Principle*.

The New Copernican Principle is an epistemic principle relating to the interpretation of constants that appear in our dynamical equations and be simply stated as an assertion that in a situation in which our epistemic access to different scales is inherently limited, the categorization of constants as constants of nature or constants of motion is always open to revision based upon the acquisition of a wider epistemic window. Articulations of forms of the principle can be found in the work of Poincaré (2017), Schlick (1980), and Lévy-Leblond (1977). Furthermore, the principle has already played an implicit role in our analysis in terms of the distinction between the energy and a constant of nature and energy as a constant of motion interpretations of Jacobi theories.

This principle is interesting and important in a purely classical context. In such a context it is however primarily an interpretative idea that allows for distinct yet empirically equivalent specifications of the structure of a physical theory. In a quantum context, by contrast, the principle is potentially transformative. This is because constants of motion and constants of nature are treated entirely differently in quantum theories. Whereas the superpositions of the first can correspond to pure states, the second are subject to superselection rules that prohibit observables having matrix values in sectors of Hilbert space corresponding to different values.

The application of the New Copernican Principle to Jacobi-type theories is then equivalent to the extension of the distinction between the energy as a constant of nature and energy as a constant of motion interpretations to the quantum theory. Crucially, however, in the quantum context this is now a distinction with a physical difference. Theories in which energy is superselected to a zero eigenvalue have different empirical consequences to those in which the quantum state can be composed of superpositions of non-zero energy eigenvalues. Wheres the latter gives a Wheeler–DeWitt-type theory, the former gives a Schrödinger-type theory, which is equivalent to standard quantum theory.

The formal implementation of relational quantization can be achieved in a number different ways. The most straightforward is to explicitly promote the time parameter and its conjugate momentum to phase space variables and then apply a standard Dirac quantization algorithm. The formal steps involved in such an approach are set out in Gryb and Thébault (2014). A similar procedure can also be applied based upon a Faddeev–Popov path integral approach. This is described in Gryb and Thébault (2011). Finally, and arguably most insightfully, one can appeal to the generalized Hamilton–Jacobi formalism described in 12.3 combined with the original heuristic for quantization discovered by Schrödinger, to directly derive a dynamical equation. This

---

there is therefore very little substantive difference between our approach and the complete and partial observable programme in terms of model class restriction. cf. Fortin, Lombardi and Pasqualini (2022).

procedure is described in Gryb and Thébault (2016*a*). Relational quantizaion thus rests on an extremely solid conceptual and formal basis. Moreover, the approach provides an explicit extension of our solution to the problem of time in classical mechanics to the quantum domain. It remains to be seen, however, the extent to which this approach generalizes to the case of gravitational theories. In the following section we will sketch the details of formal results that embody such an extension and, furthermore, provide new insights into the resolution of big bang singularities in quantum cosmology.

14.6   Resolving the Big Bang

The 'big bang' singularity and the cosmological constant are well-established features of classical cosmological models.[10] In classical cosmology, the big bang is identified in terms of both a global dynamical notion of incompleteness of inextendible causal (i.e. non-spacelike) past-directed curves and a local notion of the existence of a curvature pathology.[11] The standard understanding of the cosmological constant is then as additional term in the Einstein equation which plays the functional role of an accelerative 'dark energy' as determined by an additional, extremely small, constant of nature, Λ. In the context of quantum cosmology, there is the long standing hope, generically unfulfilled in Wheeler–DeWitt approaches, that quantization will resolve singular behaviour in the sense of restoring well-posed global dynamics and removing the local pathological features, for example, in terms of bounded expectation values for curvature (Thébault, 2023). The standard mechanism for achieving quantum singularity resolution is the introduction of a Planck-scale cut-off, applied in tandem with the complete and partial observables scheme (Ashtekar, Corichi and Singh, 2008). There is a well-developed research programme devoted to consideration of the phenomenology of such models in the context of Loop Quantum Cosmology (Ashtekar and Singh, 2011; Ashtekar and Bianchi, 2021; Bojowald, 2020).

In contrast to the differing fate of singularities in classical and quantum cosmological models, in standard approaches to quantum cosmology the cosmological constant receives very much the same treatment in classical and quantum cosmological models: it is a constant of nature classically, and thus quantum solutions are supersselected to eigenstates labelled by its classical value. It is in this context that relational quantization offers a new approach to quantum cosmology. Remarkably, when applied to simple cosmological models with a single Hamiltonian constraint it is found that relational quantization generically resolves the big bang singularity such that unitary dynamics and finite expectation values are guaranteed within a class of models that are pathological both classically and within the Wheeler–DeWitt quantization scheme. As was discussed in the last section, the crucial formal and conceptional ingredient of relational quantization is the reinterpretation of constants according to the new copernican principle. In the context of cosmology the relevant constant is the cosmological constant and thus relational quantization is built around treating the cosmological constant as a constant of motion. This leads to a quantum cosmological formalism

[10] Here and below we are following the discussion provided in Gryb and Thébault (2018).

[11] See Curiel (2019) for an excellent overview and Penrose 1965; Hawking and Penrose 1970; Hawking and Ellis 1973; Senovilla 1998; Curiel 1999 for more details.

where the wavefunction of the universe can be in superpositions of eigenstates of Λ. Relational quantization in this sense leads to a formalism that is closely connected to unimodular gravity (Unruh and Wald, 1989) and certain quantum bounce scenarios (Gielen and Turok, 2016).

Significantly, bouncing models derived via relational quantization are physically distinct from those derived via loop quantization precisely because they can feature superpositions of the cosmological constant. Characteristic features of the cosmological bounce persist into a 'super-inflation' regime that contains universal phenomenology that can be rendered insensitive to the underlying Planck-scale physics in very nature way. In particular, the model displays a 'cosmic beat' phenomenon and associated 'bouncing envelope'. The cosmic beats can be identified with Planck-scale effects and, under certain parameter constraints, are negligible compared with the effective envelope physics. Under these same constraints, the bouncing envelope persists into the super-inflation regime where it is insensitive to the beat effects in a manner that is closely analogous to *Rayleigh scattering*. Significantly, this 'Rayleigh' limit is only available when superpositions of the cosmological constant are allowed. This behaviour constitutes a remarkable unique feature of the bouncing unitary cosmologies identified.

Our novel approach to the problem of time thus leads to a formally distinct and potentially physically insightful approach to the early universe. An overview of the approach detailing the key results for the minisuperspace class of models can be found in Gryb and Thébault (2018). For an extended discussion of the general solution space of the model including the crucial formal conditions that guarantee of unitarity see Gryb and Thébault (2019a). For explicit construction of the solutions including numerical exploration see Gryb and Thébault (2019b). Further study of this model in the context of unitarity and the comparison with de-parametrized models can be found in Gielen and Menéendez-Pidal (2020, 2022).

References

Abraham, R. and Marsden, J. E. (1978). *Foundations of Mechanics.* Benjamin/Cummings.

Albert, D. (2009). *Time and Chance.* Harvard University Press.

Alexander, H. G. (1998). *The Leibniz–Clarke Correspondence: Together with Extracts from Newton's Principia and Optics.* Manchester University Press.

Alliluev, S. (1958). On the relation between 'accidental' degeneracy and 'hidden' symmetry of a system. *Soviet Journal of Experimental and Theoretical Physics*, **6**, 156.

Anderson, E. (2014). Beables/observables in classical and quantum gravity. *SIGMA*, **10**(092), 092.

Anderson, E. (2017). *The Problem of Time: Quantum Mechanics Versus General Relativity.* Vol. 190. Springer.

Anderson, I. M. and Torre, C. G. (1996). Classification of local generalized symmetries for the vacuum Einstein equations. *Communications in Mathematical Physics*, **176**(3), 479–539.

Anderson, J. L. and Bergmann, P. G. (1951). Constraints in covariant field theories. *Physical Review*, **83**(5), 1018.

Ariew, R. (1994). *G. W. Leibniz, Life and Works.* Cambridge University Press.

Arnold, V. I. (1990). *Singularities of Caustics and Wave Fronts.* Vol. 62. Springer Science & Business Media.

Arnold, V. I. (1992). *Ordinary Differential Equations.* Springer Science & Business Media.

Arnold, V. I. (2013). *Mathematical Methods of Classical Mechanics.* Vol. 60. Springer Science & Business Media.

Arnold, V. I. and Givental, A. B. (2001). *Dynamical Systems IV: Symplectic Geometry and Its Applications.* Vol. 4. Springer Science & Business Media.

Arnowitt, R., Deser, S., and Misner, C. W. (1960). Canonical variables for general relativity. *Physical Review*, **117**, 1595–1602.

Arthur, R. T. W. (1985). Leibniz's theory of time. In *The Natural Philosophy of Leibniz* (eds K. Okruhlik and J. R. Brown), pp. 263–313. Springer.

Arthur, R. T. W. (1995). Newton's fluxions and equably flowing time. *Studies in History and Philosophy of Science Part A*, **26**(2), 323–351.

Arthur, R. T. W. (2014). *Leibniz.* Polity Press.

Arthur, R. T. W. (2019). *The Reality of Time Flow.* Springer.

Arthur, R. T. W. (2021). *Leibniz on Time, Space and Relativity.* Oxford University Press.

Ashtekar, A. and Bianchi, E. (2021). A short review of loop quantum gravity. *Reports on Progress in Physics*, **84**, 042001.

Ashtekar, A., Corichi, A., and Singh, P. (2008). Robustness of key features of loop quantum cosmology. *Physical Review D*, **77**(2), 024046.

Ashtekar, A. and Singh, P. (2011). Loop quantum cosmology: A status report. *Classical and Quantum Gravity*, **28**(21), 213001.

Banks, E. C. (2003). *Ernst Mach's World Elements: A Study in Natural Philosophy*. Vol. 68. Springer Science & Business Media.

Banks, E. C. (2014). *The Realistic Empiricism of Mach, James, and Russell: Neutral Monism Reconceived*. Cambridge University Press.

Banks, E. C. (2021). The Case for Mach's Neutral Monism. In *Interpreting Mach* (ed J. Preston), pp. 258–279, Cambridge University Press.

Barbour, J. B. (1981). Mach's Mach's principles, especially the second. In *Grundlagen-Probleme der Modernen Physik* (eds J. Nitsch and E.-W. Stachow), pp. 41–65. Bibliographisches Institut, Mannheim/Wien/Züric.

Barbour, J. B. (1995). General relativity as a perfectly machian theory. In *Mach's Principle: From Newton's Bucket to Quantum Gravity* (eds J. B. Barbour and H. Pfister), pp. 214–236. Birkhäuser.

Barbour, J. B. (2003). Scale-Invariant Gravity: Particle Dynamics. *Classical and Quantum Gravity*, **20**, 1543–1570.

Barbour, J. B. (2001). *The Discovery of Dynamics: A Study from a Machian Point of View of the Discovery and the Structure of Dynamical Theories*. Oxford University Press.

Barbour, J. B. (2009a). Mach's Principle, General Relativity and Gauge Theory (unpublished).

Barbour, J. B. (2009b). The nature of time. *arXiv preprint*, 0903.3489.

Barbour, J. B. (2012). Shape dynamics: An introduction. In *Quantum Field Theory and Gravity: Conceptual and Mathematical Advances in the Search for a Unified Framework* (eds F. Finster, O. Müller, M. Nardmann, J. Tolksdorf, E. Zeidler), pp. 257–297. Springer.

Barbour, J. B. and Foster, B. Z. (2008). Constraints and gauge transformations: Dirac's theorem is not always valid. *arXiv preprint*. 808.1223.

Barbour, J., Koslowski, T., and Mercati, F. (2014). Identification of a gravitational arrow of time. *Physical Review Letters*, **113**(18), 181101.

Barbour, J. B. and Bertotti, B. (1982). Mach's principle and the structure of dynamical theories. *Proceedings of the Royal Society of London. A. Mathematical and Physical Sciences*, **382**(1783), 295–306.

Baron, S. and Evans, P. W. (2021). What's so spatial about time anyway? *British Journal for the Philosophy of Science*, **72**(1), 159–183.

Barrett, T. W. and Halvorson, H. (2016). Glymour and Quine on theoretical equivalence. *Journal of Philosophical Logic*, **45**(5), 467–483.

Belot, G. (1999). Rehabilitating relationalism. *International Studies in the Philosophy of Science*, **13**, 35–52.

Belot, G. (2003). Symmetry and gauge freedom. *Studies In History and Philosophy of Modern Physics*, **34**, 189–225.

Belot, G. (2007). The representation of time and change in mechanics. In *Handbook of Philosophy of Physics* (eds J. Butterfield and J. Earman), Chapter 2. Elsevier.

Belot, G. (2013). Symmetry and equivalence. In *The Oxford Handbook of Philosophy of Physics* (ed R. Batterman). Chapter 9. Oxford University Press.

Belot, G. (2018). Fifty million Elvis fans can't be wrong. *Noûs*, **52**(4), 946–981.

Belot, G. and Earman, J. (2001). Pre-Socratic quantum gravity. In *Physics Meets Philosophy at the Planck Scale* (eds C. Callender and N. Hugget), Chapter 10. Cambridge University Press.

Bergmann, P. G. and Komar, A. (1972). The coordinate group symmetries of general relativity. *International Journal of Theoretical Physics*, **5**(1), 15–28.

Black, M. (1959). The 'direction' of time. *Analysis*, **19**(3), 54–63.

Blackmore, J. T. (ed). (1992). *Ernst Mach—A Deeper Look: Documents and New Perspectives*. Springer Science & Business Media.

Bojowald, M. (2020). Critical evaluation of common claims in loop quantum cosmology. *Universe*, **6**(3), 36.

Bojowald, M., Höhn, P. A., and Tsobanjan, A. (2011*a*). An effective approach to the problem of time. *Classical and Quantum Gravity*, **28**(3), 035006.

Bojowald, M., Höhn, P. A., and Tsobanjan, A. (2011*b*). Effective approach to the problem of time: general features and examples. *Physical Review D*, **83**(12), 125023.

Borzeszkowski, H. and Wahsner, R. (1995). Mach's criticism of Newton and Einstein's reading of Mach: The stimulating role of two misunderstandings. *Einstein Studies*, **6**, 58–66.

Brading, K. (2016). Time for empiricist metaphysics. In *Metaphysics and the Philosophy of Science New Essays* (eds M. Slater and Z. Yurdell), pp. 13–40. Oxford University Press.

Brading, K. and Brown, H. R. (2003). Symmetries and Noether's theorems. In *Symmetries in Physics: Philosophical Reflections*, (eds K. Brading and E. Castellani), pp. 89–109. Cambridge University Press.

Brading, K. and Brown, H. R. (2004). Are gauge symmetry transformations observable? *The British Journal for the Philosophy of Science*, **55**(4), 645–665.

Bradley, C. and Weatherall, J. O. (2019). On representational redundancy, surplus structure, and the hole argument. *arXiv preprint arXiv:1904.04439*.

Bradley, C. and Weatherall, J. O. (2020). On representational redundancy, surplus structure, and the hole argument. *Foundations of Physics*, 1–24.

Bravetti, A. (2018). Contact geometry and thermodynamics. *International Journal of Geometric Methods in Modern Physics*, **16**(supp01), 1940003.

Bravetti, A. and Garcia-Chung, A. (2021). A geometric approach to the generalized Noether theorem. *Journal of Physics A: Mathematical and Theoretical*, **54**(9), 095205.

Brighouse, C. (1994). Spacetime and holes. In *PSA: Proceedings of the Biennial Meeting of the Philosophy of Science Association*, pp. 117–125. Cambridge University Press.

Brown, H. R. (2005). *Physical Relativity: Space-Time Structure from a Dynamical Perspective*. Oxford University Press.

Brush, S. G. (ed) (1966). *Kinetic Theory: Irreversible Processes*. Pergamon Press.

Brush, S. G. (1968). Mach and atomism. *Synthese*, **18**(2-3), 192–215.

Bunge, M. (1966). Mach's critique of Newtonian mechanics. *American Journal of Physics*, **34**(7), 585–596.

Butterfield, J. (1989). The hole truth. *British Journal for the Philosophy of Science*, **40**(1), 1–28.

Butterfield, J. (2007). On symplectic reduction in classical mechanics. In *Philosophy of Physics* (eds J. Butterfield and J. Earman), pp. 1–131. Elsevier.

Callender, C. (2017). *What Makes Time Special?* Oxford University Press.

Callender, C. (2021). Thermodynamic asymmetry in time. In *The Stanford Encyclopaedia of Philosophy* (Summer 2021 edn) (ed E. N. Zalta). Metaphysics Research Lab, Stanford University.

Carnap, R. (2019). *Space: A Contribution to the Theory of Science*. (trans. A. Carus and M. Friedmann). In *Rudolf Carnap: Early Writings: The Collected Works of Rudolf Carnap, Volume 1*, (eds A.W. Carus, M. Friedman, W. Kienzler, A. Richardson and S. Schlotter). Oxford University Press.

Castellani, L. (1982). Symmetries in constrained Hamiltonian systems. *Annals of Physics*, **143**(2), 357–371.

Caulton, A. (2015). The role of symmetry in the interpretation of physical theories. *Studies in History and Philosophy of Science Part B: Studies in History and Philosophy of Modern Physics*, **52**, 153–162.

Chaichian, M. and Martinez, D. L. (1994). On the Noether identities for a class of systems with singular Lagrangians. *Journal of Mathematical Physics*, **35**(12), 6536–6545.

Corichi, A. (2008). On the geometry of quantum constrained systems. *Classical and Quantum Gravity*, **25**(13), 135013.

Cover, J. (1997). Non-basic time and reductive strategies: Leibniz's theory of time. *Studies in History and Philosophy of Science Part A*, **28**(2), 289–318.

Curiel, E. (1999). The analysis of singular spacetimes. *Philosophy of Science*, S119–S145.

Curiel, E. (2010). The Geometry of the Euler-Lagrange Equation *Unpublished*, http://strangebeautiful.com/papers/curiel-geom-ele.pdf.

Curiel, E. (2015). Measure, topology and probabilistic reasoning in cosmology *preprint arXiv*, 1509.01878.

Curiel, E. (2018). On the existence of spacetime structure. *The British Journal for the Philosophy of Science*, **69**(2), 447–483.

Curiel, E. (2019). On geometric objects, the non-existence of a gravitational stress-energy tensor, and the uniqueness of the Einstein field equation. *Studies in History and Philosophy of Science Part B: Studies in History and Philosophy of Modern Physics*, **66**, 90–102.

Curiel, E. (2019). Singularities and black hole. *Stanford Encyclopedia of Philosophy*. In *The Stanford Encyclopaedia of Philosophy* (Summer 2019 edn) (ed E. N. Zalta). Metaphysics Research Lab, Stanford University.

Dasgupta, S. (2015). Substantivalism vs relationalism about space in classical physics. *Philosophy Compass*, **10**(9), 601–624.

Dasgupta, S. (2016). Symmetry as an epistemic notion (twice over). *The British Journal for the Philosophy of Science*, **67**(3), 837–878.

De Gandt, F. (2014). *Force and Geometry in Newton's Principia*. Princeton University Press.

de Leon, M. and Lainz Valcázar, M. (2019). Contact Hamiltonian systems. *Journal of Mathematical Physics*, **60**(10), 102902.

De Risi, V. (2007). *Geometry and Monadology: Leibniz's Analysis Situs and Philosophy of Space*. Vol. 33. Springer Science & Business Media.

Descartes, R. (1984). *The Philosophical Writings of Descartes, vols. 1 and 2, (trans. J. Cottingham, R. Stoothoff, and D. Murdoch*. Cambridge University Press.

Dewar, N. (2019). Sophistication about symmetries. *The British Journal for the Philosophy of Science*, **70**(2), 485–521.

Díaz, B., Higuita, D., and Montesinos, M. (2014). Lagrangian approach to the physical degree of freedom count. *Journal of Mathematical Physics*, **55**(12), 122901.

Diaz, B. and Montesinos, M. (2018). Geometric Lagrangian approach to the physical degree of freedom count in field theory. *Journal of Mathematical Physics*, **59**(5), 052901.

Díez, V. E., Maier, M., Méndez-Zavaleta, J. A., and Tehrani, M. T. (2020). Lagrangian constraint analysis of first-order classical field theories with an application to gravity. *Physical Review D*, **102**(6), 065015.

Dirac, P. A. (1959, May). Fixation of coordinates in the Hamiltonian theory of gravitation. *Physical Review*, **114**, 924–930.

Dirac, P. A. M. (1950). Generalized Hamiltonian dynamics. *Canadian Journal of Mathematics*, **2**, 129–148.

Dirac, P. A. M. (1958). The theory of gravitation in hamiltonian form. *Proceedings of the Royal Society of London. Series A, Mathematical and Physical Sciences*, **246**, 333–343.

Dirac, P. A. M. (1964). *Lectures on Quantum Mechanics*. Dover Publications.

DiSalle, R. (1990). Gereon Wolters' Mach I, Mach II, Einstein, und Die Relativitätstheorie. Eine Fälschung und Ihre Folgen. *Philosophy of Science*, **57**(4), 712–723.

DiSalle, R. (2002). Reconsidering Ernst Mach on space, time, and motion. In *Reading Natural Philosophy: Essays in the History and Philosophy of Science and Mathematics*, (ed D. B. Malament), pp. 167–191. Open Court,.

DiSalle, R. (2006). *Understanding Space-Time: The Philosophical Development of Physics from Newton to Einstein*. Cambridge University Press.

DiSalle, R. (2016). *Newton's Philosophical Analysis of Space and Time* (2nd)., pp. 34–60. Cambridge University Press.

DiSalle, R. (2020). Space and time: Inertial frames. In *The Stanford Encyclopaedia of Philosophy* (Winter 2016 edn) (ed E. N. Zalta). Metaphysics Research Lab, Stanford University.

Dittrich, B. (2006). Partial and complete observables for canonical general relativity. *Classical and Quantum Gravity*, **23**, 6155.

Dittrich, B. (2007). Partial and complete observables for hamiltonian constrained systems. *General Relativity and Gravitation*, **39**, 1891.

Dittrich, B., Höhn, P. A., Koslowski, T. A., and Nelson, M. I. (2017). Can chaos be observed in quantum gravity? *Physics Letters B*, **769**, 554–560.

Ducheyne, S. (2006). Newton's secularized onto-theology versus Descartes' and Leibniz', or on the importance of unifying tendencies in the secularization-process. *Theology and Science*, **4**(1), 71–85.

Earman, J. (1989). *World Enough and Spacetime*. MIT Press.

Earman, J. (2003). Tracking down gauge: An ode to the constrained Hamiltonian formalism. In *Symmetries in physics* (eds K. Brading and E. Castellani), pp. 150–162. Cambridge University Press.

Earman, J. (2006). The 'Past Hypothesis': Not even false. *Studies in History and Philosophy of Science Part B: Studies in History and Philosophy of Modern Physics*, **37**(3), 399–430.

Euler, L. (1748). Réflexions sur l'espace et le temps. *Histoire de l'Academie Royale des sciences et belles lettres*, **4**, 324–33.

Evans, P. W. and Thébault, K. P. Y. (2020). On the limits of experimental knowledge. *Philosophical Transactions of the Royal Society A: Mathematical, Physical and Engineering Sciences*, **378**, 20190235.

Farr, M. (2012). *Towards a C Theory of Time*. Ph.D. thesis, University of Bristol.

Farr, M. (2016). Causation and time reversal. *The British Journal for the Philosophy of Science*, **71**(1)

Farr, M. (2020). C-theories of time: On the adirectionality of time. *Philosophy Compass*, **15**(12), e12714.

Feyerabend, P. K. (1984). Mach's theory of research and its relation to Einstein. *Studies in History and Philosophy of Science Part A*, **15**(1), 1–22.

Fletcher, S. C. (2018). On representational capacities, with an application to general relativity. *Foundations of Physics*, 1–22.

Fortin, S., Lombardi, O., and Pasqualini, M. (2022). Relational event-time in quantum mechanics. *Foundations of Physics*, **52**(1), 10.

Friederich, S. (2015). Symmetry, empirical equivalence, and identity. *British Journal for the Philosophy of Science*, **66**(3), 537–559.

Friedman, M. (1983). *Foundations of Space-time Theories: Relativistic Physics and Philosophy of Science*. Princeton University Press.

Friedman, M. (2001). *Dynamics of Reason*. Publications Stanford.

Friedman, M. (2019). *Editorial Notes to Space: A Contribution to the Theory of Science*. In *Rudolf Carnap: Early Writings: The Collected Works of Rudolf Carnap, Volume 1*, (eds A.W. Carus, M. Friedman, W. Kienzler, A. Richardson and S. Schlotter). Oxford University Press.

Gambini, R. and Porto, R. A. (2001). Relational time in generally covariant quantum systems: four models. *Physical Review D*, **63**(10), 105014.

Gambini, R., Porto, R. A., Pullin, J., and Torterolo, S. (2009). Conditional probabilities with dirac observables and the problem of time in quantum gravity. *Physical Review D*, **79**(4), 041501.

Garber, D. (1994). Leibniz: Physics and philosophy. In *The Cambridge Companion to Leibniz* (ed N. Jolley), pp. 270–352. Cambridge University Press.

Gaukroger, S. (2002). *Descartes' System of Natural Philosophy*. Cambridge University Press.

Gaukroger, S. (2006). *The emergence of a Scientific Culture: Science and the Shaping of Modernity 1210–1685*. Oxford University Press.

Gielen, S. and Menéndez-Pidal, L. (2020). Singularity resolution depends on the clock. *Classical and Quantum Gravity*, **37**(20), 205018.

Gielen, S. and Menéndez-Pidal, L. (2022). Unitarity, clock dependence and quantum recollapse in quantum cosmology. *Classical and Quantum Gravity*, **39** (7), 075011

Gielen, S. and Turok, N. (2016). Perfect quantum cosmological bounce. *Physical Review Letters*, **117**(2), 021301.

Gitman, D. and Tyutin, I. V. (1990). *Quantization of fields with constraints*. Springer.

Giulini, D. and Marolf, D. (1999*a*). A uniqueness theorem for constraint quantization. *Classical and Quantum Gravity*, **16**, 2489–2505.

Giulini, D. and Marolf, D. (1999*b*). On the generality of refined algebraic quantization. *Classical and Quantum Gravity*, **16**, 2479–2488.

Glymour, C. (1970). Theoretical realism and theoretical equivalence. In *PSA: Proceedings of the Biennial Meeting of the Philosophy of Science Association*, Vol. 1970, pp. 275–288. D. Reidel Publishing.

Glymour, C. (1980). *Theory and Evidence*. Princeton University Press.

Glymour, C. (2013). Theoretical equivalence and the semantic view of theories. *Philosophy of Science*, **80**(2), 286–297.

Goldstein, H. (1980). *Classical Mechanics*. Adison-Wesley Publishing Company.

Goldstein, H., Poole, C., and Safko, J. (2014). *Classical Mechanics*. Pearson Education Limited

Goldstein, S. (2001). Boltzmann's approach to statistical mechanics. In *Chance in Physics: Foundations and Perspectives*, (eds Bricmont, J., Dürr, D., Galavotti, M.C., Ghirardi, G., Petruccione, F. and Zanghi, N.), pp. 39–54. Springer.

Goldstein, S. and Lebowitz, J. L. (2004). On the (Boltzmann) entropy of non-equilibrium systems. *Physica D: Nonlinear Phenomena*, **193**(1-4), 53–66.

Gomes, H. (2013). A construction principle for ADM-type theories in maximal slicing gauge. *arXiv preprint arXiv:1307.1097*.

Gomes, H. (2014). A Birkhoff theorem for Shape Dynamics. *Classical and Quantum Gravity.*, **31**, 085008.

Gomes, H. (2020, March). Gauge-invariance and the empirical significance of symmetries. *PhilSci preprint 16981*

Gomes, H. and Gryb, S. (2020). Angular momentum without rotation: turbocharging relationalism. *arXiv preprint arXiv:2011.01693*.

Gomes, H. and Koslowski, T. (2012). The link between general relativity and shape dynamics. *Classical and Quantum Gravity*, **29**, 075009.

Gomes, H. and Koslowski, T. (2013). Frequently asked questions about shape dynamics. *Foundations of Physics*, **43**, 1428–1458.

Gomes, H., Gryb, S., and Koslowski, T. (2011). Einstein gravity as a 3d conformally invariant theory. *Classical and Quantum Gravity*, **28**(4), 045005.

Gori, P. (2021). Ernst Mach and Friedrich Nietzsche: On the prejudices of scientists. In *Interpreting Mach*, (ed J. Preston), pp. 123–141. Cambridge University Press.

Gray, J. (2013). *Henri Poincaré: A Scientific Biography*. Princeton University Press.

Greaves, H. and Wallace, D. (2014). Empirical consequences of symmetries. *The British Journal for the Philosophy of Science*, **65**(1), 59–89.

Gryb, S. (2021). New difficulties for the past hypothesis. *Philosophy of Science*, **88**(3), 511–532.

Gryb, S. and Sloan, D. (2021). When scale is surplus. *Synthese*, **199**(5), 14769–14820.

Gryb, S. and Thébault, K. P. Y. (2011). The role of time in relational quantum theories. *Foundations of Physics*, 1–29.

Gryb, S. and Thébault, K. P. Y. (2014). Symmetry and evolution in quantum gravity. *Foundations of Physics*, **44**(3), 305–348.

Gryb, S. and Thebault, K. P. Y. (2015). Schrödinger evolution for the universe: Reparametrization. *arXiv preprints arXiv:1502.01225*

Gryb, S. and Thébault, K. P. Y. (2016a). Schrödinger evolution for the universe: reparametrization. *Classical and Quantum Gravity*, **33**(6), 065004.

Gryb, S. and Thébault, K. P. (2016b). Time remains. *The British Journal for the Philosophy of Science*, **67**(3), 663–705.

Gryb, S. and Thébault, K. P.Y. (2018). Superpositions of the cosmological constant allow for singularity resolution and unitary evolution in quantum cosmology. *Physics Letters B*, **784**, 324–329.

Gryb, S. and Thébault, K. P. Y. (2019a). Bouncing unitary cosmology I. mini-superspace general solution. *Classical and Quantum Gravity*, **36**(3), 035009.

Gryb, S. and Thébault, K. P. Y. (2019b). Bouncing unitary cosmology II. mini-superspace phenomenology. *Classical and Quantum Gravity*, **36**(3), 035010.

Guicciardini, N. (2003). *Reading the Principia*. Cambridge University Press.

Guicciardini, N. (2016). Newtonian absolute time vs fluxional time. *Almagest*, **13**(2), pp.68–83,

Guzzardi, L. (2014). Energy, metaphysics, and space: Ernst Mach's interpretation of energy conservation as the principle of causality. *Science & Education*, **23**(6), 1269–1291.

Haddock, G. E. R. (2016). *The Young Carnap's Unknown Master: Husserl's Influence on Der Raum and Der logische Aufbau der Welt*. Routledge.

Halvorson, H. (2012). What scientific theories could not be. *Philosophy of Science*, **79**(2), 183–206.

Harper, W. L. (2011). *Isaac Newton's Scientific Method: Turning Data into Evidence about Gravity and Cosmology*. Oxford University Press.

Hartz, G. A. and Cover, J. (1988). Space and time in the Leibnizian metaphysic. *Noûs*, 493–519.

Hawking, S. W. and Ellis, G. F. R. (1973). *The Large Scale Structure of Space-time*. Cambridge University Press.

Hawking, S. W. and Penrose, R. (1970). The singularities of gravitational collapse and cosmology. In *Proceedings of the Royal Society of London A: Mathematical, Physical and Engineering Sciences*, 314, 529–548.

Healey, R. (2009). Perfect symmetries. *The British Journal for the Philosophy of Science*, **60**(4), 697–720.

Henneaux, M. and Teitelboim, C. (1992). *Quantization of Gauge Systems*. Princeton University Press.

Hintikka, J. (2001). Ernst Mach at the crossroads of of twentieth-century philosophy. *Future Pasts: The Analytic Tradition in Twentieth-century Philosophy*, 81.

Höhn, P. A. (2019). Switching internal times and a new perspective on the 'wave function of the universe'. *Universe*, **5**(5), 116.

Hooft, G. t. (2010). The Conformal Constraint in Canonical Quantum Gravity. *arXiv preprint arXiv:1011.0061.*

Huggett, N. (2008). Why the parts of absolute space are immobile. *The British Journal for the Philosophy of Science*, **59**(3), 391–407.

Huggett, N. (2012). What did Newton mean by 'Absolute Motion'? In *Interpreting Newton: Critical Essays* (eds A. Janiak and E. Schliesser), pp. 196-218. Cambridge University Press.

Isham, C. and Kakas, A. (1984*a*). A group theoretical approach to the canonical quantisation of gravity. i. construction of the canonical group. *Classical and Quantum Gravity*, **1**(6), 621.

Isham, C. and Kakas, A. (1984*b*). A group theoretical approach to the canonical quantisation of gravity. ii. unitary representations of the canonical group. *Classical and Quantum Gravity*, **1**(6), 633.

Ismael, J. and Van Fraassen, B. C. (2003). Symmetry as a guide to superfluous theoretical structure. In *Symmetries in Physics* (eds K. Brading and E. Castellani), Chapter 23, pp. 371–392. Cambridge University Press.

James, L. (2022). A New Perspective on Time and Physical Laws *British Journal for the Philosophy of Science*, **73**(4): 849-877.

Janiak, A. (2008). *Newton as Philosopher.* Cambridge University Press.

Kelvin, W. T. B. and Tait, P. G. (1867). *Treatise on Natural Philosophy.* Clarendon Press.

Kiefer, C. and Zeh, H. (1995). Arrow of time in a recollapsing quantum universe. *Physical Review D*, **51**(8), 4145.

Klein, A. (2021). On the Philosophical and Scientific Relationship between Ernst Mach and William James. In *Interpreting Mach*, (ed J. Preston), pp. 103–122. Cambridge University Press.

Koslowski, T. A., Mercati, F., and Sloan, D. (2018). Through the big bang: Continuing Einstein's equations beyond a cosmological singularity. *Physics Letters B*, **778**, 339–343.

Kosmann-Schwarzbach, Y. (2010). *The Noether Theorems: Invariance and Conservation Laws in the Twentieth Century.* Springer Science & Business Media.

Kosso, P. (2000). The empirical status of symmetries in physics. *The British Journal for the Philosophy of Science*, **51**(1), 81–98.

Kostant, B. (1970). Quantization and unitary representations. *Lectures Notes in Mathematics*, 170, 87–208.

Kuchař, K. (1991). The problem of time in canonical quantization of relativistic systems. In *Conceptual Problems of Quantum Gravity* (eds A. Ashtekar and J. Stachel), p. 141. Boston University Press.

Kuchař, K. (1992). Time and interpretations of quantum gravity. In *4th Canadian Conference on General Relativity and Relativistic Astrophysics held 16-18 May, 1991 at University of Winnipeg* (eds G. Kunstatter, D.E. Vincent and J.G. Williams), p. 211. World Scientific.

Kuchař K. (1999). The problem of time in quantum geometrodynamics. *The Arguments of Time* (ed J. Butterfield), 169–196. Oxford University Press.

Ladyman, J. and Presnell, S. (2019). The hole argument in homotopy type theory. *Foundations of Physics*, 1–11.

Lange, L. (1885). Ueber das beharrungsgesetz. *Berichte der Königlichen Sachsischen Gesellschaft der Wissenschaften zu Leipzig, Mathematisch-physische*, **37**, 333–51.

Lebowitz, J. L. (1993). Boltzmann's entropy and time's arrow. *Physics Today*, **46**, 32–32.

Leibniz, G. W. (1998). *Philosophical Texts*. Oxford University Press.

Lévy-Leblond, J.-M. (1977). On the conceptual nature of the physical constants. *La Rivista del Nuovo Cimento*, **7**(2), 187–214.

Locke, J. (2008). *An Essay Concerning Human Understanding*. Oxford University Press.

Loemker, L. E. (1969). *Gottfried Wilhelm Leibniz: Philosophical Papers and Letters*. Kluwer.

Logan, J. D. (1977). *Invariant Variational Principles*. Academic Press.

Lusanna, L. (1990). An enlarged phase space for finite-dimensional constrained systems, unifying their Lagrangian, phase-and velocity-space descriptions. *Physics Reports*, **185**(1), 1–54.

Lusanna, L. (1991). The second Noether theorem as the basis of the theory of singular Lagrangians and Hamiltonians constraints. *La Rivista del Nuovo Cimento (1978–1999)*, **14**(3), 1–75.

Lutz, S. (2017). What was the syntax-semantics debate in the philosophy of science about? *Philosophy and Phenomenological Research*, **95**(2), 319–352.

Mach, E. (1895). *Popular Scientific Lectures* (1st English, 1st German edn). Open Court Publishing

Mach, E. (1911). *History and Root of the Principle of the Conservation of Energy* (1st English, 2nd German edn). Open Court Publishing.

Mach, E. (1914). *The Analysis of Sensations, and the Relation of the Physical to the Psychical*. Open Court Publishing.

Mach, E. (1919). *The Science of Mechanics: A Critical and Historical Account of its Development* (2nd English, 4th German edn). Open Court Publishing Company.

Mach, E. (1976). *Knowledge and Error*. Springer.

Mach, E. (2012*a*). *Die Mechanik in ihrer Entwicklung: historisch-kritisch dargestellt*. Xenomoi Verl.

Mach, E. (2012*b*). *Principles of the Theory of Heat: Historically and Critically Elucidated*. Springer Science & Business Media.

Malament, D. B. (2012). *Topics in the Foundations of General Relativity and Newtonian Gravitation Theory*. University of Chicago Press.

Mannheim, P. D. (2012). Making the case for conformal gravity. *Foundations of Physics*, **42**, 388–420.

Martens, N. and Read, J. (2021). Sophistry about symmetries? *Synthese*, **199** 315–344

Massimi, M. (2007). Saving unobservable phenomena. *British Journal for the Philosophy of Science*, **58**, 235–262.

Maudlin, T. (1988). The essence of space-time. In *PSA: Proceedings of the Biennial Meeting of the Philosophy of Science Association*, pp. 82–91. Cambridge University Press

Maudlin, T. (2012). *Philosophy of Physics: Space and Time*. Princeton University Press.

McGuire, J. E. (1976). 'Labyrinth continui' Leibniz on substance, activity and matter. In *Motion and Time, Space and Matter*. (ed R. T. P. Machamer), pp. 290–327. Ohio State University Press.

McLaughlin, B. and Bennett, K. (2018). Supervenience. In *The Stanford Encyclopedia of Philosophy* (Winter 2018 edn) (ed E. N. Zalta). Metaphysics Research Lab, Stanford University.

McTaggart, J. E. (1908). The unreality of time. *Mind*, 457–474.

Mercati, F. (2018). *Shape Dynamics: Relativity and Relationalism*. Oxford University Press.

Mercati, F. (2019). Through the Big Bang in inflationary cosmology. *Journal of Cosmology and Astroparticle Physics*, , **10**, 025.

Meyer, U. (2013). *The Nature of Time*. Oxford University Press.

Miller, W., Post, S., and Winternitz, P. (2013). Classical and quantum superintegrability with applications. *Journal of Physics A: Mathematical and Theoretical*, **46**(42), 423001.

Mittelstaedt, P. (1980). *Der Zeitbegriff in der Physik*. Spektrum Akademischer Verlag.

Mugani, M. (2012). Leibniz's theory of relations: a last word? In *Oxford Studies in Early Modern Philosophy*, Vol. 6, Chapter pp. 171–208. Oxford University Press.

Mukunda, N. (1980). Generators of symmetry transformations for constrained hamiltonian systems. *Physica Scripta*, **21**(6), 783.

Nakahara, M. (2003). *Geometry, Topology and Physics*. CRC Press.

Neuenschwander, D. E. (2017). *Emmy Noether's Wonderful Theorem*. JHU Press.

Neumann, C. (1870). *Ueber die Principien der Galilei-Newton'schen Theorie*. Teubner.

Newton, I. (1952). *Opticks, Or A Treatise of the Reflections, Refractions, Inflections & Colours of Light*. Dove.

Newton, I. (1962). *Principia, Vol. I: The Motion of Bodies* (trans Motte, Cajori). *University of California Press*.

Newton, I. (1999). *The Principia: Mathematical Principles of Natural Philosophy* (trans Cohen, Whitman). University of California Press.

Newton-Smith, W. (1982). *The Structure of Time*. Routledge & Kegan Paul.

Nguyen, J. (2017). Scientific representation and theoretical equivalence. *Philosophy of Science*, **84**(5), 982–995.

Nguyen, J., Teh, N. J., and Wells, L. (2020). Why surplus structure is not superfluous. *The British Journal for the Philosophy of Science*, **71**(2), 665–695.

Nietzsche, F. (1998). *On the Genealogy of Morality*. Hackett Publishing.

Norton, J. (1995). Mach's principle before Einstein. In *Mach's Principle: From Newton's Bucket to Quantum Gravity* (eds J. B. Barbour and H. Pfister), pp. 9–57. Birkhäuser.

Norton, J. D. (2019). The hole argument. In *The Stanford Encyclopedia of Philosophy* (Summer 2019 edn) (ed E. N. Zalta). Metaphysics Research Lab, Stanford University.

Okruhlik, K. (2014). Bas van Fraassen's philosophy of science and his epistemic voluntarism. *Philosophy Compass*, **9**(9), 653–661.

Olver, P. J. (1991). *Applications of Lie groups to differential equations.* Vol. 107. Springer Science & Business Media.

Pashby, T. (2015). Taking times out: Tense logic as a theory of time. *Studies in History and Philosophy of Science Part B: Studies in History and Philosophy of Modern Physics*, **50**, 13–18.

Patton, L. (2021). *Abstraction, Pragmatism, and History in Mach's Economy of Science*, pp. 142–163. Cambridge University Press.

Penrose, R. (1965). Gravitational collapse and space-time singularities. *Physical Review Letters*, **14**(3), 57.

Penrose, R. (1979). Singularities and time-asymmetry. In *General Relativity: An Einstein Centenary Survey* (eds S. Hawking and W. Israel), pp. 581–638.

Penrose, R. (1994). On the second law of thermodynamics. *Journal of Statistical Physics*, **77**(1-2), 217–221.

Pfister, H. (2014). Ludwig Lange on the law of inertia. *The European Physical Journal H*, **39**(2), 245–250.

Pfister, H. and King, M. (2015). *Inertia and Gravitation: The Fundamental Nature and Structure of Space-time.* Springer.

Pitts, J. B. (2013). A first class constraint generates not a gauge transformation, but a bad physical change: The case of electromagnetism. *arXiv preprint arXiv:1310.2756*.

Pitts, J. B. (2014). Change in hamiltonian general relativity from the lack of a time-like killing vector field. *Studies In History and Philosophy of Science Part B: Studies In History and Philosophy of Modern Physics*, 47 68–89.

Poincaré, H. (2017). *Science and Hypothesis: The Complete Text.* Bloomsbury Publishing.

Pojman, P. (2010). From Mach to Carnap: A tale of confusion. *Discourse on a New Method: Reinvigorating the Marriage of History and Philosophy of Science* (eds M Domski and M. Dickson), 295–310.

Pojman, P. (2011). The influence of biology and psychology upon physics: Ernst mach revisited *Perspectives on Science*, **19**(2), 121–135.

Pons, J. (2005). On Dirac's incomplete analysis of gauge transformations. *Studies In History and Philosophy of Science Part B: Studies In History and Philosophy of Modern Physics*, **36**, 491.

Pons, J., Salisbury, D., and Shepley, L. (1997). Gauge transformations in the Lagrangian and Hamiltonian formalisms of generally covariant theories. *Physical Review D*, **55**(2), 658–668.

Pons, J., Salisbury, D., and Sundermeyer, K. A. (2010). Observables in classical canonical gravity: folklore demystified *Journal of Physics A: Mathematical and General*, **222**, 12018.

Pooley, O. (2001). Relationism rehabilitated? ii: Relativity. *PhilSci preprint 221*.

Pooley, O. (2004). Comments on Sklar's "Barbour's relationist metric of time". This is a corrected version of a paper that was originally printed in *Chronos: Proceedings of the Philosophy of Time Society*, 6 (2003–4): 77–86.

Pooley, O. (2013). Substantivalist and relationalist approaches to spacetime. In *The Oxford Handbook of Philosophy of Physics* (ed R. Batterman), 522–586. Oxford University Press.

Pooley, O. (2017). Background independence, diffeomorphism invariance and the meaning of coordinates. In *Towards a Theory of Spacetime Theories*, (eds D. Lehmkuhl, G. Schiemann, E. Scholz), pp. 105–143. Springer.

Pooley, O. and Read, J. (2021). On the mathematics and metaphysics of the hole argument *The British Journal for the Philosophy of Science*, DOI: 10.1086/718274.

Pooley, O. and Wallace, D. (2022). First-class constraints generate gauge transformations in electromagnetism (reply to Pitts). *arXiv preprint arXiv:2210.09063*.

Preston, J. (ed) (2021*a*). *Interpreting Mach*. Cambridge University Press.

Preston, J. (2021*b*). Phenomenalism, or Neutral Monism, in Mach's Analysis of Sensations? In *Interpreting Mach*, (ed J. Preston), pp. 235–257.. Cambridge University Press.

Price, H. (1997). *Time's Arrow & Archimedes' Point: New Directions for the Physics of Time*. Oxford University Press.

Price, H. (2002). Boltzmann's time bomb. *The British Journal for the Philosophy of Science*, **53**(1), 83–119.

Price, H. (2004). On the origins of the arrow of time: Why there is still a puzzle about the low-entropy past? In *Contemporary Debates in Philosophy of Science* (ed C. Hitchcock), pp. 219–239 Blackwell.

Price, H. (2011). The flow of time. In *The Oxford Handbook of Philosophy of Time* (ed C. Callender), pp. 276–311. Oxford University Press.

Prince, G. and Eliezer, C. (1981). On the Lie symmetries of the classical Kepler problem. *Journal of Physics A: Mathematical and General*, **14**(3), 587.

Quine, W. V. (1975). On empirically equivalent systems of the world. *Erkenntnis*, 313–328.

Reichenbach, H. (1956). *The Direction of Time*. University of California Press.

Rescher, N. (1979). *Leibniz An Introduction to His Philosophy*. Blackwell Publishing.

Rickles, D. (2004). Symmetry and possibility: To reduce or not reduce? *PhiSci preprint 1846*

Rickles, D. (2007). *Symmetry, Structure, and Spacetime*. Elsevier.

Rickles, D. (2020). *Covered in Deep Mist: The Development of Quantum Gravity 1916–1956*. Oxford University Press.

Roberts, B. W. (2014*a*). Disregarding the 'hole argument'. *arXiv preprint arXiv:1412.5289*.

Roberts, B. W. (2014*b*). A general perspective on time observables. *Studies in History and Philosophy of Science Part B: Studies in History and Philosophy of Modern Physics*, **47**, 50–54.

Roberts, B. W. (2020). Regarding 'Leibniz equivalence'. *Foundations of Physics*, 1–20.

Roberts, B. W. (2022). *Reversing the Arrow of Time*. Cambridge University Press.

Roberts, J. T. (2008). A puzzle about laws, symmetries and measurability. *The British Journal for the Philosophy of Science*, **59**(2), 143–168.

Rovelli, C. (2001). A note on the foundation of relativistic mechanics. i: Relativistic observables and relativistic states. *arXiv preprint gr-qc/0111037*.

Rovelli, C. (2002). Partial observables. *Physical Review D*, **65**, 124013.

Rovelli, C. (2004). *Quantum Gravity*. Cambridge University Press.

Rovelli, C. (2007). Comment on 'Are the spectra of geometrical operators in loop quantum gravity really discrete?' by B. Dittrich and T. Thiemann. *arXiv preprint arXiv:0708.2481*

Russell, B. (1900). *A Critical Exposition of the Philosophy of Leibniz, with an Appendix of Leading Passages*. Cambridge University Press.

Russell, B. (1903). *Principles of Mathematics*. Allen & Unwin.

Russell, B. (1927). *The Analysis of Matter*. Kegan Paul.

Rynasiewicz, R. (1995*a*). By their properties, causes and effects: Newton's scholium on time, space, place and motion—i. the text. *Studies In History and Philosophy of Science Part A*, **26**(1), 133–153.

Rynasiewicz, R. (1995*b*). By their properties, causes and effects: Newton's scholium on time, space, place and motion—ii. the context. *Studies In History and Philosophy of Science Part A*, **26**(2), 295–321.

Saunders, S. (2013). Rethinking Newton's Principia. *Philosophy of Science*, **80**(1), 22–48.

Schiffrin, J. S. and Wald, R. M. (2012). Measure and probability in cosmology. *Physical Review D*, **86**, 023521.

Schlick, M. (1980). Reflections on the causal principle. In *Philosophical Papers. Vol. 1 (1909-1922)* (ed H. L. Mulder, B. F. van de Velde-Schlick, and P. Heath), pp. 295 – 321. D. Reidel Publishing Company.

Schliesser, E. (2013). Newton's philosophy of time. *A Companion to the Philosophy of Time* (ed H. Dyke, A. Bardon), 87–101. Wiley

Schmaltz, T. M. (2009). Descartes on the extensions of space and time. *Analytica. Revista de Filosofia*, **13**(2), 113–147.

Senovilla, J. M. (1998). Singularity theorems and their consequences. *General Relativity and Gravitation*, **30**(5), 701–848.

Sklar, L. (1974). *Space, Time, and Spacetime*. University of California Press.

Sklar, L. (2004). Barbour's relationalist metric of time. *Chronos: Proceedings of the Philosophy of Time Society*, **6**, 64–76.

Sloan, D. (2018). Dynamical similarity. *Physical Review D*, **97**(12), 123541.

Sloan, D. (2019). Scalar fields and the FLRW singularity. *Classical Quantum Gravity*, **36**(23), 235004.

Sloan, D. (2021). New action for cosmology. *Physical Review D*, **103**(4), 043524.

Slowik, E. (2009). Newton's metaphysics of space: A 'tertium quid' betwixt substantivalism and relationism, or merely a 'god of the (rational mechanical) gaps'? *Perspectives on Science*, **17**(4), 429–456.

Slowik, E. (2011). Newton, the parts of space, and the holism of spatial ontology. *HOPOS: The Journal of the International Society for the History of Philosophy of Science*, **1**(2), 249–272.

Slowik, E. (2016). *The Deep Metaphysics of Space*. Springer.

Souriau, J. (1997). *Structure of Dynamical Systems: A Symplectic View of Physics*. Birkhauser.

Souriau, J.-M. (1974). Sur la variété de Kepler. *Symposia Mathmatica*, 14, 343–260.

Stadler, F. (ed) (2019). *Ernst Mach—Life, Work, Influence*. Springer.

Stadler, F. (2021). Ernst Mach and the Vienna Circle: A Re-evaluation of the Reception and Influence of His Work. In *Interpreting Mach* (ed J. Preston), pp. 184–207. Cambridge University Press.

Staley, R. (2021). Mother's Milk and More: On the Role of Ernst Mach's Relational Physics in the Development of Einstein's Theory of Relativity In *Interpreting Mach* (ed J. Preston), pp. 28–47. Cambridge University Press.

Stein, H. (1967). Newtonian space-time. *Texas Quarterly*, **10**, 174-200.

Stein, H. (2002). Newton's metaphysics. In *The Cambridge Companion to Newton* (eds R. Iliffe and G. E. Smith), pp. 257–307. Cambridge University Press.

Stöltzner, M. (1999). Vienna indeterminism: Mach, Boltzmann, Exner. *Synthese*, **119**(1-2), 85–111.

Stöltzner, M. (2003). The principle of least action as the logical empiricist's shibboleth. *Studies in History and Philosophy of Science Part B: Studies in History and Philosophy of Modern Physics*, **34**(2), 285–318.

Stöltzner, M. (2021). Narratives Divided: The Austrian and the German Mach. In *Interpreting Mach* (ed J. Preston), pp. 208–234. Cambridge University Press.

Streintz, H. (1883). *Die physikalischen Grundlagen der Mechanik*. BG Teubner.

Sudarshan, E. C. G. and Mukunda, N. (1974). *Classical Dynamics: A Modern Perspective*. World Scientific.

Sundermeyer, K. (1982). Constrained dynamics with applications to Yang–Mills theory, general relativity, classical spin, dual string model.

Tait, P. G. (1884). Note on reference frames. *Proceedings of the Royal Society of Edinburgh*, 743-45.

Tal, E. (2016). Making time: A study in the epistemology of measurement. *The British Journal for the Philosophy of Science*, **67**(1), 297–335.

Tambornino, J. (2012). Relational observables in gravity: a review. *SIGMA. Symmetry, Integrability and Geometry: Methods and Applications*, **8**, 017.

Teitelboim, C. (1973). How commutators of constraints reflect the spacetime structure. *Annals of Physics*, **79**(2), 542–557.

Thébault, K. P. Y. (2023). Big bang singularity resolution in quantum cosmology. *Classical and Quantum Gravity*, **40** 055007.

Thébault, K. P. Y. (2021a). On Mach on time. *Studies in History and Philosophy of Science Part A*, **89**, 84–102.

Thébault, K. P. Y. (2021*b*). The problem of time. In *Routledge Companion to Philosophy of Physics* (ed A. Wilson and E. Knox), Chapter 26. Routledge.

Thiemann, T. (2007). *Modern Canonical Quantum General Relativity*. Cambridge University Press.

Thomas, E. (2018). *Absolute Time: Rifts in Early Modern British Metaphysics*. Oxford University Press.

Thomson, J. (1884). On the law of inertia; the principle of chronometry; and the principle of absolute clinural rest, and of absolute rotation. *Proceedings of the Royal Society of Edinburgh*, **12**, 568–578.

Torretti, R. (1996). *Relativity and Geometry*. Dover.

Uebel, T. (2021). Ernst Mach's Enlightenment Pragmatism: History and Economy in Scientific Cognition In *Interpreting Mach* (ed J. Preston), pp. 84–102. Cambridge University Press.

Unruh, W. G. and Wald, R. M. (1989). Time and the interpretation of canonical quantum gravity. *Physical Review D*, **40**(8), 2598.

Vailati, E. (1997). *Leibniz and Clarke: A Study of their Correspondence*. Oxford University Press.

van Fraassen, B. C. (1980). *The Scientific Image*. Oxford University Press.

van Fraassen, B. C. (1989). *Laws and Symmetry*. Oxford University Press.

van Fraassen, B. C. (2008). *The Empirical Stance*. Yale University Press.

Wallace, D. (2022). Observability, redundancy and modality for dynamical symmetry transformations. In *The Philosophy and Physics of Noether's Theorems* (eds J. Read and N. J. Teh), Chapter 13. Cambridge University Press.

Weatherall, J. O. (2016). Understanding gauge. *Philosophy of Science*, **83**(5), 1039–1049.

Weatherall, J. O. (2018). Regarding the 'hole argument'. *The British Journal for the Philosophy of Science*, **69**(2), 329–350.

Weatherall, J. O. (2019*a*). Part 1: Theoretical equivalence in physics. *Philosophy Compass*, **14**(5), e12592. e12592 10.1111/phc3.12592.

Weatherall, J. O. (2019*b*). Part 2: Theoretical equivalence in physics. *Philosophy Compass*, **14**(5), e12591. e12591 10.1111/phc3.12591.

Weyl, H. (1922). *Space-Time-Matter* (trans H. L. Brose) Dover.

Willard, S. (2004). *General Topology*. Dover.

Winterbourne, A. (1982). On the metaphysics of Leibnizian space and time. *Gottfried Wilhelm Leibniz. Critical Assessments*, **3**, 62–75.

Woodhouse, N. M. J. (1997). *Geometric Quantization*. Oxford University Press.

Zahar, E. (1977). Mach, Einstein, and the rise of modern science. *The British Journal for the Philosophy of Science*, **28**(3), 195–213.

Zahar, E. (1981). Second thoughts about machian positivism: A reply to feyerabend. *The British Journal for the Philosophy of Science*, **32**(3), 267–276.

Zwart, P. (1972). The flow of time. *Synthese*, **24**(1-2), 133–158.

Index